Recent Adva

P. Parvatha Reddy

Recent Advances in Crop Protection

Springer

P. Parvatha Reddy
Former Director
Indian Institute of Horticultural Research
Bangalore, Karnataka, India

ISBN 978-81-322-2847-9 ISBN 978-81-322-0723-8 (eBook)
DOI 10.1007/978-81-322-0723-8
Springer India Heidelberg New York Dordrecht London

Springer is part of Springer Science+Business Media (www.springer.com)

Preface

In the recent years, the need to increase food production to meet the demands of rapidly increasing population from a limited land resource necessitated the use of intensive farming systems, with the inputs like narrow genetic base, high fertilisers, irrigation, multiple cropping, etc., which favour disease and pest development. The introduction of new high-yielding genotypes susceptible to the pests and the pathogens and changing cropping patterns including cultivation in non-traditional areas have resulted in a spurt of pests and diseases in crop pathosystems, remarkably changing the scenario of biotic stresses. It is not only that those new pest problems emerged but also the minor pests assumed major status and vice versa. The effect of changing global climate, particularly of sharp increase in CO_2 concentration from the current level of 330–670 ppm, increases the susceptibility of plants to pathogens and pests.

Presently practised crop protection measures mainly orient towards chemical control. The excessive dependence on chemical pesticides leads to the development of resistance in insect pests and pathogens, outbreaks of secondary pests and pathogens/biotypes and occurrence of residues in food chain. The chemical control also has other limitations such as high cost, low cost–benefit ratio; poor availability; selectivity; temporary effect; efficacy affected by physico-chemical and biological factors; resurgence of pests; health hazards; toxicity towards plants, animals and natural enemies; and environmental pollution. Because of high intensive agricultural practices and chemicalisation of agriculture, the age-old environment-friendly pest management practices like sanitation, crop rotation, mixed cropping, adjustment of date of planting, fallowing, summer ploughing, green manuring, composting, etc., to combat plant pests are not being practised in Indian agriculture. The pace of development and durability of resistant varieties had been slow and unreliable.

Considering these limitations, there has been a growing awareness and increasing demand for novel and improved crop protection approaches to guarantee effective and sustainable food production which offers new opportunities for crop protection research. The information on recent advances in crop protection (bacteria, fungi, nematodes, insects, mites and weeds) is very much scattered. There is no book at present which comprehensively and exclusively deals with the above aspects. The present book deals with the most recent advances in crop protection such as avermectins, bacteriophages, biofumigation, biotechnological approaches, bio-priming of seeds, disguising the leaf surface, non-pathogenic strains, plant defence activators, plant growth-promoting rhizobacteria, pathogenesis-related proteins, soil solarisation, strobilurin fungicides, variety mixtures/cultivar mixtures/multilines, biointensive integrated pest management and other recent advances (RNA interference, fusion protein-based biopesticides, seed mat technology and environmental methods). The book is illustrated with excellent quality photographs enhancing the quality of publication. The book is written in lucid style and easy to understand language along with adoptable recommendations for pest management.

This book can serve as a useful reference to policymakers, research and extension workers and students. The material can also be used for teaching postgraduate courses. Suggestions to improve the contents of the book are most welcome (e-mail: reddy_parvatha@yahoo.com). The publisher, Springer (India) Pvt. Ltd., New Delhi, deserves commendation for their professional contribution.

Bangalore P. Parvatha Reddy

About the Author

Dr. P. Parvatha Reddy obtained his Ph.D. degree in plant pathology jointly from the University of Florida, USA, and the University of Agricultural Sciences, Bangalore. He served as the director of the prestigious Indian Institute of Horticultural Research (IIHR) at Bangalore from 1999 to 2002 during which period the institute was honoured with 'ICAR Best Institution Award'. He also served as the Head, Division of Entomology and Nematology at IIHR and gave tremendous impetus and direction to research, extension and education in developing biointensive integrated pest management strategies in horticultural crops. Dr. Reddy has about 34 years of experience working with horticultural crops and is involved in developing an F1 tomato hybrid 'Arka Varadan' resistant to root-knot nematodes. He has over 237 scientific publications to his credit, which also include 22 books. Dr. Reddy has been honoured with the following prestigious awards: 'Association for Advancement Pest Management in Horticultural Ecosystems Award', 'Dr. GI D'souza Memorial Lecture Award', 'Prof. H. M. Shah Memorial Award' and 'Hexamar Agricultural Research and Development Foundation Award' for his unstinted efforts in developing sustainable, biointensive and eco-friendly integrated pest management strategies in horticultural crops.

Contents

1 Introduction

Abstract

Because of high intensive agricultural practices and chemicalisation of agriculture, the age-old environment-friendly pest management practices like sanitation, crop rotation, mixed cropping, adjustment of date of planting, fallowing, summer ploughing, green manuring, composting, etc., to combat plant pests are not being practised in Indian agriculture. The pace of development and durability of resistant varieties had been slow and unreliable.

Considering these limitations, there have been growing awareness and increasing demand for novel and improved crop protection approaches to guarantee effective and sustainable food production which offers new opportunities for crop protection research. The most recent advances in crop protection include avermectins, bacteriophages, biofumigation, biotechnological approaches, bio-priming of seeds, disguising the leaf surface, non-pathogenic strains, plant growth-promoting rhizobacteria, soil solarisation, strobilurin fungicides, variety mixtures/cultivar mixtures/multilines, biointensive integrated pest management, pathogenesis-related proteins, RNA interference, fusion protein-based biopesticides, seed mat technology and environmental methods.

In the recent years, the need to increase food production to meet the demands of rapidly increasing population from a limited land resource necessitated the use of intensive farming systems, with the inputs like narrow genetic base, high fertilisers, irrigation, multiple cropping, etc., which favour disease and pest development. The intensive agriculture, especially the introduction of new high-yielding genotypes susceptible to the pests and the pathogens, and changing cropping patterns including cultivation in nontraditional areas have resulted in a spurt of pests and diseases in crop pathosystems, remarkably changing the scenario of biotic stresses. It is not only that those new pest problems emerged but also the minor pests assumed major status and vice versa. The intensification of cropping systems has led to increase in biotic stresses on account of introduction of new pest problems (e.g. cotton leaf curl, type B of whitefly, spiralling whitefly), increased intensity of the existing pests (e.g. white rust of mustard, leaf blight of wheat, sheath blight of rice) and development of resistance to pesticides (e.g. American bollworm, whitefly).

P.P. Reddy, *Recent Advances in Crop Protection*,
DOI 10.1007/978-81-322-0723-8_1, © Springer India 2013

1.1 Emerging Pest Scenario

1.1.1 Emerging Insect Pest Scenario

The change of insect pest scenario in Indian agriculture was more influenced by the introduction of new agricultural production technology based on high-yielding varieties and increased use of inputs like irrigation water, fertilisers and pesticides during the last four decades. Many ecological attributes, which are favourable for plant growth, also favour multiplication of pests. Basic principles of herbivore were highlighted in several pestilence episodes. Most of these intense pestilences in various crops could be mitigated through regulated plant nutrition system. The quality of plant nutrition is the determinant of the intensity and build-up of agricultural pests, which are capable of adapting to large changes in the environment. Consequently, there is a change in the insect pest scenario in almost all the crops due to changes in agroecosystem.

The *desi* tall varieties of rice were found to suffer from only five insect pests, namely, yellow stem borer, hispa, rice bug, rice grasshopper and surface grasshopper. Following the introduction of high-yielding dwarf varieties and many river command area projects, and also the associated chemical intensive technology, there has been a considerable increase in area under rice. Extensive sole cropping of rice is very common in many river and tank command areas, which favoured the development and multiplication of multitude of pests. Similarly, leaf folder has emerged as a serious pest in Kerala, Tamil Nadu, Andhra Pradesh, Karnataka, Gujarat, Punjab and Haryana. Gall midge has been observed to occur during *rabi* in Godavari delta of Andhra Pradesh and coastal Maharashtra. Gundhi bug, which was earlier confined to north and north-eastern states of India, has been frequently causing serious damage to rice in parts of Tamil Nadu and Andhra Pradesh. There is a need to cultivate mosaics of varieties of crop plants in monocropping systems in large geographic tracts to prevent generation and invasions of various pests as a tool to avoid enormous biotic stresses.

1.1.2 Emerging Disease Scenario

Plant disease epiphytotics have been a major cause to change agricultural patterns and even food habits in many parts of the world. For example, cereal rye was replaced by potato due to ergot (*Claviceps purpurea*) disease, and potato was replaced by wheat due to late blight (*Phytophthora infestans*) in Europe. Some other similar devastating diseases which had far-reaching impact on agriculture were wheat rust in Mexico, coffee rust in Sri Lanka and Brazil, southern corn blight in the USA, cassava mosaic and maize rust in Africa, Panama disease (*Fusarium oxysporum* f. sp. *cubense)* of banana in South America and Liberia, bunchy top of banana in Australia, swollen shoot of cocoa in Ghana, bacterial rot (*Pseudomonas solanacearum*) of potato in Kenya and many others. In India, the brown spot of rice (*Cochliobolus miyabeanus*) caused the great Bengal famine in 1943; red rot of sugar cane caused severe epiphytotics in Uttar Pradesh and Bihar during 1938–1942; wheat rust in Uttar Pradesh and Madhya Pradesh in 1946–1947; *Helminthosporium* blight of wheat and barley in Uttar Pradesh and Bihar during 1979–1981; leaf curl of cotton in Punjab and Rajasthan during 1994–1995; tungro disease of rice in Punjab during 1998–1999; and necrosis disease of sunflower in Karnataka are just a few examples of the serious problems caused by plant diseases. Ironically, most of these diseases have caused destruction in crops grown under rain-fed agriculture, indicating vulnerability of such cropping systems.

In wheat, rusts were the most serious problems until the mid-1970s, but currently with the wide use of rust-resistant varieties (although there is overwhelming evidence of the yellow rust-resistant wheat variety, PB 343 appears to be susceptible to this dreadful disease in the 'green revolution area' of the country), a minor disease of the past, namely, Karnal bunt, has assumed serious proportions. Karnal bunt in some wheat-growing states and breakdown of resistance in PB 343 variety for yellow rust race, 78S 84, have emerged as serious disease problems. The rice tungro virus and the bacterial leaf blight of rice are the most devastating diseases in the new varieties.

Sheath blight has become serious on rice in the nonconventional areas. Maize and millets are now devastated by downy mildews. Blight of cotton became a major problem when the indigenous diploid cottons were replaced by exotic cotton tetraploid cottons. The new exotic cotton varieties are also highly susceptible to a new disease known as 'parawilt of cotton' of unknown aetiology. During the past few years, the viruses of cotton (particularly whitefly-transmitted leaf curl) have become dangerous and required special efforts to check the spread to different areas. Such examples of changing disease scenario are available for pulses, oilseeds, vegetables, fruits, etc., that is, the crops in which productivity has increased tremendously. Certain diseases of complex or unknown aetiology, for example, parawilt of cotton, coconut root wilt, dieback in citrus, mango malformation, crown rot of oil palms, and brown bast disease of rubber need special efforts to develop management practices to minimise the losses.

1.1.3 Emerging Plant-Parasitic Nematode Scenario

Plant-parasitic nematodes, the unseen enemies of the farmer, have been recognised as serious perpetual problems in agricultural production all over the world. Favourable weather and almost continuous availability of host crops in the tropical and subtropical regions including India favour their build-up. Being soil-borne and with wide host range, nematodes are one of the toughest pests to control. International estimates have put crop losses due to nematodes at over $100 billion (Sasser and Freckman 1987). The root-knot nematodes cause intense pestilence problem in vegetables, fruit crops, rice, pulses, fibre and oilseed crops.

Extensive nematode quarantine and control efforts in developed countries have paid rich dividends. India suffers heavy quantitative and qualitative losses in various food, fibre and commercial crops due to nematodes. A number of new nematode problems have emerged in the intensive cropping systems.

Rice revolution in the country has brought about many second-generation nematode problems such as a new root-knot nematode species *Meloidogyne triticoryzae*, which damage the rice and wheat in the north-western plains with rice–wheat cropping system. *Meloidogyne graminicola* infestation was intense in rice crop of Mandya district in Karnataka. *Molya* disease in wheat caused by the cereal cyst nematode, *Heterodera avenae,* in Haryana and Bihar has emerged as a serious disease problem.

1.2 Impact of Climate Change

The effect of changing global climate, particularly of sharp increase in temperature, through the last century on the intensity of pests and diseases is largely unknown. It now appears that the southern Asia will become warmer and unseasonably much wetter. The major changes in global temperature and climate are expected to be mainly due to atmospheric CO_2, methane and chlorofluorocarbons. Initial experiments have already shown that an increase in atmospheric CO_2 concentration from the current level of 330–670 ppm increases the susceptibility of plants to pathogens and pests. Practically, there is no data on the effect of ongoing climatic variability on diseases and pests. Most of the studies have been more concerned with the influence of day-to-day weather conditions rather than with year-to-year climatic variability on pest and disease appearance and build-up. Climatic variability can affect any part of the life cycle of the pathogen, insect, nematode as well as the interaction between or amongst these organisms. The integrated pest management (IPM) strategies strongly rely on natural controlling factors such as weather and natural enemies of various pest organisms. Based on physiography, soil, climate and growing period, 21 agroecological regions (AERs) have been identified. Most of the major disease problems in rain-fed agriculture occur in sub-humid and semi-arid conditions which represent 72% of the total geographical area (329 m ha) of the country. The other AERs have relatively fewer serious disease problems. Utilising the current scientific tools, appropriate information on the biological transportations that the herbivorous organisms undergo in crop fields should be studied.

There is a strong need to revamp the direction of research in plant protection in order to acquire better understanding of the internal processes of each crop species alone and in combination with others of a cropping system and in sequence. This could be best addressed through discovering a new paradigm of crop health management. Agrarian husbandry has to be made more meaningful with crop health management in which defined health attributes of the crops have to be pursued. Having currently a national pool of around 1,000 plant protection scientists (in Entomology, Plant Pathology, Nematology, Acarology, Apiculture, vertebrate organisms, etc.) in DARE (ICAR and SAU systems), reorganisation of resources has to be rationalised to involve effectively in the prioritised research programmes in plant protection.

1.3 Need for Novel Approaches to Crop Protection

Presently practised crop protection measures mainly orient towards chemical control. The excessive dependence on chemical pesticides leads to the development of resistance in insect pests and pathogens, outbreaks of secondary pests and pathogens/biotypes and occurrence of residues in food chain. The chemical control also has other limitations such as high cost, low cost to benefit ratio; poor availability; selectivity; temporary effect; efficacy affected by physico-chemical and biological factors; resurgence of pests; health hazards; toxicity towards plants, animals and natural enemies; environmental pollution; etc. Because of high intensive agricultural practices and chemicalisation of agriculture, the age-old environment-friendly pest management practices like sanitation, crop rotation, mixed cropping, adjustment of date of planting, fallowing, summer ploughing, green manuring, composting, etc., to combat plant pests are not being practised in Indian agriculture. The pace of development and durability of resistant varieties had been slow and unreliable.

Considering these limitations, there have been growing awareness and increasing demand for novel and improved crop protection approaches to guarantee effective and sustainable food production which offers new opportunities for crop protection research. The most recent advances in crop protection include avermectins, bacteriophages, biofumigation, biotechnological approaches, bio-priming of seeds, disguising the leaf surface, non-pathogenic strains, plant growth-promoting rhizobacteria, soil solarisation, strobilurin fungicides, variety mixtures/cultivar mixtures/multilines, biointensive integrated pest management, pathogenesis-related proteins, RNA interference, fusion protein-based biopesticides, seed mat technology and environmental methods.

1.4 Recent Advances in Crop Protection

1.4.1 Avermectins

The avermectins are a new class of macrocyclic lactones derived from mycelia of the soil actinomycete, *Streptomyces avermitilis* (soil inhabiting which is ubiquitous in nature). These compounds were reported to be possessing insecticidal, acaricidal and nematicidal properties (Putter et al. 1981). They are commonly distributed in most of the cultivated soils and are in widespread use, especially as agents affecting plant-parasitic nematodes, mites and insect pests. The water solubility of avermectin B1 is approximately 6–8 ppb, and its leaching potential through many types of soil is extremely low. These physical properties also confer many advantages upon the use of avermectins as pesticides. Their rapid degradation in soil and poor leaching potential suggest that field applications would not result in persistent residues or contamination of ground water.

1.4.2 Bacteriophages

A bacteriophage (from 'bacteria' and Greek *phagein* 'to eat') is any one of a number of viruses that infect bacteria. Bacteriophages are amongst the most common biological entities on earth. Kotila and Coons (1925) isolated bacteriophages from soil samples that were active against the causal agent of blackleg disease of potato, *Erwinia carotovora* subsp. *atroseptica*. They demonstrated in growth chamber experiments that co-inoculation of *E. carotovora* subsp. *atroseptica* with phage successfully inhibited the pathogen and prevented rotting of tubers. These workers also isolated phages against *Erwinia carotovora* subsp. *carotovora* and *Agrobacterium tumefaciens* from various sources such as soil, rotting carrots and river water (Coons and Kotila 1925). Thomas (1935) reported that treatment of corn seed infected with *Pantoea stewartii*, the causal agent of Stewart's wilt of corn, with bacteriophage isolated from diseased plant material reduced the disease incidence from 18 to 1.4%.

Despite the promising early work, phage therapy did not prove to be a reliable and effective means of controlling phytobacteria. Since the early 1990s, various approaches were attempted to improve the competitive advantage of phages in the environment in order to improve their efficacy including prevention of development of phage-resistant mutants, proper selection of efficient phages, timing of phage application, maximising chances for interaction between phage and target bacterium, overcoming adverse factors in phyllosphere on phage persistence, development of solar protectants to increase phase bioefficacy and delivery of phages in the presence of phage-sensitive bacterium.

1.4.3 Biofumigation

Biofumigation is an agronomic technique that makes use of some plants' defensive systems. The main plant species in which this is found are the Brassicaceae (cabbage, cauliflower, kale and mustard), Capparidaceae (cleome) and Moringaceae (horseradish). In suitable conditions, the biofumigation technique is able to efficiently produce a number of important substances. In the above plant families, one of the most important enzymatic defensive systems is the myrosinase–glucosinolate system. With this system, tissues of these plants can be used as a soft, eco-compatible alternative to chemical fumigants and sterilants. In a number of countries over the past few years, several experiments have been carried out to evaluate the effectiveness of the myrosinase–glucosinolate system, in particular using the glucosinolate-containing plants as a biologically active rotation and green manure crop for controlling several soil-borne pathogens and diseases. The use of this technique is growing, and it is studied in several countries at a full-field scale (the USA, Australia, Italy, the Netherlands and South Africa), thus triggering the interest of some seed companies, with a positive effect on the 'biofumigation' seed market, which is significantly growing year after year. New potential has also been found for the dehydrated plant tissues and/or for defatted meal pellets production and use. An intense discussion amongst researchers of this topic in various countries seems to be of fundamental importance particularly to define and develop future common strategies. The current strategies include biofumigation as part of an integrated approach to methyl bromide replacement in agriculture.

1.4.4 Biotechnological Approaches

Biotechnology offers many opportunities for agriculture and provides the means to address many of the constraints placed to productivity. It uses the conceptual framework and technical approaches of molecular biology and plant cell culture systems to develop commercial processes and products. With the rapid development of biotechnology, agriculture has moved from a resource-based to a science-based industry, with plant breeding being dramatically augmented by the introduction of recombinant DNA technology based on knowledge of gene structure and function. The concept of utilising a transgenic approach to host plant resistance was realised in the mid-1990s with the commercial introduction of transgenic maize, potato and cotton plants expressing genes encoding the insecticidal δ-endotoxin from *Bacillus thuringiensis*. Similarly, the role of herbicides in agriculture entered a new era with the introduction of glyphosate-resistant soybeans in 1995. Currently, the commercial area planted to transgenic crops is in excess of 90 million hectares with approximately 77% expressing herbicide tolerance, 15% expressing insect resistance genes and approximately 8% expressing both traits. Despite the increasing disquiet over the growing of such crops in Europe and Africa (at least by the media and certain NGOs) in recent years, the latest figures available demonstrate that the market is increasing, with an 11% increase between 2004 and 2005.

1.4.5 Bio-priming of Seeds

Bio-priming is a new technique of seed treatment that integrates biological (inoculation of seed with beneficial organism to protect seed) and physiological aspects (seed hydration) of disease control. It is recently used as an alternative method for controlling many seed and soil-borne pathogens. It is an ecological approach using selected fungal antagonists against the soil and seed-borne pathogens. Biological seed treatments may provide an alternative to chemical control. Seed priming, osmo-priming and solid matrix priming were used commercially in many horticultural crops, as a tool to increase speed and uniformity of germination and improve final stand. However, if seeds are infected or contaminated with pathogens, fungal growth can be enhanced during priming, thus resulting in undesirable effects on plants. Therefore, seed priming alone or in combination with low dosage of fungicides and/or biocontrol agents have been used to improve the rate and uniformity emergence of seed and reduce damping-off disease.

1.4.6 Disguising the Leaf Surface

The leaf surface provides the first barrier that fungi must overcome in order to gain access to the leaf, but it also provides chemical and physical cues that are necessary for the development of infection structures for many fungal pathogens. Film-forming polymers can coat the leaf surface, acting not just as an extra barrier to infection but also disguising the cues necessary for germling development. Kaolin particle films can envelop the leaf in a hydrophobic particle film barrier that prevents spores or water from directly contacting the leaf surface and, as a result, can suppress infection. Adhesion of fungal spores to the leaf surface, which is important to keep spores on the leaf surface and for appropriate development of the fungus on the leaf surface, can be inhibited, leading to reduced infection and lesion development. Polymer and particle films have been shown to provide disease control in the field, whilst research on agents that inhibit spore adhesion on leaf surfaces is still in its infancy. There is an urgent need for research on the practicality of using these innovative methods under field conditions and on ways of integrating them into current crop protection programmes.

1.4.7 Non-pathogenic Strains

Non-pathogenic (avirulent) or low virulent (hypovirulent) strains are capable of colonising infection site niches on the plants' surfaces and protecting susceptible plants against their respective pathogens. Such phenomena have been demonstrated for a considerable number of plant pathogens (*Agrobacterium* spp., *Rhizoctonia* spp., *Fusarium* spp. and *Pythium* spp.). Non-pathogenic strains of various pathogens are potential candidates for development of biocontrol preparations. Some strains are already used in agriculture. The modes of protection differ amongst the non-pathogenic strains, and one strain can protect by more than one mechanism. Competition for infection sites, or for nutrients (such as carbon, iron) as well as induction of the host plant resistance, has been demonstrated for several pathogens such as *Rhizoctonia* spp., *Fusarium* spp. and *Pythium* spp. Mycoparasitism was shown for *Pythium* spp. The non-pathogenic *F. oxysporum* are easy to mass produce and formulate, but application conditions for biocontrol efficacy under field conditions have still to be determined.

1.4.8 Plant Growth-Promoting Rhizobacteria

Plant growth-promoting rhizobacteria (PGPR) were first defined by Kloepper and Schroth (1978) to describe soil bacteria that colonise the roots of plants following inoculation onto seed and that enhance plant growth and/or reduce disease or insect damage. There has been much research interest in PGPR, and there is now an increasing number of PGPR being commercialised for crops. Organic growers may have been promoting these bacteria without knowing it. The addition of compost and compost teas promote existing PGPR and may introduce additional helpful bacteria to the field. The absence of pesticides and the more complex organic rotations are likely to promote existing populations of these beneficial bacteria. However, it is also possible to inoculate seeds with bacteria that increase the availability of nutrients, including solubilising phosphate, potassium, oxidising sulphur, fixing nitrogen, chelating iron and copper.

PGPR such as *Pseudomonas* and *Bacillus* species have attracted much attention for their role in reducing plant diseases. The work to date is very promising and may offer organic growers with some of their first effective control of serious plant diseases. Some PGPR, especially if they are inoculated on the seed before planting, are able to establish themselves on the crop roots. They use scarce resources and thereby prevent or limit the growth of pathogenic microorganisms. Even if nutrients are not limiting, the establishment of benign or beneficial organisms on the roots limits the chance that a pathogenic organism that arrives later will find space to become established. Numerous rhizosphere organisms are capable of producing compounds that are toxic to pathogens like HCN.

1.4.9 Soil Solarisation

The use of clear polyethylene film to cover moistened soil and trap lethal amounts of heat from solar radiation was first reported by Katan and colleagues in Israel in the mid-1970s (Katan et al. 1976). DeVay and associates at University of California, Davis, began an intensive research programme on the promising technique shortly thereafter, and the term 'soil solarisation' was soon coined to describe the process by cooperators in the San Joaquin Valley. Researchers found that solarisation could be a useful soil disinfestation method, especially in areas with hot and arid conditions during the summer months, such as the Central Valley and southern deserts. In certain cases, the treatment has also been effective, primarily for weed management, in cooler coastal areas (Elmore et al. 1993). The pesticidal activity of solarisation was found to stem from a combination of physical, chemical and biological effects, as described in several comprehensive reviews.

Solarisation is a technique that elevates soil temperatures beneath a clear plastic layer to reduce soil-borne pests. The capacity of soil solarisation to suppress propagule numbers of soil-borne pathogens relies on many factors. The temperatures obtained in the moistened soil covered by the transparent sheeting and the exposure time of the organisms to these elevated temperatures are both important characteristics of this pre-plant soil treatment. Solarisation has been effective in disease control in many geographical locations around the world. It is most successful in regions with the appropriate meteorological parameters such as high air temperatures and extended periods of high radiation.

1.4.10 Strobilurin Fungicides

Natural strobilurins are produced by certain forest mushrooms and secreted into the decaying wood on which they grow. The powerful fungicidal activity of this secretion prevents invasion by other fungi, so protecting the nutrient source of the original mushroom. This fungal antibiotic fights infections of the plants. After German scientists first discovered strobilurins in 1977, it didn't take long for people to realise its potential for use as a fungicide. Thus, the development of what would become one of the most important classes of fungicides began.

The fungus *Strobilurus tenacellus*, which grows on fallen pine cones, produces strobilurin A. This rather insignificant grey to yellowish-brown mushroom grows to a height of 5–7 cm and is edible, with a mild, slightly bitter taste, but it is remarkable for its fungitoxic activity. Through the production of strobilurin, it is capable of keeping other fungi at bay that might otherwise compete for nutrients. All fungi need to produce their own energy supply in order to grow and produce new spores. This supply is especially important during the early establishment phase of the disease life cycle. It is produced by a complex series of chemical processes in the mitochondria that are part of every living fungal cell. Strobilurins work by blocking electron transfer within this chemical process, thus denying the fungus energy and preventing development, even at the earliest stages of the life cycle, the spore germination stage.

1.4.11 Cultivar Mixtures

Wolfe (1985) defined cultivar mixtures as 'mixtures of cultivars that vary for many characters including disease resistance, but have sufficient similarity to be grown together'. Cultivar mixtures do not cause major changes to the agricultural system, generally increase yield stability and in some cases can reduce pesticide use. They are also quicker and cheaper to formulate and modify than 'multilines', which are defined as mixtures of genetically uniform lines of a crop species (near-isogenic lines) that differ only in a specific disease or pest resistance (Browning and Frey 1981).

Cultivars used in the mixture must possess good agronomic characteristics and may be phenotypically similar for important traits including maturity, height, quality and grain type, depending on the agronomic practices and intended use. Cultivar mixtures in barley for the control of powdery mildew are an example of phenotypically similar mixtures, whereas red- and white-grained sorghum mixtures used in Africa are an example of phenotypically different mixtures.

The principles driving use of variety mixtures for disease control are soundly based on ecology. Epidemics are the exception in natural and semi-natural ecosystems, reflecting the balance derived from the co-evolution of hosts and pathogens. However, in modern agriculture in particular, this balance is far from equilibrium, and epidemics would be frequent were it not for highly effective pesticides and a plant breeding industry which introduces new cultivars to the market with new or different resistance genes. Such a situation is generally profitable when commodity prices are high, but it is costly and rates very poorly on sustainability and ecological or environmental parameter scales.

1.4.12 Biointensive Integrated Pest Management (BIPM)

Biointensive IPM is defined as 'A systems approach to pest management based on an understanding of pest ecology. It begins with steps to accurately diagnose the nature and source of pest problems, and then relies on a range of preventive tactics and biological controls to keep pest populations within acceptable limits. Reduced-risk pesticides are used if other tactics have not been adequately effective, as a last resort, and with care to minimize risks' (Benbrook 1996).

Biointensive IPM incorporates ecological and economic factors into agricultural system design and decision-making and addresses public concerns about environmental quality and food safety. The benefits of implementing biointensive IPM can include reduced chemical input costs, reduced on-farm and off-farm environmental impacts and more effective and sustainable pest management. An ecology-based IPM has the potential of decreasing inputs of fuel, machinery and synthetic chemicals – all of which are energy intensive and increasingly costly in terms of financial and environmental impact. Such reductions will benefit the grower and society.

The primary goal of biointensive IPM is to provide guidelines and options for the effective management of pests and beneficial organisms in an ecological context. The flexibility and environmental compatibility of a biointensive IPM strategy make it useful in all types of cropping systems. Biointensive IPM would likely decrease chemical use and costs even further.

1.4.13 Plant Defence Activators

1.4.13.1 Chemical Plant Defence Activators

A number of natural and synthetic compounds induce plant defences against pathogens and herbivores and act at different points in plant defence pathways (Karban and Baldwin 1997; Gozzo 2004). The non-protein amino acid DL-β-aminobutyric acid (BABA) is a potent inducer of plant resistance and is effective against a wide range of biotic and abiotic stresses. BABA is rarely found naturally in plants, but, when applied as a root drench or foliar spray, it has been shown to protect against viruses, bacteria, oomycetes, fungi and phytopathogenic nematodes, as well as abiotic stresses such as drought and extreme temperatures (Jakab et al. 2001). BABA-induced resistance (BABA-IR) can provide effective protection for crop plants in many botanical families, including legumes, cereals, Brassicas and Solanaceae (Jakab et al. 2001). Unlike other chemical inducers (e.g. INA and BTH), BABA does not directly activate the plant's defence arsenal and therefore does not cause direct trade-off effects on plant growth due to energetically demanding investment in defence mechanisms. Instead, BABA appears to condition the plant for a faster and stronger activation of defence responses once the induced plant is exposed to stress, a process known as 'sensitisation' or 'priming' (Conrath et al. 2002).

BTH [benzo (1, 2, 3) thiadiazole-7-carbothioic acid S-methyl ester] is strongly effective against *Peronospora tabacina*, causative agent of blue mould, the most important worldwide distributed tobacco disease. Applied in minimal amounts (around 50 g ha^{-1}), BTH provides field protection lasting until flowering without negative influence on growth, development and yield of tobacco. BTH appears more efficient than metalaxyl, the commonly used blue mould fungicide. It ensures 90% disease reduction on the 17th day after its application versus only 46% for metalaxyl (Tally et al. 1999). It is noteworthy that BTH is an effective inducer of resistance in tobacco not only against fungal pathogens but also against viruses and bacteria (Tally et al. 1999). BTH was also found to be effective in inducing SAR in wheat (Görlach et al. 1996), pea (Dann and Deverall 2000), potato (Bokshi et al. 2003), cotton against *Alternaria* leaf spot, bacterial blight and *Verticillium* wilt (Colson-Hanks et al. 2000), tomato against bacterial canker (*Clavibacter michiganensis* subsp. *michiganensis*) (Soylu et al. 2003).

1.4.13.2 Biological Plant Defence Activators

Plants are endowed with several defence genes which are involved in synthesis of antifungal, antibacterial and antiviral compounds like pathogenesis-related proteins (PRs), phenolics, phytoalexins, lignin, callose and terpenoids conferring resistance against plant pathogens. Most of the defence genes are sleeping genes (quiescent in healthy plants) which require specific signals to activate them. Several antagonistic organisms have been shown to provide signals, which activate the defence genes, and they are called 'biological plant defence activators' or 'biological plant activators'. Several elicitors have been isolated from these antagonistic organisms. Elicitors are the primary signal molecules of the antagonists, which elicit host defence mechanisms. The elicitor provides necessary signal for activation of the defence genes. Lipopolysaccharides, chitin oligomers and glucans, siderophores, some enzymes (xylanases), a low molecular weight protein (oligandrin) and salicylic acid are some of the plant defence activators produced by antagonistic organisms.

1.4.14 Pathogenesis-Related Proteins (PRs)

The defence strategy of plants against stress factors involves a multitude of tools, including various types of stress proteins with putative protective functions. A group of plant-coded proteins induced by different stress stimuli, named 'pathogenesis-related proteins' (PRs), is assigned an important role in plant defence against pathogenic constraints and in general adaptation to stressful environment. A large body of experimental data has been accumulated, and changing views and concepts on this hot topic have been evolved.

Pathogenesis-related proteins (initially named 'b' proteins) have focused an increasing research interest in view of their possible involvement in plant resistance to pathogens. This assumption flowed from initial findings that these proteins are commonly induced in resistant plants, expressing a hypersensitive necrotic response (HR) to pathogens of viral, fungal and bacterial origin. Thus, Antoniw et al. (1980) coined the term 'pathogenesis-related proteins' (PRs), which have been defined as 'proteins encoded by the host plant but induced only in pathological or related situations', the latter implying situations of non-pathogenic origin.

Experimental evidences substantiated the utility of PR genes to develop disease resistance in transgenic plants. This practical aspect of PR gene research resulted in the release of agronomically important crops resistant to various diseases of economical interest. One promising strategy is based on the exploitation of the genes encoding antifungal hydrolases, such as ß-1, 3-glucanase and chitinase, which are associated with SAR response in plants.

1.4.15 Other Recent Advances

1.4.15.1 RNA Interference (RNAi)

RNA interference (RNAi) is a technology that allows for the specific down-regulation of genes and is a powerful tool for the identification of new targets for crop protection compounds. Whilst a crop protection compound inhibits its target protein, often an enzyme catalysing a specific metabolic step, RNAi targets the messenger RNA which encodes this enzyme and in consequence reduces the amount of the enzyme itself. Thus, RNAi is capable of mimicking the action of crop protection compounds. The availability of genome sequences for several model organisms made gene sequences available and provided the necessary information to target any gene of interest by using RNAi.

RNA interference was introduced in crop protection research in the mid-1990s by antisense silencing of genes encoding herbicide targets and candidate target genes. In the late 1990s, the approach was widened with the goal to identify new targets by *Arabidopsis thaliana* genome-wide down-regulation of genes based on controlled antisense expression of target gene sequences.

1.4.15.2 Fusion Protein-Based Biopesticides

The biopesticide is based on fusion protein technology invented and developed collaboratively by the Food and Environment Research Agency (FERA), York and Durham University. This allows selected toxins from arthropods, which have no toxicity towards higher animals, to be combined with a carrier protein that makes them orally toxic to invertebrates, whereas they would normally only be effective when injected into a prey organism by a predator. The fusion protein, containing both the toxin and the carrier, is produced as a recombinant protein in a microbial expression system, which can be scaled up for industrial production.

1.4.15.3 Seed Mat Technology

Seed mat technology is 'advancing crop technology' replacing agrochemicals for weed, pest and disease management by the use of advanced seed mat systems that also reduce water and labour requirement whilst improving food safety, quality and shelf life. It is a novel and innovative technology that has been developed by Terraseed Ltd. (commercial product 'Terraseed') primarily as a non-chemical approach to weed control in salad and vegetable production. It has taken several years and significant investment to develop the system, and the product is now being licensed to commercial growers in the UK and overseas as an alternative for effective weed control in situations where key herbicides have been withdrawn or are due to be revoked in the near future.

It is intended to further develop and extend the technology for the control of important pests and pathogens in horticultural crops and hence further reduce, and potentially eliminate, insecticide and fungicide inputs. The technology provides the opportunity to improve yield and quality and also reduce fertiliser, water and labour inputs through improved mechanisation. The focus will be on the control of economically important pathogens and pests such as *Bremia, Sclerotinia, Rhizoctonia, Pythium*, carrot fly and aphids.

1.4.15.4 Environmental Methods

There are many ways the environment can be altered or managed to reduce plant diseases. Some of them include temperature, irrigation, humidity and host nutrition (fertiliser).

All diseases have a specific range of temperatures under which they are the worst.

If you grow plants in a greenhouse, you can alter the temperature of that structure to levels that are not optimal for the pathogen. This can greatly reduce disease severity.

Other elements, which greatly affect disease severity, are the irrigation method and the humidity of the growing environment. Many diseases are less severe under lower humidity. One of the best ways to alter the humidity around the plants is to space them farther apart. This can increase air movement between plants and thus reduce relative humidity and disease severity.

Plant nutrition can influence the feeding, longevity and fecundity of phytophagous pests; the common fertiliser elements (nitrogen, phosphorous and potassium) can have direct and indirect effects on pest suppression. In general, nitrogen in high concentrations has the reputation of increasing pest incidence, particularly of sucking pests such as mites and aphids. On the other hand, phosphorous and potassium additions are known to reduce the incidence of certain pests; for example, in low-phosphorous soils, wireworm populations often tend to increase.

References

Antoniw JF, Ritter CE, Pierpoint WS, Van Loon LC (1980) Comparison of three pathogenesis-related proteins from plants of two cultivars of tobacco infected with TMV. J Gen Virol 47:79–87

Benbrook CM (1996) Pest management at the crossroads. Consumers Union, Yonkers, 272 pp

Bokshi AI, Morris SC, Deverall BJ (2003) Effects of benzothiadiazole and acetylsalicylic acid on α-1,3-glucanase activity and disease resistance in potato. Plant Pathol 52:22–27

Browning JA, Frey KJ (1981) The multiline concept in theory and practice. In: Jenkyn JF, Plumb RT (eds) Strategies for the control of cereal disease. Blackwell Scientific, London, pp 37–36

Colson-Hanks ES, Allen SJ, Deverall BJ (2000) Effect of 2, 6- dichloroisonicotinic acid or benzothiadiazole on Alternaria leaf spot, bacterial blight and Verticillium wilt in cotton under field conditions. Austr Plant Pathol 29:170–177

Conrath U, Pieterse CMJ, Mauch-Mani B (2002) Priming in plant-pathogen interactions. Trends Plant Sci 7:210–216

Coons GH, Kotila JE (1925) The transmissible lytic principle (bacteriophage) in relation to plant pathogens. Phytopathology 15:357–370

Dann EK, Deverall BJ (2000) Activation of systemic disease resistance in pea by an avirulent bacterium or a benzothiadiazole, but not by a fungal leaf spot pathogen. Plant Pathol 49:324–332

Elmore CL, Roncaroni JA, Giraud DD (1993) Perennial weeds respond to control by soil solarization. Calif Agric 47(1):19–22

Görlach J, Volrath S, Knauf Beiter G, Hengy G, Beckhove U, Kogel KH, Oostendorp M, Staub T, Ward E, Kessman H, Ryals J (1996) Benzothiadiazole, a novel class of inducers of systemic acquired resistance, activates gene expression and disease resistance in wheat. Plant Cell 8:629–643

Gozzo F (2004) Systemic acquired resistance in crop protection. Outlook Pest Manage 2004:20–23

Jakab G, Cottier V, Touquin V et al (2001) Beta-aminobutyric acid- induced resistance in plants. Eur J Plant Pathol 107:29–37

Karban R, Baldwin IT (1997) Induced responses to herbivory. University of Chicago Press, London, 31 pp

Katan J, Greenburger A, Alon H, Grinstein A (1976) Solar heating by polyethylene mulching for control of diseases caused by soil-borne pathogens. Phytopathology 66:683–688

Kloepper JW, Schroth MN (1978) Plant growth-promoting rhizobacteria on radishes. In: Proceedings of the 4th international conference on plant pathogenic bacteria, vol. 2. Station de Pathologie Vegetale et Phytobacteriologie, INRA, Angers, France, pp 879–882

Kotila JE, Coons GH (1925) Investigations on the black leg disease of potato. Michigan Agric Expt Stn Tech Bull 67:3–29

Putter JG, Mac Connell FA, Preiser FA, Haidri AA, Rishich SS, Dybas RA (1981) Avermectins: novel class of insecticides, acaricides and nematicides from a soil microorganism. Experientia 37:963–964

Sasser JN, Freckman DW (1987) A world perspective on nematology: the role of the society. In: Vetch JA, Dickson DW (eds) Vistas in nematology. Society of Nematologists Inc., Hyattsville, pp 7–14

Soylu S, Baysal O, Soylu EM (2003) Induction of disease resistance by the plant activator, acibenzolar-S methyl (ASM) against bacterial canker (*Clavibacter michiganensis* sub sp *michiganensis*) in tomato seedlings. Plant Sci 165:1069–1076

Tally A, Oostendorp M, Lawton K, Staub T, Bassy B (1999) Commercial development of elicitors of induced resistance to pathogens. In: Agrawal AA, Tuzun S, Bent E (eds) Inducible plant defenses against pathogens and herbivores: biochemistry, ecology, and agriculture. American Phytopathological Society Press, St Paul, pp 357–369

Thomas RC (1935) A bacteriophage in relation to Stewart's disease of corn. Phytopathology 25:371–372

Wolfe MS (1985) The current status and prospects of multiline cultivars and variety mixtures for disease control. Annu Rev Phytopathol 23:251–273

2 Avermectins

Abstract

The avermectins are a new class of macrocyclic lactones derived from mycelia of the soil actinomycete, *Streptomyces avermitilis* (soil inhabiting which is ubiquitous in nature). These compounds were reported to be possessing insecticidal, acaricidal and nematicidal properties (Putter JG, Mac Connell FA, Preiser FA, Haidri AA, Rishich SS, Dybas RA. Experientia 37:963–964, 1981). They are commonly distributed in most of the cultivated soils and are in widespread use, especially as agents affecting plant-parasitic nematodes, mites and insect pests. The water solubility of avermectin B1 is approximately 6–8 ppb, and its leaching potential through many types of soil is extremely low. These physical properties also confer many advantages upon the use of avermectins as pesticides. Their rapid degradation in soil and poor leaching potential suggest that field applications would not result in persistent residues or contamination of ground water.

Avermectins offer an outstanding alternative to any of the available synthetic pesticides. Their novel mode of action, high potency and specific physico-chemical properties makes the avermectins excellent candidates for further insecticidal, acaricidal and nematicidal studies.

2.1 Distinguishing Characteristics of *Streptomyces avermitilis*

- Brownish-grey spore colour
- Smooth spore surface
- Spiral sporophore structure (Fig. 2.1)
- Spores in chain (Fig. 2.1)
- Production of melanoid pigments
- Cultural and carbon utilisation patterns
- Preferred temperatures of 28 and 37°C for growth

2.2 Chemical Structure of Avermectins

The avermectin complex contains four closely related major components – A1a, A2a, B1a and B2a – in varying proportions and four minor components – A1b, A2b, B1b and B2b (Fig. 2.2). Since these compounds are of biological origin, they have no residual problems and other environmental side effects.

P.P. Reddy, *Recent Advances in Crop Protection*,
DOI 10.1007/978-81-322-0723-8_2, © Springer India 2013

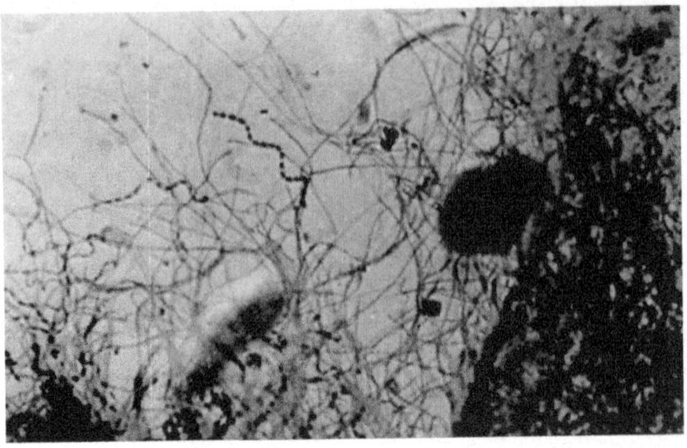

Fig. 2.1 Spiral sporophores of *Streptomyces avermitilis* and spores in chain

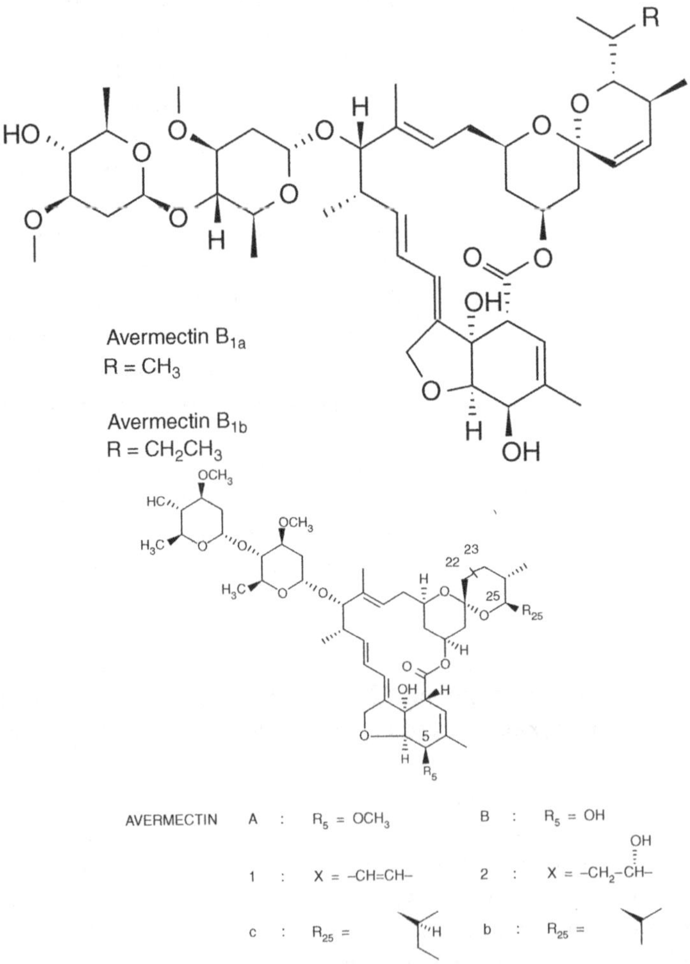

Fig. 2.2 Chemical structure of avermectins

2.3 Abamectin

Abamectin is a blend of B_{1a} and B_{1b} avermectins (Fig. 2.3). Avermectin B1 (abamectin), the major component of the fermentation, also showed potent activity against arthropods in preliminary laboratory evaluations and was subsequently selected for development to control phytophagous mites and insect pests on a variety of agricultural and horticultural crops worldwide. Major applications for which abamectin is currently registered include uses on ornamental plants, citrus, cotton, pears and vegetable crops at rates in the range of 5–27 g abamectin per hectare as a foliar spray. Abamectin has shown low toxicity to non-target beneficial arthropods which has accelerated its acceptance into integrated pest management programmes. Extensive studies have been conducted to support the safety of agricultural uses of abamectin to man and the environment. Abamectin is highly unstable to light and has been shown to photodegrade rapidly on plant and soil surfaces and in water following agricultural applications. Abamectin was also found to be degraded readily by soil microorganisms. Abamectin residues in or on crops are very low, typically less than 0.025 ppm, resulting in minimal exposure to man from harvesting or consumption of treated crops. In addition, abamectin does not persist or accumulate in the environment. Its instability as well as its low water solubility and tight binding to soil limits abamectin's bioavailability in non-target organisms and, furthermore, prevents it from leaching into groundwater or entering the aquatic environment.

Abamectin, the active ingredient in Avicta seed treatment nematicide, is composed of two molecules that are produced by the soil microorganism, *Streptomyces avermitilis*. Abamectin interferes with the signal transmission between nerve cells inside the nematode at a novel target site, the GABA receptor protein. Avicta is a contact nematicide, in that the active ingredient kills the nematode immediately upon contact and does not allow the nematode to feed or reproduce.

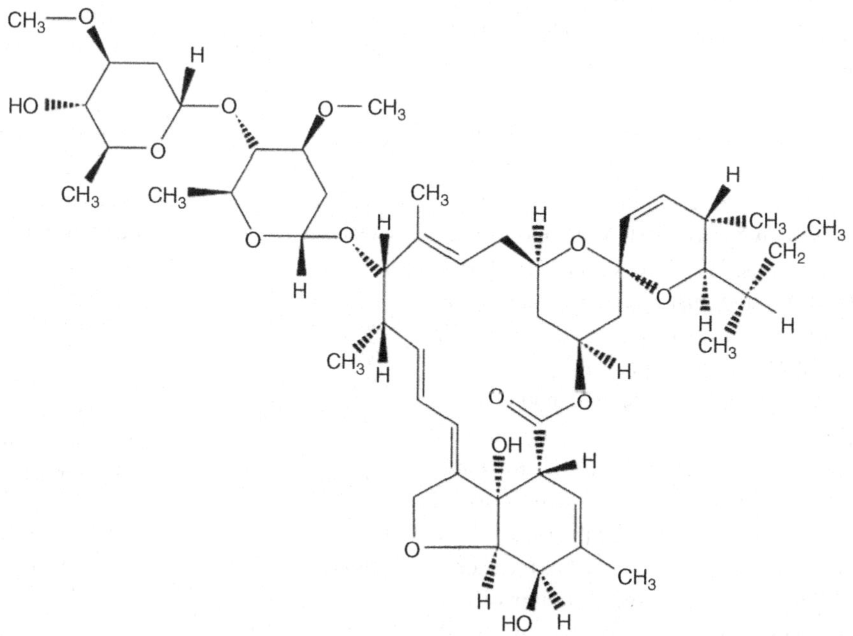

Fig. 2.3 Structure of abamectin (avermectin B_1: mixture of avermectins containing >80% avermectin B_{1a} and <20% avermectin B_{1b})

2.4 Mode of Action

2.4.1 GABA Antagonists

The avermectins affect invertebrates by potentiating the ability of neurotransmitters, such as glutamate and/or GABA, to stimulate and influx of chloride ions into nerve cells. The chloride ion flux produced by the opening of the channel into neurons results in loss of cell function and disruption of nerve impulses. Consequently, invertebrates are paralysed irreversibly and stop feeding. The avermectins do not exhibit rapid knock-down effect. The safety of avermectin to mammals is due to (i) the lack of glutamate-gated chloride channels in mammals, (ii) the low affinity of avermectins for other mammalian ligand-gated chloride channels and (iii) their inability to readily cross the blood–brain barrier.

The mode of action of avermectins has been studied on plant-parasitic nematodes in terms of their gross effects on the movement and infective behaviour of the parasites. Juveniles of *Meloidogyne incognita* exposed to a 120 nM aqueous solution of avermectin B2a-23-ketone (i) initially lost movement within 10 min whilst being responsive to touch, (ii) partially recovered within 30 min of exposure and (iii) irreversibly lost movement after 120 min (Wright et al. 1984a). A similar triphasic response was also seen in *M. incognita* juveniles exposed to avermectin B1. The initial loss of movement in *M. incognita* may be reflective of avermectin's activity as GABA antagonists as inhibitory synapses. Wright et al. (1984b) found that the GABA antagonist picrotoxin and bicuculline counteract the effects of avermectins on the locomotion of *M. incognita* juveniles.

Avermectins potentiate glutamate- and GABA-gated chloride-channel opening. A number of total syntheses of simpler analogues of the natural products were undertaken with the hope to get access to constructs containing the pharmacophore which would be biologically active but more economical to prepare. None of these attempts led to a compound with high activity. On the other hand, the search for derivatives of the natural products with improved biological properties turned out to be very successful.

2.5 Commercial Products (Tables 2.1, 2.2, 2.3, and 2.4)

The commercial products of abamectin are as follows:

Table 2.1 Commercial products of abamectin

Commercial products	Pests controlled	Application/ comments
Avid	Spider mites, leaf miners	Many beneficials can be released one week after use
Avicta	Root-knot and reniform nematodes on cotton and vegetables	Seed treatment
Agri-Mek	Apples and pears as an acaricide/insecticide against European red mite and pear psylla and spotted tentiform leaf miner	Foliar spray

Table 2.2 Suppliers and branded products of avermectins

Branded products	Suppliers
Akomectin	AAKO BV
Mectinide	AgriGuard Ltd.
Avermk	Agro-Care Chemical Industry Group Limited
Greyhound	ArborSystems
AbaMecK, Transact	Astra Industrial Complex Co. Ltd. (Astrachem)
Zoro	Cheminova A/S
Temprano	Chemtura AgroSolutions
Abacin	Crystal Phosphates Ltd.
Denka-Flylure	Denka International B.V.
Fertimectin	Fertiagro Pvt. Ltd.
Gilmectin	Gilmore Marketing & Development Inc.
Satin	Hubei Sanonda Co. Ltd.
Laotta	Lainco S.A.
Abba	Makhteshim Agan Group
Abba, Abba Ultra	MANA – Makhteshim Agan North America Inc.
Pilarmectin	Pilar AgriScience (Canada) Corp.
Romectin	Rotam CropSciences Inc.
Eagrow	Shandong Kesai Eagrow Co. Ltd.
Agri-Mek, Avicta, Avid, Clinch, Dynamec, Epi-Mek, Varsity, Vertimec, Zephyr	Syngenta
Alba	Wangs Crop-Science Co. Ltd.
Zabamec	Zagro Singapore Pte. Ltd.

Table 2.3 Formulators and trade names of avermectins

Trade name/s	Formulators
Artig	Agroquimicos Versa S.A. de C.V.
Cam-Mek	Cam for Agrochemicals
Bermectin	Chema Industries
Minx	Cleary Chemical LLC
Bentar, Crysabamet, Crysmectin	Dupocsa Protectores Quimicos para el Campo SA
Torpedo	Hektas Ticaret TAS
Eminence, Instar AD, Noflye	Ingenieria Industrial S.A. de C.V.
Inimectin	Insecticidas Internacionales, C.A.
Saddle	Ladda Co. Ltd.
Reaper	Loveland Products Inc.
Medamec	Medmac Agrochemicals
Nagmectin	Multiplex Fertilizers Pvt. Ltd.
Bermectine	Probelte, S.A.
Arvilmec	SAFA TARIM AS
ManChongGai	Shanghai Nong Le Biological Products Co. Ltd.
Wopro-abamectin	BV Industrie & Handelsonderneming Simonis
Frog	Sinochem Ningbo Chemicals Co. Ltd.
Ieungaechoong	SM-BT Co. Ltd.
Acaritina	Stockton Agrimor AG
Odin, Tinamex	TRAGUSA (Tratamientos Guadalquivir S.L.)
Arrow, Aviat, Biok	United Phosphorus Ltd.
Abamex, Maysamectin, Vapcomic	VAPCO, Veterinary and Agricultural Products Mfg Co. Ltd.
Virbamec	Vietnam Pesticide Joint Stock Co. (VIPESCO)
Abamine, Akirmactin, Armada, Bitech, Verkotin	Willowood Ltd.

Table 2.4 Premix products of avermectins

Premix products	Trade name/s	Suppliers
Abamectin + *beta*-cypermethrin	Awei Gaolv	Shanghai Agro-Chemical Industry Co. Ltd.
Abamectin + bifenazate	Sirocco	OHP Inc.
Abamectin + bifenthrin	Athena	FMC Corp.
Abamectin + chlorpyrifos	Paragon	BMC Group Co. Ltd.
	Slamfast	
	Vibafos	Vietnam Pesticide Joint Stock Co (VIPESCO)
	Vinc	Wangs Crop-Science Co. Ltd.
Abamectin + cyromazine	Locking	Wangs Crop-Science Co. Ltd.
Abamectin + imidacloprid	Extra Power	Wangs Crop-Science Co. Ltd.
Abamectin + propargite	Choice	Wangs Crop-Science Co. Ltd.
Abamectin + thiamethoxam	Acceleron INT-210	Monsanto Co
	Agri-Flex	Syngenta
	Avicta Duo Corn	Syngenta
	Avicta Duo Cotton	Syngenta
Abamectin + azoxystrobin + fludioxonil + metalaxyl-M + thiamethoxam	Avicta Complete Corn	Syngenta
Abamectin + azoxystrobin + thiamethoxam	Avicta Complete Cotton	Syngenta
Abamectin + fludioxonil + thiamethoxam	Avicta Complete Beans	Syngenta

2.6 Pest Management Using Avermectins

2.6.1 Insect Pests

Abamectin (Agri-Mek) is a natural fermentation product containing a macrocyclic glycoside, used on apples and pears as an insecticide. When used as currently recommended, it controls pear psylla and aids in the control of spotted tentiform leaf miner. Abamectin is toxic to bees and predator mites on contact, but the foliar residue dissipates quickly, making it essentially non-toxic to these species after a few hours (low bee-poisoning hazard).

Abamectin is also effective against dipterous (*Liriomyza*) and lepidopterous (Gracillanidae) leaf miners, lepidopterous tomato pinworm (*Keiferia lycopersicella*), citrus thrips (*Scirtothrips citri*), fire ants in turf and lepidopterous pear psylla (*Cacopsylla pyricola*).

A new class of insecticidal and antiparasitic agents, 4′-amino-4′-deoxy avermectins, has been developed by chemical modification of avermectin B1. The most effective of these compounds are 1,500-fold more potent than avermectin B1 (abamectin) against the beet army worm, *Spodoptera exigua,* and show similar potency against other lepidopteron larvae (Mrozik et al. 1989).

2.6.1.1 Potato Leaf Miner, *Liriomyza huidobrensis*

Abamectin applied once at recommended field rates, early in the potato-growing season significantly reduced leaf miner, *Liriomyza huidobrensis,* as compared to non-treated control. Eulophid parasitoid populations (*Diglyphus isaea*) from abamectin treatments were significantly reduced as compared to the non-treated control. However, parasitoid populations from abamectin-treated plots recovered sooner in treated plots (Weintraub 2001).

2.6.1.2 Chilli Thrips, *Scirtothrips dorsalis*

Spraying of Vertimec at 0.3–0.4 ml/l was found effective against thrips for 45 days.

2.6.1.3 Cabbage Diamondback Moth, *Plutella xylostella*

The percentage reduction in population and the percentage reduction in feeding damage (% activity) by *Plutella xylostella* was more than 80% in cabbage by avermectin B1.

2.6.1.4 Bean Leaf Miner, *Liriomyza huidobrensis*

The evaluation of abamectin applied alone or mixed with plant oil on leaf miner fly, *Liriomyza huidobrensis,* on bean plants revealed that the addition of plant oil to abamectin sprays to get a 1% oil spray concentration increased the effectiveness of the insecticide to the extent that the active ingredient of the insecticide could be reduced by one-half to three-fourths of the normal dosage (0.15%). Both eggs and fly larvae were affected. The synergistic effect shown by the mixture of abamectin and plant oil allows a reduction in the commercially recommended concentration of abamectin without any loss in effectiveness. As a result, treatment costs can be reduced by 60%, and more farmers will be able to use abamectin to control leaf miner fly (Mujica et al. 2000).

2.6.1.5 Rose Thrips, *Rhipiphorothrips cruentatus, Scirtothrips dorsalis*

Spraying Vertimec (biopesticide of *Streptomyces avermitilis*) was found effective against rose thrips.

2.6.1.6 Poinsettia Whitefly, *Trialeurodes vaporariorum*

In the greenhouse experiment, the combined treatment of abamectin and the parasitoid, *Encarsia formosa,* maintained significantly lower densities of greenhouse whiteflies (*Trialeurodes vaporariorum*) on poinsettia, throughout the season with fewer abamectin applications than did abamectin alone. Moreover, the percentage of parasitism did not differ significantly amongst plants treated with and without abamectin. Abamectin might be used to reduce whitefly numbers on poinsettia without eliminating the parasitoid population when releases of *E. formosa* are not satisfactory (Zchori-Fein et al. 1994).

2.6.1.7 Soybean *Spodoptera littoralis*

The percentage reduction in population and the percentage reduction in feeding damage (% activity) by *Spodoptera littoralis* was more than 80% in soybean by avermectin B1.

2.6.1.8 Maize *Spodoptera littoralis* and *Diabrotica balteata*

The percentage reduction in population of *Spodoptera littoralis* and *Diabrotica balteata* was more than 80% in maize by avermectin B1.

2.6.1.9 Cotton Whitefly, *Bemisia tabaci*

The residual toxicity of abamectin applied at 1 mg a.i./l to cotton seedlings, under laboratory conditions without exposure to sunlight, resulted in adult whitefly, *Bemisia tabaci,* mortalities that declined from 86 to 39% over 28 days. A mixture of 1 mg a.i./l abamectin with 0.5% mineral oil resulted in higher mortalities, 100 and 88% at 0 and 28 days after application, respectively. When the seedlings were placed outdoors daily for 3 h, a low mortality of 20% was obtained 2 days after application with 1 mg a.i./l abamectin. A mixture of 1 mg a.i./l abamectin with 0.5% mineral oil resulted in high adult mortality (93–100%) 20 days after treatment. In field trials, a mixture of 18 g a i/ha abamectin with 1% mineral oil decreased larval population levels throughout the experiment to a greater extent than the insecticide applied alone, resulting in 2.9 larvae per leaf at day 27 in the abamectin–oil mixture as compared with 9.6 and 14.6 larvae per leaf in the abamectin or mineral oil, respectively. Abamectin in combination with mineral oil is a potential agent for controlling *B. tabaci* and may be used in alternation with other effective novel compounds in insecticide resistance management strategies, especially when whiteflies and spider mites are both present in the field.

2.6.1.10 Wheat Storage Insect Pests

Avermectin B1 caused 100% mortality in parent adult *Sitophilus oryzae, Rhyzopertha dominica* and *Oryzaephilus surinamensis* exposed to a dose of 320 ppb on wheat. *Tribolium castaneum* was more tolerant; at a dose of 2.6 ppm, only 36% mortality occurred although at 160 ppb, the insects appeared sluggish.

Suppression of F1 progeny was achieved at doses of 10 ppb in *Sitotroga cerealella*, 20 ppb in *R. dominica*, 160 ppb in *S. oryzae* and *O. surinamensis* and 640 ppb in *Plodia interpunctella*. The half-life decay for avermectin B1 on wheat at 26.7°C and 60% RH was 3–6 months.

2.6.2 Mite Pests

Abamectin (Agri-Mek) is a natural fermentation product containing a macrocyclic glycoside, used on apples and pears as an acaricide. When used as currently recommended, it controls European red mite.

Abamectin penetrates leaf tissue and provides long-term (3–5 weeks) control of various mite species (Fig. 2.4) in field-grown roses and other ornamentals, strawberry, citrus, cotton and pears.

2.6.2.1 Chilli Yellow Mite, *Polyphagotarsonemus latus*

Spraying of Vertimec at 0.3–0.4 ml/l was found effective against yellow mites on chilli for 45 days.

2.6.2.2 Bean Spider Mite, *Tetranychus urticae*

The percentage reduction in population (% activity) by *Tetranychus urticae* was more than 80% in bean by avermectin B1.

2.6.2.3 Rose Red Spider Mite, *Tetranychus urticae*

Five applications of abamectin applied at a concentration of 12 ppm and at 3- to 5-days intervals as full canopy sprays provided effective control of *Tetranychus urticae* on greenhouse roses, but also eliminated a biological control agent, *Phytoseiulus persimilis*. However, abamectin applications confined to the upper portion of the canopy (from where the marketed portion of the crop is taken) provided chemical control of spider mites in the upper canopy. The

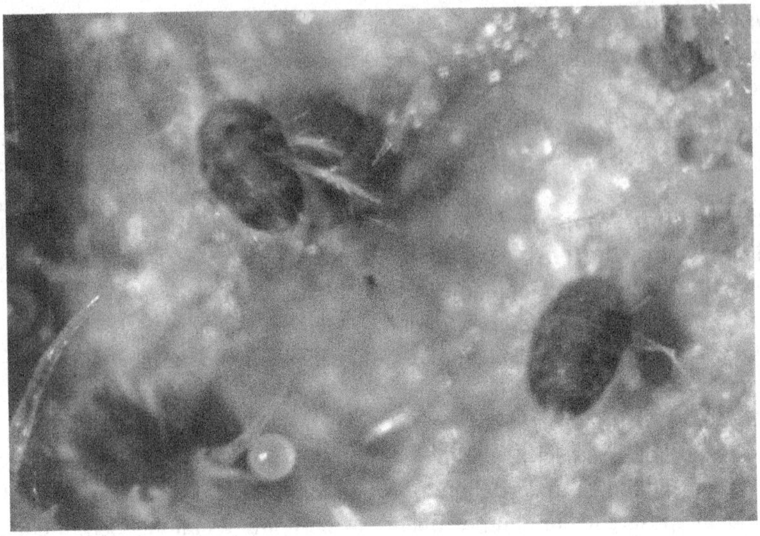

Fig. 2.4 Red spider mites on rose leaf

lower canopy supported a rapidly decreasing spider mite population and a slightly increasing population of *P. persimilis*. These results show the potential for integration of biological control of *T. urticae* with chemical control of mite and other pests of greenhouse roses (Sanderson and Zhang 1995).

New miticides like Vertimec at 0.0025% offer very good solution for the mite problem on rose (Jhansi Rani and Jagan Mohan 1997).

2.6.2.4 Carnation and Gerbera Red Spider Mite, *Tetranychus urticae*

In a polyhouse, application of avermectins at 0.001% achieved 92% mortality of red spider mites in carnation and gerbera (Parvatha Reddy and Nagesh 2002).

Recent acaricides like 0.0025% Vertimec or abamectin 1.9 EC at 0.5 ml/l provide highly effective control of the mite (Jhansi Rani and Jagan Mohan 1997).

2.6.3 Nematode Pests

The available literature on avermectins indicates that *Meloidogyne incognita*, *M. javanica*, *M. arenaria* and *Tylenchulus semipenetrans* were the commonly used test organisms to assess the nematicidal activity of these metabolites on tomato and citrus (Garabedian and Van Gundy 1983).

Whilst avermectins B1 and B2a have shown high toxicity against *M. incognita* in greenhouse conditions when incorporated into soil, avermectin B2a was found to be more potent than B1 (Putter et al. 1981).

In a microplot study, the toxicity of avermectin B1 at 0.17 kg a i/ha was equal to ethoprop, aldicarb, fenamiphos, oxamyl and carbofuran at 6.7 kg a i/ha (Nordmeyer and Dickson 1981).

Abamectin is being used as a seed treatment to control plant-parasitic nematodes (*M. incognita* and *Rotylenchulus reniformis*) on cotton and some vegetable crops. Using an assay of nematode mobility, LD_{50} values of 1.56 and 32.9 μg/ml were calculated based on 2 h exposure for *M. incognita* and *R. reniformis*, respectively. There was no recovery of either nematode after exposure for 1 h. Mortality of *M. incognita* continued to increase following a 1 h exposure, whereas *R. reniformis* mortality remained unchanged at 24 h after the nematodes were removed from the abamectin solution. Sublethal concentrations of 1.56–0.39 μg/ml for *M. incognita* and 32.9–8.2 μg/ml for *R. reniformis* reduced infectivity of each nematode on tomato roots. The toxicity of abamectin to these nematodes was comparable to that of aldicarb (Faske and Starr 2006).

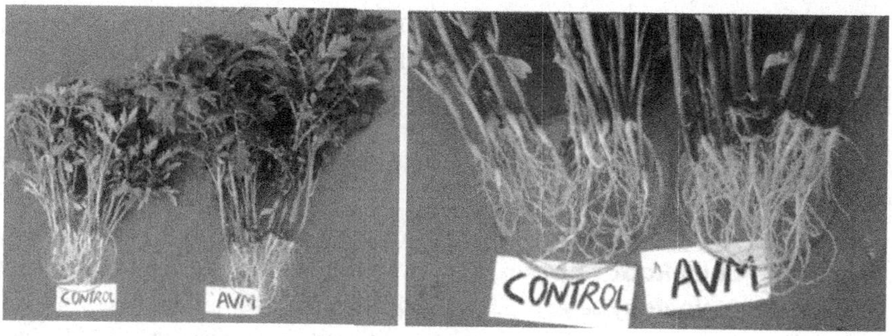

Fig. 2.5 Tomato seedlings from avermectin treated and untreated nursery beds

2.6.3.1 Banana Nematodes, *Meloidogyne javanica, Radopholus similis*

Injections (1 ml) of ≥100 μg a.i./plant of abamectin into banana (*Musa acuminata* cv. Cavendish) pseudostems were effective in controlling *M. javanica* and *R. similis* and were comparable to control achieved with a conventional chemical nematicide, fenamiphos, in a protectant assay. Abamectin injections of 250 and 500 μg a.i./plant were effective at reducing nematode infections 28–56 days after inoculation. Abamectin was more effective than ivermectin in controlling nematodes after their establishment in banana roots. Injections of between 100 and 1,000 μg a.i./plant were effective in controlling nematodes for at least 56 days after treatment. These studies demonstrated that abamectin has potential for controlling nematode parasites on banana when injected into the pseudostem (Jansson and Rabatin 1997).

2.6.3.2 Citrus Nematode, *Tylenchulus semipenetrans*

In citrus, a monthly rate of 1.1 kg a i/ha of avermectins for 7 months gave maximum increase in yield and reduction of *T. semipenetrans* population (Garabedian and Van Gundy 1983).

2.6.3.3 Tomato Root-Knot Nematode, *Meloidogyne incognita*

The per cent root infection by *M. incognita* juveniles in tomato seedlings raised in seed pans treated with aqueous solution of avermectins (100 ml of 0.001% solution w/v) was lowest as compared to carbofuran (2 g a.i./seed pan) or neem cake (250 g/seed pan) (Parvatha Reddy and Nagesh 2002).

The aqueous solution of avermectins (250 ml of 0.001%/m^2 nursery bed) significantly reduced root galls of tomato seedlings raised in root-knot-infested nursery beds as compared to carbofuran (10 g a i/m^2 nursery bed) or neem cake (1 kg/m^2 nursery bed). The treated tomato nursery beds yielded robust and healthy seedlings with no root-knot nematode infection (Fig. 2.5) (Parvatha Reddy and Nagesh 2002).

Talc and charcoal formulations of *S. avermitilis* at 100 g/m^2 nursery bed significantly controlled *M. incognita* infection in tomato and were superior to carbofuran. The above treatments also gave vigorous and root-knot-free tomato seedlings (Parvatha Reddy and Nagesh 2002).

In a root-knot-infested microplot study, avermectins at 0.015 kg/ha effectively controlled *M. incognita* on tomato and increased yield by 11 and 8% compared to the untreated and carbofuran-treated plots, respectively.

All sublethal concentrations greater than 0.39 μg abamectin/ml inhibited ($P<0.05$) infection of tomato roots by *M. incognita*. No reduction of root galls occurred with abamectin concentration less than 0.15 μg/ml (Faske and Starr 2006).

Avermectins B1 and B2a applied to soil through drip irrigation systems at 0.093–0.34 kg a i/ha applied as a single dose or 0.24 kg a i/ha applied as three doses each at 0.08 kg a i/ha on tomatoes against *M. incognita* were as effective as oxamyl and aldicarb at 3.36 kg a i/ha. There was no significant difference in efficacy between B1 and B2a (Garabedian and Van Gundy 1983).

2.6.3.4 Tomato Reniform Nematode, *Rotylenchulus reniformis*

Sublethal concentrations greater than 8.2 μg abamectin/ml lowered ($P<0.05$) the number of *R. reniformis* females observed per root (Faske and Starr 2006).

2.6.3.5 Brinjal and Chilli Root-Knot Nematode, *Meloidogyne incognita*

Avermectins effectively reduced root-knot nematode infection in chilli and brinjal under field conditions (Parvatha Reddy and Nagesh 2002).

2.6.3.6 Cucumber Root-Knot Nematode, *Meloidogyne incognita*

Under soil-free conditions, avermectin B2a-23-ketone reduced the invasion of cucumber roots by *M. incognita* larvae and their further development at concentrations much lower than were needed to immobilise the juveniles. Wright et al. (1984a) proposed that avermectins might affect the behavioural sequence preceding invasion, a mode of action also suggested for organophosphorus and carbamate nematicides.

2.6.3.7 Carnation and Gerbera Root-Knot Nematode, *Meloidogyne incognita*

Avermectins (0.001%) applied as post-plant treatment at 250 ml/m^2 at two intervals (6 and 12 months after planting) effectively controlled root-knot nematodes (*M. incognita*) in carnation and gerbera in commercial polyhouses (Parvatha Reddy and Nagesh 2002).

2.6.3.8 Garlic Stem and Bulb nematode, *Ditylenchus dipsaci*

Abamectin at 10–20 ppm as the 20-min hot dip (49°C) or as a 20-min cool dip (18°C) following a 20-min hot-water dip was highly effective in controlling *D. dipsaci* and was non-injurious to garlic seed cloves. This treatment was not as effective as a hot-water-formalin dip and was non-eradicative, but showed high efficacy on heavily infected seed cloves relative to non-treated controls. Abamectin was most effective as a cool dip. These abamectin cool-dip (following hot-water dip) treatments can be considered as effective alternatives to replace formalin as a dip additive for control of clove-borne *D. dipsaci* (Roberts and Matthews 1995).

2.6.3.9 Tobacco Root-Knot Nematode, *Meloidogyne incognita*

In a small-plot field trial, soil incorporation of the avermectins B1, B2a and B2a-23-ketone were equally effective in inhibiting root-gall development and reproduction of *M. incognita* on tobacco. At application rates ranging from 0.168 to 1.52 kg a i/ha, these avermectins were as effective as ethoprop and fenamiphos at 6.73 kg a i/ha (Sasser et al. 1982).

In another microplot study, the toxicity of avermectins B1a at 0.17 kg a i/ha was equal to ethoprop, aldicarb, fenamiphos, oxamyl and carbofuran at 6.7 kg a i/ha in significantly increasing yields of tobacco infested with *M. incognita* or *M. arenaria* (Nordmeyer and Dickson 1985).

2.6.3.10 Cotton Root-Knot Nematode, *Meloidogyne incognita*

In greenhouse tests 35 days after planting (DAP), plants from seed treated with abamectin were taller than plants from non-treated seed, and root galling severity and nematode reproduction were lower where treated seed was used. The number of second-stage juveniles that had entered the roots of plants from seed treated with 100 g abamectin/kg seed was lower during the first 14 DAP than with non-treated seed. In microplot tests, seed treatment with abamectin and soil application of aldicarb at 840 g/kg of soil reduced the numbers of juveniles penetrating seedling roots during the first 14 DAP compared to the non-treated seedlings. In field plots, population densities of *M. incognita* were lower 14 DAP in plots that received seed treated with abamectin at 100 g/kg seed than where aldicarb (5.6 kg/ha) was applied at planting (Monfort et al. 2006).

The number of galls caused by *M. incognita* race 3 in cotton was also reduced more than 80% with abamectin at 0.1 mg a i seed^{-1} (Cabrera et al. 2009).

Based on the position of initial root-gall formation along the developing taproot from 21 to 35 DAP, infection by *M. incognita* was reduced by abamectin seed treatment. Penetration of

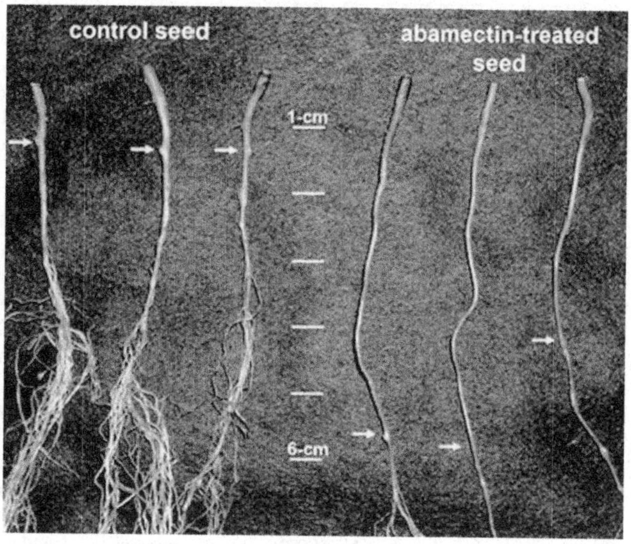

Fig. 2.6 Effect of abamectin seed treatment on the position of initial gall formation by *Meloidogyne incognita* on cotton taproots. To better visualise galling, secondary roots were removed 2 cm past the initial gall for both control and abamectin-treated seed

developing taproots by nematode species was suppressed at taproot length of 5 cm by abamectin-treated seed, but root penetration increased rapidly with taproot development. Based on an assay of nematode mobility to measure abamectin toxicity, the mortality of *M. incognita* associated with 2-day-old emerging cotton radical was lower than mortality associated with the seed coat, indicating that more abamectin was on the seed coat than on the radical. Thus, the limited protection of early stage root development suggested that only a small portion of abamectin applied to the seed was transferred to the developing root system (Fig. 2.6) (Faske and Starr 2007).

2.6.3.11 Sugar Beet Cyst Nematode, *Heterodera schachtii*

Penetration of *Heterodera schachtii* in sugar beets was reduced over 60% when sugar beet seeds were treated with abamectin at a concentration of 0.3 mg a i seed^{-1} (Cabrera et al. 2009).

2.6.3.12 Maize Lesion Nematode, *Pratylenchus zeae*

Penetration of *Pratylenchus zeae* was reduced more than 80% in maize with abamectin at a dose of 1.0 mg a i seed^{-1}.

2.7 Conclusions

Avermectins offer an outstanding alternative to any of the available synthetic pesticides as they showed excellent insecticidal, nematicidal and acaricidal action. Their novel mode of action, high potency and specific physicochemical properties make the avermectins excellent candidates for further insecticidal, acaricidal and nematicidal studies. Mobility of avermectins may be increased by formulating with suitable surfactants which compete for adsorption sites (Morton 1986) or by forming micelles around the avermectin molecules which will reduce adsorption. Some other approaches such as using slow release and encapsulated formulations (Morton 1986) can also be applied to advance research with the avermectins.

Viable and effective formulations need to be developed for wide-scale application. The talc, charcoal and coir pith formulations examined were effective for short-term/immediate use for pest control. Keeping the cost of commercial production, they can be recommended for high value and polyhouse-grown crops.

References

Cabrera JA, Kiewnick S, Grimm C, Dababat AA, Sikora RA (2009) Efficacy of abamectin seed treatment on *Pratylenchus zeae*, *Meloidogyne incognita* and *Heterodera schachtii*. J Plant Dis Prot 116:124–128

Faske TR, Starr JL (2006) Sensitivity of *Meloidogyne incognita* and *Rotylenchulus reniformis* to Abamectin. J Nematol 38:240–244

Faske TR, Starr JL (2007) Cotton root protection from plant-parasitic nematodes by abamectin-treated seed. J Nematol 39:27–30

Garabedian S, Van Gundy SD (1983) Use of avermectins for the control of *Meloidogyne incognita* on tomatoes. J Nematol 15:503–510

Jansson RK, Rabatin S (1997) Curative and residual efficacy of injection applications of avermectins for control of plant-parasitic nematodes on banana. J Nematol 29:695–702

Jhansi Rani B, Jagan Mohan N (1997) Pest management in ornamental crops. In: Yadav IS, Choudhary ML (eds) Progresssive floriculture. House of Sarpan, Bangalore, pp 169–181

Monfort WS, Kirkpatrick TL, Long DL, Rideout S (2006) Efficacy of a novel nematicidal seed treatment against *Meloidogyne incognita* on cotton. J Nematol 38:245–249

Morton HV (1986) Modification of proprietary chemicals for increasing efficacy. J Nematol 18:123–128

Mrozik H, Eskola P, Linn BO, Lusi A, Shih M, Tischler M, Waksmunski FS, Wyvratt MJ, Hilton NJ, Anderson TE, Babu JR, Dybas RA, Preiser FA, Fisher MH (1989) Discovery of novel avermectins with unprecedented insecticidal activity. Experientia 45:315–316

Mujica N, Pravatiner M, Cisneros F (2000) Effectiveness of abamectin and plant-oil mixtures on eggs and larvae of the leaf miner fly, *Liriomyza huidobrensis* Blanchard. CIP Prog Rep 1999–2000:161–166

Nordmeyer D, Dickson DW (1981) Effect of oximecarbamates and organophosphates and one avermectin on the oxygen uptake of three *Meloidogyne* spp. J Nematol 13:452–453

Nordmeyer D, Dickson DW (1985) Management of *Meloidogyne javanica, M. arenaria* and *M. incognita* on flue-cured tobacco with organophosphate, carbamate and avermectin nematicides. Plant Dis 69:67–69

Parvatha Reddy P, Nagesh M (2002) Avermectins: isolation, fermentation, preliminary characterization and screening for nematicidal activity, Technical Bulletin 17. Indian Institute of Horticulture Research, Bangalore, 28 pp

Putter JG, Mac Connell FA, Preiser FA, Haidri AA, Rishich SS, Dybas RA (1981) Avermectins: novel class of insecticides, acaricides and nematicides from a soil microorganism. Experientia 37:963–964

Roberts PA, Matthews WC (1995) Disinfection alternatives for control of *Ditylenchus dipsaci* in garlic seed cloves. J Nematol 27:448–456

Sanderson JP, Zhang ZQ (1995) Dispersion, sampling, and potential for integrated control of two spotted spider mite (Acari: Tetranychidae) on greenhouse roses. J Econ Entomol 88:343–351

Sasser JN, Kirkpatrick TL, Dybas RA (1982) Efficacy of avermectins for root-knot control in tobacco. Plant Dis 66:691–693

Weintraub PG (2001) Effects of cyromazine and abamectin on the pea leaf miner *Liriomyza huidobrensis* (Diptera: Agromyzidae) and its parasitoid *Diglyphus isaea* (Hymenoptera: Eulophidae) in potatoes. Crop Prot 3:207–213

Wright DJ, Birtle AJ, Corps AE, Dybas RA (1984a) Efficacy of avermectins against a plant parasitic nematode *Meloidogyne incognita*. Ann Appl Biol 103:465–470

Wright DJ, Birtle AJ, Roberts TJ (1984b) Triphasic locomotor response of a plant-parasitic nematode to avermectin: inhibition by the GABA antagonists bicuculline and picrotoxin. Parasitology 88:375–382

Zchori-Fein E, Roush RT, Sanderson JP (1994) Potential for integration of biological and chemical control of greenhouse whitefly (Homoptera: Aleyrodidae) using *Encarsia formosa* (Hymenoptera: Aphelinidae) and abamectin. Environ Entomol 23:1277–1282

Bacteriophages

3

Abstract

Bacteriophage-based control of bacterial plant diseases is a fast developing field of research. A wide range of strategies (prevention of development of phage-resistant mutants, proper selection of efficient phages, timing of phage application, maximising chances for interaction between phage and target bacterium, overcoming adverse factors in phyllosphere on phage persistence, development of solar protectants to increase phase bioefficacy and delivery of phages in the presence of phage-sensitive bacterium) have been utilised to increase control efficacy. Phages are utilised as a component in developing integrated disease management strategies along with SAR inducers, avirulent strains and copper-mancozeb. Phage treatment is presently used in greenhouse and fields in Florida, USA, as a part of standard integrated management programme for tomato bacterial spot control (Momol et al., Integrated management of bacterial spot on tomato in Florida. EDIS, Institute of Food and Agricultural Sciences, University of Florida, Sept 2002, Report no. 110, 2002). Owing to their increasing efficacy and contribution to sustainable agriculture, phage-based products are likely to gain a bigger share in the bactericide market in the future.

3.1 What Is a Bacteriophage?

A bacteriophage (from 'bacteria' and Greek *phagein* 'to eat') is any one of a number of viruses that infect bacteria. Bacteriophages are amongst the most common biological entities on Earth. The term is commonly used in its shortened form, phage.

Typically, bacteriophages consist of an outer protein capsid enclosing genetic material. The genetic material can be ssRNA, dsRNA, ssDNA or dsDNA ('ss-' or 'ds-' prefix denotes single strand or double strand) along with either circular or linear arrangement (Figs. 3.1 and 3.2).

Phages are estimated to be the most widely distributed and diverse entities in the biosphere. Phages are ubiquitous and can be found in all reservoirs populated by bacterial hosts, such as soil or the intestines of animals. One of the densest natural sources for phages and other viruses is seawater, where up to 9×10^8 virions per millilitre have been found in microbial mats at the surface and up to 70% of marine bacteria may be infected by phages. They have been used for over 90 years

P.P. Reddy, *Recent Advances in Crop Protection*,
DOI 10.1007/978-81-322-0723-8_3, © Springer India 2013

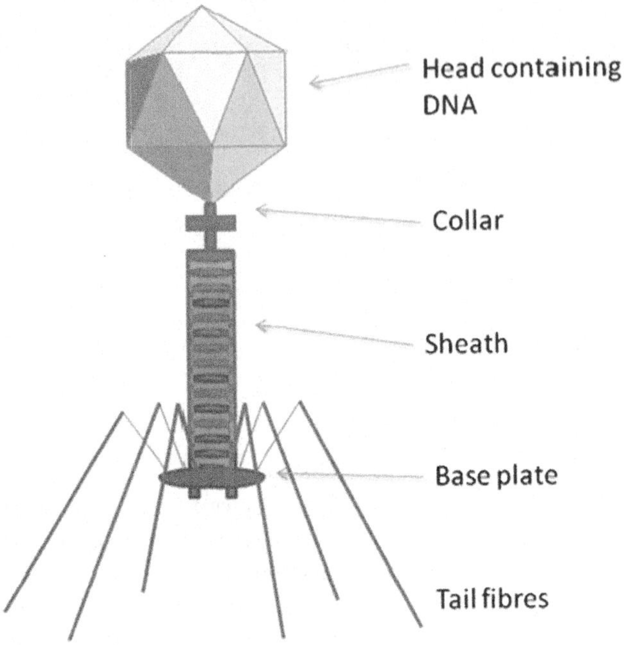

Fig. 3.1 The structure of a typical bacteriophage

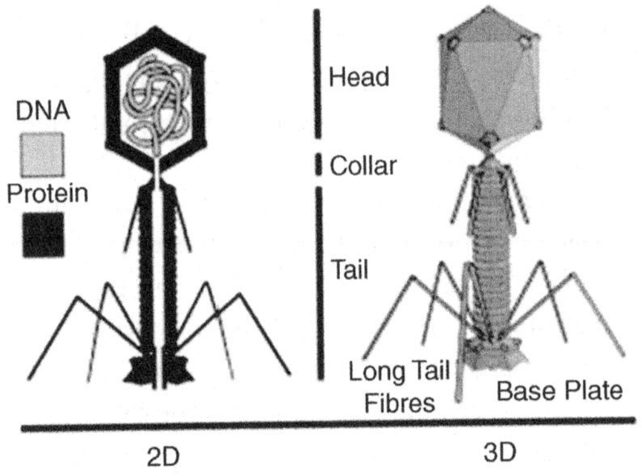

Fig. 3.2 Diagram of a typical tailed bacteriophage structure

as an alternative to antibiotics in the former Soviet Union and Eastern Europe as well as France. They are seen as a possible therapy against multidrug-resistant strains of many bacteria.

The most striking form of phage infection is that in which all of the infected bacteria are destroyed in the process of the formation of new phage particles. This results in the clearing of a turbid liquid culture as the infected cells lyse. When lysis occurs in cells fixed as a lawn of bacteria growing on a solid medium, it produces holes, or areas of clearing, called plaques. These represent colonies of bacteriophage. The size and other properties of the plaque vary with individual viruses and host cells.

3.2 Control of Bacterial Plant Diseases

Diseases incited by bacterial plant pathogens are responsible for major economic losses to agricultural production. Disease control is challenging for many diseases incited by bacteria. Chemical control with bactericides has been extremely difficult because few effective bactericides are available. Copper has been used more extensively than any other chemical for control of bacterial plant diseases. However, copper resistance is present in many plant pathogenic bacteria and is associated with plasmid-borne and chromosomal resistance. Antibiotics (streptomycin) have also been used as part of a management strategy for various bacterial diseases (Thayer and Stall 1961). As a result of extensive use, streptomycin-resistant strains became prevalent, which resulted in reduced disease control efficacy of bacterial spot of tomato and bell pepper (Thayer and Stall 1961), fire blight of apple and pear (Manulis et al. 1998) and many other bacterial plant pathogens. More recently, systemic acquired resistance (SAR) plant inducers have been used and have shown some success against bacterial diseases of tomato and bell pepper, *Xanthomonas* leaf blight of onion and fire blight of apple. These inducers also may have some negative effects on yield in certain plant species. Plant inducers have been ineffective for disease control in some pathosystems.

3.3 Biological Control

Biological control has gained recent interest for controlling bacterial plant diseases. Various strategies for using biological control for bacterial diseases include the use of non-pathogenic or pathogenically attenuated strains of the pathogen, saprophytic bacteria and plant growth-promoting rhizobacteria (PGPR) to suppress pathogen populations or induce a reaction in the plant such that the pathogen is reduced in its ability to colonise the plant and cause disease. Disease control using these approaches has been variable.

3.4 Early Use of Phages in Agriculture

Soon after the first medical (Summers 2005) and veterinary (d'Herelle 1921) applications, phages were evaluated for the control of plant diseases. Mallman and Hemstreet (1924) isolated the 'cabbage-rot organism', *Xanthomonas campestris* pv. *campestris*, from rotting cabbage and demonstrated that the filtrate of the liquid collected from the decomposed cabbage inhibited in vitro growth of the pathogen. The following year, Kotila and Coons (1925) isolated bacteriophages from soil samples that were active against the causal agent of blackleg disease of potato, *Erwinia carotovora* subsp. *atroseptica*. They demonstrated in growth chamber experiments that co-inoculation of *E. carotovora* subsp. *atroseptica* with phage successfully inhibited the pathogen and prevented rotting of tubers (Kotila and Coons 1925). These workers also isolated phages against *E. carotovora* subsp. *carotovora* and *Agrobacterium tumefaciens* from various sources such as soil, rotting carrots and river water (Coons and Kotila 1925). Thomas (1935) treated corn seeds that were infected with *Pantoea stewartii*, the causal agent of Stewart's wilt of corn, with bacteriophage isolated from diseased plant material. The seed treatment reduced disease incidence from 18 to 1.4%.

Despite the promising early work, phage therapy did not prove to be a reliable and effective means of controlling phytobacteria. Several workers questioned if positive results were possible. In 1963, Okabe stated, 'in general, the phage seems to be ineffective for [controlling] the disease development' (Okabe and Goto 1963). Three decades later, Goto (1992) concluded that 'practical use of phages for control of bacterial plant disease in the field has not been successful'. Chemical control with antibiotics and copper compounds became the standard for controlling bacterial plant diseases (Duckworth and Gulig 2002; Marco and Stall 1983).

3.5 Advantages and Disadvantages About Phage Therapy

3.5.1 Advantages

Greer (2005) and Kutter (1997) identified the several advantages of using phages for disease control.

- Phages are self-replicating and self-limiting; they replicate only as long as the host bacterium is present in the environment but are quickly degraded in its absence (Kutter 1997).
- Phages are natural components of the biosphere; they can readily be isolated from wherever bacteria are present, including soil, water, plants, animals (Adams 1959; Goyal et al. 1987; Woods et al. 1981) and the human body (Osawa et al. 1981).
- Phages could be targeted against bacterial receptors that are essential for pathogenesis, so resistant mutants would be attenuated in virulence (Kutter 1997).

- Phages are non-toxic to the eukaryotic cell (Greer 2005). Thus, they can be used in situations where chemical control is not allowed due to legal regulations, such as for treatment of peach fruit before harvest (Zaccardelli et al. 1992) or for control of human pathogens in fresh-cut produce (Leverentz et al. 2001, 2003).
- Phages are specific or highly discriminatory, eliminating only target bacteria without damaging other possibly beneficial members of the indigenous flora. Thus, their use can also be coupled with the application of antagonistic bacteria for increased pressure on the pathogen (Tanaka et al. 1990), or they can be used to promote a desired strain against other members of the indigenous flora (Basit et al. 1992).
- Phage preparations are fairly easy and inexpensive to produce and can be stored at 4°C in complete darkness for months without significant reduction in titer (Greer 2005). Application can be carried out with standard farm equipment, and since phages are not inhibited by the majority of agrochemicals (Balogh et al. 2004; Zaccardelli et al. 1992), they can be tank mixed with them without significant loss in titer. Copper-containing bactericides have been shown to inactivate phages (Balogh et al. 2004; Alvarez et al. 1991), and inhibition was eliminated if phages were applied at least 3 days after copper (Iriarte et al. 2007).

3.5.2 Disadvantages

A number of disadvantages and concerns have been raised in relation to phage therapy (Greer 2005; Kutter 1997; Vidaver 1976); moreover, additional problems specific to agricultural applications have surfaced.

- Limited host range can be a disadvantage, as often there is diversity in phage types of the target bacterium (Greer 2005). Several approaches have been tried for addressing this problem: using broad host-range phages (Saccardi et al. 1993; Svircev et al. 2006), using host-range mutant phages (Flaherty et al. 2000, 2001; Obradovic et al. 2004), applying phages in mixtures (Flaherty et al. 2000) or even to breed them (Hibma et al. 1997).
- The requirement of threshold numbers of bacteria (10^4–10^6 cfu/ml) may limit the impact of phages (Wiggins and Alexander 1985).
- Emergence of phage-resistant mutants can render phage treatment ineffective. However, using mixtures of phages that utilise distinct cell receptors can suppress the emergence of resistance (Tanji et al. 2004). Also, phage resistance often comes at some metabolic cost to the bacteria. Loss of virulence was observed with phage-resistant mutants of *Ralstonia solanacearum* (Hendrick and Sequeira 1984), *Xanthomonas campestris* pv. *pruni* (Randhawa and Civerolo 1986) and *Pantoea stewartii* (Thomas 1935).
- Environmental effects, such as temperature, pH and physiology of bacterium can hinder control. Civerolo and Maas Geesteranus (1972) observed that *Xanthomonas phaseoli* phages attacked *Xanthomonas* and *Pseudomonas* species only at temperatures above 20°C. Vidaver and Schuster (1960) suggested that *P. syringae* and *P. phaseolicola*, causal agents of halo blight and brown spot of bean, may be more prevalent below 22°C because of phage resistance. Leverentz et al. (2003) noted that phage treatment caused a significant population reduction of the *Listeria monocytogenes* on melons but not on apples, because phages were unstable on apple slices, possibly due to low pH (4.37 in apple vs. 5.77 in melon) (Leverentz et al. 2003).
- Unavailability of target organism can hinder control. Plant pathogenic bacteria often occur in non-homogeneous masses surrounded by extracellular polysaccharides that protect them from phage attachment (Goto 1992; Okabe and Goto 1963) or reside in protected spaces on the surface or inside the plant and unavailable for the control agents (Civerolo and Maas Geesteranus 1972).
- There is a concern that phages have the potential of transducing undesirable characteristics, such as virulence factors, between bacteria (Vidaver 1976).

- Lysogenic conversion, alteration of phenotypic characteristics of lysogenised bacteria by their prophages, has been found to have undesirable consequences, such as resistance to bacteriophages, toxin production or even increased virulence. When *Xanthomonas axonopodis* pv. *citri* strain XCJ19 was lysogenised with temperate phage PXC7, it became resistant to phage CP2 (Wu 1972). Phage-associated toxin production has not been documented amongst phytobacteria, but such cases are known amongst human pathogens (Wagner and Waldor 2002) and in bacteria of plant-associated nematodes (Ophel et al. 1993). Goto (1992) reported that *Xanthomonas campestris* pv. *oryzae* strains lysogenised by phages Xf or Xf-2 became more virulent on rice.
- Consumer perception of adding viruses to food products also could become an issue (Greer 2005).
- Vidaver (1976) raised the concern that 'transducing phages can introduce active prokaryotic genes into plant and animal cells'.
- Despite the generally narrow host ranges of phages, negative side effects due to inhibition of beneficial bacteria are possible. Examples for negative phage impact in agriculture include studies in which the phage-incited reduction in symbiotic nitrogen-fixing bacteria reduced growth and nitrogen content of cowpea (Ahmad and Morgan 1994) and in which biocontrol ability of *Pseudomonas fluorescens* was abolished by a lytic bacteriophage (Keel et al. 2002).

3.6 Return of Phage-Based Disease Management

Several factors have contributed to the re-evaluation of phage therapy for plant disease control. The use of antibiotics has been largely discontinued in agriculture due to the emergence of antibiotic-resistant bacteria in the field (Manulis et al. 1998; Minsavage et al. 1990; Thayer and Stall 1961) and because of concerns of possible transfer of antibiotic resistance from plant pathogens to human pathogens. The feasibility of reliance on copper compounds is questioned because of the emergence of copper-tolerant strains amongst phytobacteria (Marco and Stall 1983; Voloudakis et al. 2005), phytotoxicity caused by ionic copper (Momol et al. 2002; Stein et al. 2005) and soil contamination from extended heavy use (Koller 1998). Additionally, concerns about food safety and environmental protection and the goal of achieving sustainable agriculture necessitated development of safer, more specific and environment-friendly pesticides (Reichelderfer and Barry 1995). These factors, together with the expanding knowledge base about phage application in medicine (Barrow 2001; Duckworth and Gulig 2002; Kutter 1997), led to renewed interest in bacteriophage-based disease control in modern agriculture.

3.7 Recent Approaches for Using Phages on Bacterial Diseases

Since early 1990s, various approaches were attempted to improve the competitive advantage of phages in the environment in order to improve their efficacy including prevention of development of phage-resistant mutants, proper selection of efficient phages, timing of phage application, maximising chances for interaction between phage and target bacterium, overcoming adverse factors in phyllosphere on phage persistence, development of solar protectants to increase phase bioefficacy and delivery of phages in the presence of phage-sensitive bacterium.

3.7.1 Prevention of Development of Phage-Resistant Mutants

A major factor limiting the use of phages for control of plant diseases was the probability of developing bacterial strains resistant to the phage. Jackson (1989) developed a strategy to prevent occurrence of phage-resistant mutants. This involved preparing mixtures of host-range mutant phages (h-mutants) that lyse bacterial strains that are resistant to the parent phage, whilst maintaining the ability to lyse the wild-type bacterium. Using this strategy, a mixture of four phages including wild-type and h-mutant phages were applied twice weekly and provided significantly better tomato bacterial spot disease control and produced greater yield of extra-large fruits than the standard copper-mancozeb (Flaherty et al. 2000).

3.7.2 Proper Assay for Efficient Phage Selection

An important and often neglected aspect of phage-based biological control is identifying specific phages with particular characteristics that may prove effective in control rather than arbitrarily selecting them based strictly on lytic activity for disease control. The proper assay for phage selection is critical. Saccardi and co-workers selected from a collection of eight phages a lytic phage with the broadest host range to use in studies to control bacterial spot of peaches caused by *Xanthomonas campestris* pv. *pruni* (Saccardi et al. 1993). Balogh (2006) found no correlation between in vitro characteristics, such as antibacterial activity or phage multiplication rate, and disease control efficacy. On the other hand, he found that phages which multiplied more efficiently on their host phyllosphere were also better in disease control.

3.7.3 Timing of Phage Application

Timing of phage applications relative to the arrival of the pathogen influenced efficacy of disease control. Civerolo and Keil (1969) achieved marked reduction of peach bacterial spot only if phage treatment was applied 1 h or 1 day before inoculation of the pathogen (*X campestris* pv. *pruni*). There was a slight disease reduction when phage was applied 1 h after inoculation and no effect if applied 1 day later. Civerolo and Maas Geesteranus (1972) suggested that bacteria were inaccessible to phage in intercellular spaces, or there were not enough phages reaching the pathogen. Schnabel et al. (1999) achieved a significant reduction of fire blight on apple blossoms when the phage mixture was applied at the same time as the pathogen, *Erwinia amylovora*. In contrast, disease reduction was not significant when phages were applied a day before inoculation. On cabbage, significant reduction of black rot disease (*Xanthomonas campestris* pv. *campestris*) was achieved if the phage treatment was applied 3 days before to 1 day after inoculation, whereas on bell pepper from 3 days before to the day of inoculation of *X. campestris* pv. *vesicatoria* (bacterial spot) (Bergamin Filho and Kimati 1981). The greatest disease reduction occurred with application of phages on the same day of inoculation in both pathosystems.

3.7.4 Maximising Chances for Interaction Between Phage and Target Bacterium

Maximising chances for an interaction between phage and target bacterium is of critical importance. In the case of phage therapy, there is a need for high populations of both phage and bacterium, in order to start the 'chain reaction' of bacterial lysis. Therefore, a threshold concentration should exist above which phages provide good control regardless of the applied concentration, but below those threshold phages will not exert a pronounced effect on the bacterial population. Balogh (2002) found that a phage mixture provided similar levels of control of tomato bacterial spot if applied at 10^6 or 10^8 PFU/ml concentrations but was ineffective at 10^4 PFU/ml.

3.7.5 Overcoming Adverse Factors in Phyllosphere on Phage Persistence

Phages encounter a number of adverse factors in the phyllosphere (sunlight irradiation, especially UVA and UVB regions; desiccation and exposure to certain chemical pesticides, such as copper compounds), which substantially reduce their persistence, rapidly diminishing their populations under the threshold level, thus reducing their residual activity. Additionally, phase persistence varies depending on the ambient temperature. Under field conditions, sunlight irradiation is the single major factor hindering phage persistence (Iriarte et al. 2007). Phage populations declined sharply during the early afternoon hours but persisted at much higher levels when applied in the early evening and were highly correlated with the intensity of sunlight UV irradiation.

3.7.6 Development of Solar Protectants to Increase Phase Bioefficacy

Several strategies have been evaluated for increasing phage persistence, including the use of protective formulations, application scheduling for sunlight avoidance and co-application of bacterial hosts for in vivo phage propagation. Development of solar protectants for increasing biocontrol efficacy of phages has been the focus of considerable research. Balogh (2002) identified compounds that, when mixed with phage, extended the ability of phage to survive on leaf surfaces. Furthermore, Balogh et al. (2003) enhanced the efficacy of phage treatment with protective formulations that increased phage persistence on tomato foliage. These formulations enhanced the phages' ability to persist in the presence of UV and fluorescent light (Iriarte et al. 2007). The use of these formulations led to increased phage residual activity and, consequently, enhanced disease control efficacy (Balogh et al. 2003). Obradovic et al. (2004) determined that application of formulated phages resulted in reduced bacterial spot disease and increased tomato yield.

Flaherty et al. (2000) effectively controlled the tomato bacterial spot pathogen in field experiments by applying phages in the early morning hours prior to sunrise and no control if phages were applied during the day. Iriarte et al. (2007) showed that phages persisted better if applied in the evening rather than in the morning. Balogh et al. (2003) more definitively demonstrated that sunlight avoidance during phage application led to increased control by showing that phages applied to tomato plants in the field in the evening significantly reduced disease compared to morning applications, resulting in 26.9 and 13.1% disease reduction, respectively.

3.7.7 Delivery of Phages in the Presence of Phage-Sensitive Bacterium

The ability to increase phage numbers can be exploited if phages are applied into an environment where a phage-sensitive bacterium is present, or, alternatively, they are delivered together with the host. In an environment where high host populations are present and conditions are favourable, phages persist much better than without the host. Tanaka et al. (1990) used an avirulent strain of *Ralstonia solanacearum* and its phage that was active against both the virulent and avirulent strains to reduce tobacco bacterial wilt incited by *R. solanacearum*. Whilst the application of avirulent strain alone caused a significant 59% reduction in the number of wilted plants, the co-application of phage with the avirulent strain increased control significantly to 82%. When phages selected based on the ability to lyse both the target organism, the pathogen *Erwinia amylovora*, and also a closely related antagonistic phyllosphere bacterium, *Pantoea agglomerans* together with the phage, *P. agglomerans* served as biological control agent as well as phage carrier, a vehicle of delivery and medium of propagation on the leaf surface (Svircev et al. 2006). Whilst *P. agglomerans* alone significantly reduced disease, its combination with phage resulted in significantly better disease control which was comparable to streptomycin treatment.

3.8 Disease Management Using Phages

3.8.1 Bacterial Spot of Peach, *Xanthomonas campestris* pv. *pruni*

There has been considerable amount of work on the use of phages for control of bacterial spot of peach, caused by *Xanthomonas campestris* pv.

pruni. Civerolo and Keil (1969) reduced bacterial spot severity on peach leaves under greenhouse conditions with a single application of a single-phage suspension. Zaccardelli et al. (1992) isolated eight phages active against the pathogen, screened them for host range and lytic ability and selected a lytic phage with the broadest host range for disease control trials. Biweekly spray applications of the phage suspension in producing orchards significantly reduced bacterial spot incidence on fruits (Saccardi et al. 1993).

3.8.2 Fire Blight Pathogen of Apple, Pear and Raspberry, *Erwinia amylovora*

Control of *Erwinia amylovora*, the fire blight pathogen of apple, pear and raspberry, with bacteriophages was investigated in Canada and the USA. Schnabel et al. (1999) used a mixture of three phages for controlling fire blight on apple blossoms and achieved significant (37%) disease reduction. Gill et al. (2003) isolated 47 phages capable of lysing *E amylovora* and categorised them based on plaque morphology and host range. Later, the phages were evaluated for disease control ability in pear blossom bioassays, and the ones with broad host ranges and best disease control ability were selected for subsequent orchard trials (Svircev et al. 2006). *Pantoea agglomerans*, a bacterial antagonist that was also sensitive to the phages, was used to deliver and propagate them on the leaf surface. Disease control comparable to streptomycin was achieved (Svircev et al. 2006).

3.8.3 Citrus Canker, *Xanthomonas axonopodis* pv. *citri*

Bacteriophages reduced citrus canker disease (*Xanthomonas axonopodis* pv. *citri*) severity both in greenhouse and field trials (Balogh 2006). The level of control was inferior to chemical control with copper bactericides. The combination of bacteriophage and copper treatments did not result in increased control. The efficacy of phage treatment on a similar bacterial citrus disease, citrus bacterial spot, incited by *X. axonopodis* pv. *citrumelo* was evaluated. Application of a mixture of three phages applied twice weekly in skim milk formulation contributed to a significant bacterial spot disease (*X. axonopodis* pv. *citrumelo*) reduction in Valencia orange (moderately sensitive to bacterial spot disease) in a commercial citrus nursery in Florida. Some disease suppression also occurred on grapefruit (highly susceptible to the disease), but it was not significant ($p=0.1585$) (Balogh 2006). In summary, bacteriophages show significant promise as part of an integrated management strategy for controlling citrus canker.

The application of a mixture of four bacteriophages (CP2, CP31, ccΦ7 and ccΦ13-2) without skim milk significantly reduced grapefruit citrus canker (*X. axonopodis* pv. *citri* strain Xac65) disease severity. No phages could be recovered 2 days after application from plants if they were applied without skim milk formulation, whereas if applied with the formulation, phage populations ranged from 10^4 to 10^7 PFU/ml. However, interestingly, formulated phage treatment did not decrease disease severity (Balogh 2006).

3.8.4 Tomato Bacterial Spot, *Xanthomonas campestris* pv. *vesicatoria*

There has been extensive research on suppressing tomato bacterial spot with phage. Flaherty et al. (2000) effectively controlled the disease in greenhouse and field experiments with a mixture of four host-range mutant phages active against the two predominant races of the pathogen, *X. campestris* pv. *vesicatoria*. Balogh et al. (2003) enhanced the efficacy of phage treatment with protective formulations that increased phage persistence on tomato foliage. Obradovic et al. (2004, 2005) used formulated phages in combination with other biological control agents and systemic acquired resistance inducers, as a part of integrated disease management approach. Phage-based integrated management of tomato bacterial spot is now officially recommended to tomato growers in Florida (Momol et al. 2002), and bacteriophage mixtures against the pathogen are commercially available (Agriphage from OmniLytics Inc., Salt Lake City, UT, EPA Registration # 67986-1).

3.8.5 Tobacco Bacterial Wilt, *Ralstonia solanacearum*

Tanaka et al. (1990) treated tobacco bacterial wilt, caused by *Ralstonia solanacearum*, by co-application of an antagonistic avirulent *R. solanacearum* strain and a bacteriophage that was active against both the pathogen and the antagonist. The avirulent strain alone reduced the ratio of wilted plants from 95.8 to 39.5%, whereas the co-application of the avirulent strain and the phage resulted in 17.6% wilted plants.

3.8.6 Other Diseases

Other important work includes the reduction of incidence of bacterial blight of geranium with foliar applications of a mixture of host-range mutant phages (Flaherty et al. 2001), disinfection of *Streptomyces scabies*-infected seed potatoes using a wide host-range phage (McKenna et al. 2001) and a reduction in the loss of cultivated mushrooms caused by bacterial blotch with phage applications (Munsch and Olivier 1995; Munsch et al. 1991).

3.9 Phages in Integrated Disease Management Strategy

Tanaka et al. (1990) reduced tobacco bacterial wilt by co-application of an avirulent strain of *R. solanacearum* with a phage that was active against both the virulent and avirulent strains. Svircev et al. (2006) reduced fire blight of pear with co-application of an antagonistic epiphyte, *Pantoea agglomerans*, and a phage that lysed both the antagonist and the pathogen, *Erwinia amylovora*.

Obradovic et al. (2004) combined phage treatment with SAR inducers, PGPR and antagonistic bacteria to control bacterial spot of tomato. They achieved better and more reliable disease control when combining phages with SAR inducers. However, integration with bacterial biocontrol agents did not improve control efficacy, as compared to phage alone.

Phage treatment in combination with acibenzolar-S-methyl (SAR inducer) or with copper-mancozeb resulted in enhanced control of *Xanthomonas* leaf blight of onion (Lang et al. 2007).

3.10 Other Uses of Phages in Plant Pathology

Bacteriophages still remained in use in plant pathology and have been used as tools for detection, identification, classification and enumeration of pathogenic bacteria and were also used for disease forecasting. Phage typing, as a method of differentiating different races or pathovars of the same bacterial species, became a standard method in plant epidemiological studies (Klement et al. 1990). Phages CP1 and CP2 of *Xanthomonas axonopodis* pv. *citri*, the causal agent of citrus canker, were used for species-specific identification and classification of strains of the pathogen in Japan (Goto 1992). These two phages in combination with phage CP3 were used for differentiating worldwide strains causing citrus canker (Goto et al. 1980). Wu et al. (1993) used phages CP115 and CP122 for identification of *X. axonopodis* pv. *citri* strains in Taiwan. Phages were used for detection of the host bacterium from crude samples and seed lots (Katznelson and Sutton 1951) and directly from lesions on the plant foliage (Okabe and Goto 1963) by monitoring increases in homologous phage concentration. Okabe and Goto (1963) demonstrated that phages could be used for quantifying bacterial cells based on the number of newly produced phages and average burst size. They developed a method for indirectly forecasting bacterial leaf blight by monitoring phage titers in rice fields (Okabe and Goto 1963). Reliability of this latter method was questioned later by Civerolo and Maas Geesteranus (1972).

3.11 Commercialisation

The first commercial company to produce phages specifically for control of bacterial plant diseases was AgriPhi Inc., established by Jackson (1989). To minimise development of bacterial strains resistant to the phage, mixtures of wild-type phages and h-mutants were used. Only lytic phages isolated from plant parts (leaves, stems, etc.), soil and water (field run-off, river and stream water and sewage effluents) were utilised. To ensure that temperate phages were not selected, phages were never obtained from their respective bacterial hosts. Although an h-mutant will attack and kill both its specific wild-type host and bacterial phage-resistant mutants selected from this parent, a secondary bacterial mutant resistant to the h-mutant could occur. The secondary phage-resistant mutant and its progeny will be resistant to the one h-mutant being used. Any time a phage-resistant mutant is encountered, new h-mutant phages could be developed against those strains and included in the phage mixture. However, even if h-mutants are used, they may not infect all strains depending on the bacterial diversity that may exist. Therefore, microbial diversity needs to be determined and carefully monitored. The mixture of phages needs to reflect the microbial diversity.

Recently, as a result of greenhouse and field research, OmniLytics Inc., in Salt Lake City, UT (formerly AgriPhi Inc.), received the first EPA registration (EPA registration # 67986-1) to use phages in agriculture. The registration is for using host-specific phages on tomatoes in greenhouses and production fields in Florida as a part of a standard integrated management programme to control tomato bacterial spot.

3.12 Future Outlook

Use of bacteriophages for controlling plant diseases is an emerging field with great potential. The concern about environment-friendly sustainable agriculture and the rise of organic production necessitates improvements in biological disease control methods, including the use of bacteriophages against bacterial plant pathogens. On the other hand, the lack of knowledge about the biology of phage–bacterium–plant interaction and influencing factors hinders progress in the field. Much research in these areas is needed before phages can become effective and reliable agents of plant disease management.

The use of phages for disease control is a fast expanding area of plant protection with great potential to replace the chemical control measures now prevalent. Phages can be used effectively as part of integrated disease management strategies. The relative ease of preparing phage treatments and low cost of production of these agents make them good candidates for widespread use in developing countries as well. However, the efficacy of phages, as is true of many biological control agents, depends greatly on prevailing environmental factors as well as on susceptibility of the target organism. Great care is necessary during development, production and application of phage treatments. In addition, constant monitoring for the emergence of resistant bacterial strains is essential. Phage-based disease control management is a dynamic process with a need for continuous adjustment of the phage preparation in order to effectively fight potentially adapting pathogenic bacteria.

References

Adams MH (1959) Bacteriophages. Interscience, New York

Ahmad MH, Morgan V (1994) Characterization of a cowpea (*Vigna unguiculata*) rhizobiophage and its effects on cowpea nodulation and growth. Biol Fertil Soils 18:297–301

Alvarez AM, Benedict AA, Mizumoto CY, Pollard LW, Civerolo EL (1991) Analysis of *Xanthomonas campestris* pv *citri* and *X c citrumelo* with monoclonal antibodies. Phytopathology 81:857–865

Balogh B (2002) Strategies of improving the efficacy of bacteriophages for controlling bacterial spot of tomato. MS thesis, University of Florida, Gainesville, FL, USA

Balogh B (2006) Characterization and use of bacteriophages associated with citrus bacterial pathogens for disease control. PhD thesis, University of Florida, Gainesville, FL, USA

Balogh B, Jones JB, Momol MT, Olson SM, Obradovic A, King P, Jackson LE (2003) Improved efficacy of newly formulated bacteriophages for management of bacterial spot on tomato. Plant Dis 87:949–954

Balogh B, Jones JB, Momol MT, Olson SM (2004) Persistence of bacteriophages as biocontrol agents in the tomato canopy. In: Momol MT, Ji P, Jones JB (eds) Proceedings of the 1st international symposium on tomato diseases. ISHS, Orlando, FL, pp 299–302

Barrow PA (2001) The use of bacteriophages for treatment and prevention of bacterial disease in animals and animal models of human infection. J Chem Technol Biotechnol 76:677–682

Basit HA, Angle JS, Salem S, Gewaily EM (1992) Phage coating of soybean seeds reduces nodulation by indigenous soil bradyrhizobia. Can J Microbiol 38:1264–1269

Bergamin Filho A, Kimati H (1981) Estudos sobre um bacterofago isolado de *Xanthomonas campestris*. II. Seu emprego no controle de X campestris e X vesicatoria. Summa Phytopathologica 7:35–43

Civerolo EL, Keil HL (1969) Inhibition of bacterial spot of peach foliage by *Xanthomonas pruni* bacteriophage. Phytopathology 12:1966–1967

Civerolo EL, Maas Geesteranus HP (eds) (1972) Interaction between bacteria and bacteriophages on plant surfaces and in plant tissues. In: Proceedings of third international conference of plant pathogenic bacteria, 14–21 Apr 1971. Centre for Agricultural Publishing and Documentation, Wageningen, pp 25–37

Coons GH, Kotila JE (1925) The transmissible lytic principle (bacteriophage) in relation to plant pathogens. Phytopathology 15:357–370

D'Herelle F (1921) Le bactériophage: Sone rôle dans l'immunité. Masson et Cie, Paris

Duckworth DH, Gulig PA (2002) Bacteriophages: potential treatment for bacterial infections. BioDrugs 16:57–62

Flaherty JE, Jones JB, Harbaugh BK, Somodi GC, Jackson LE (2000) Control of bacterial spot on tomato in the greenhouse and field with H-mutant bacteriophages. Hortic Sci 35:882–884

Flaherty JE, Harbaugh BK, Jones JB, Somodi GC, Jackson LE (2001) H-mutant bacteriophages as a potential biocontrol of bacterial blight of geranium. Hortic Sci 36:98–100

Gill JJ, Svircev AM, Smith R, Castle AJ (2003) Bacteriophages of *Erwinia amylovora*. Appl Environ Microbiol 69:2133–2138

Goto M (1992) Fundamentals of bacterial plant pathology. Academic, San Diego

Goto M, Takahashi T, Messina MA (1980) A comparative study of the strains of *Xanthomonas campestris* pv *citri* isolated from citrus canker in Japan and cancrosis B in Argentina. Ann Phytopathol Soc Jpn 46:329–338

Goyal SM, Gerba CP, Bitton G (eds) (1987) Phage ecology. Wiley, New York

Greer GG (2005) Bacteriophage control of food-borne bacteria. J Food Prot 68:1102–1111

Hendrick CA, Sequeira L (1984) Lipopolysaccharide-defective mutants of the wilt pathogen *Pseudomonas solanacearum*. Appl Environ Microbiol 48:94–101

Hibma AM, Jassim SAA, Griffiths MW (1997) Infection and removal of L-forms of *Listeria monocytogenes* with bred bacteriophage. Int J Food Microbiol 34:197–207

Iriarte FB, Balogh B, Momol MT, Jones JB (2007) Factors affecting survival of bacteriophage on tomato leaf surfaces. Appl Environ Microbiol 18:177–183

Jackson LE (1989) U.S. Patent No. 4828999

Katznelson H, Sutton MD (1951) A rapid phage plaque count method for the detection of bacteria as applied to the demonstration of internally borne bacterial infection of seed. J Bacteriol 61:689–701

Keel C, Ucurum Z, Michaux P, Adrian M, Haas D (2002) Deleterious impact of a virulent bacteriophage on survival and biocontrol activity of *Pseudomonas fluorescens* Strain CHA0 in natural soil. Mol Plant Microbe Interact 15:567–576

Klement Z, Rudolf K, Sands DC (eds) (1990) Methods in phytobacteriology. Akadémiai Kiadó, Budapest

Koller W (1998) Chemical approaches to managing plant pathogens. In: Ruberson JR (ed) Handbook of integrated pest management. Dekker, New York

Kotila JE, Coons GH (1925) Investigations on the black leg disease of potato. Michigan Agri Exp Stn Tech Bull 67:3–29

Kutter E (1997) Phage therapy: bacteriophages as antibiotics. The Evergreen State College, Olympia, Washington, 15 Nov 1997. Available from: http://www.evergreen.edu/phage/phagetherapy/phagetherapy.htm

Lang JM, Gent DH, Schwartz HF (2007) Management of *Xanthomonas* leaf blight of onion with bacteriophages and a plant activator. Plant Dis 91:871–878

Leverentz B, Conway WS, Alavidze Z, Janisiewicz WJ, Fuchs Y, Camp MJ, Chighladze Sulakvelidze A (2001) Examination of bacteriophage as a biocontrol method for *Salmonella* on fresh-cut fruit: a model study. J Food Prot 64:1116–1121

Leverentz B, Conway WS, Camp MJ, Janisiewicz WJ, Abuladze T, Yang M, Saftner R, Sulakvelidze A (2003) Biocontrol of *Listeria monocytogenes* on fresh-cut produce by treatment with lytic bacteriophages and a bacteriocin. Appl Environ Microbiol 69:4519–4526

Mallmann WL, Hemstreet CJ (1924) Isolation of an inhibitory substance from plants. Agri Res 28:599–602

Manulis S, Zutra D, Kleitman F, Dror O, David I, Zilberstaine M, Shabi E (1998) Distribution of streptomycin-resistant strains of *Erwinia amylovora* in Israel and occurrence of blossom blight in the autumn. Phytoparasitica 26:223–230

Marco GM, Stall RE (1983) Control of bacterial spot of pepper initiated by strains of *Xanthomonas campestris* pv *vesicatoria* that differ in sensitive to copper. Plant Dis 67:779–781

McKenna F, El-Tarabil KA, Hardy GESTJ, Dell B (2001) Novel *in vivo* use of a polyvalent *Streptomyces* phage to disinfest *Streptomyces scabies*-infected seed potatoes. Plant Pathol 50:666–675

Minsavage GV, Canteros BI, Stall RE (1990) Plasmid-mediated resistance to streptomycin in *Xanthomonas*

campestris pv *vesicatoria*. Phytopathology 80: 719–723
Momol MT, Jones JB, Olson SM, Obradovic A, Balogh B, King P (2002) Integrated management of bacterial spot on tomato in Florida. EDIS, Institute of Food and Agricultural Sciences, University of Florida, Sept 2002, Report no. 110
Munsch P, Olivier JM (1995) Biocontrol of bacterial blotch of the cultivated mushroom with lytic phages: some practical considerations. In: Elliot TJ (ed) Science and cultivation of edible fungi, Proceedings of the 14th international Congress, Oxford, vol II. Brookfield, Rotterdam, pp 595–602
Munsch P, Olivier JM, Houdeau G (1991) Experimental control of bacterial blotch by bacteriophages. In: Maher MJ (ed) Science and cultivation of edible fungi. Balkema, Rotterdam, pp 389–396
Obradovic A, Jones JB, Momol MT, Balogh B, Olson SM (2004) Management of tomato bacterial spot in the field by foliar applications of bacteriophages and SAR inducers. Plant Dis 88:736–740
Obradovic A, Jones JB, Momol MT, Olson SM, Jackson LE, Balogh B, Guven K, Iriarte FB (2005) Integration of biological control agents and systemic acquired resistance inducers against bacterial spot on tomato. Plant Dis 89:712–716
Okabe N, Goto M (1963) Bacteriophages of plant pathogens. Annu Rev Phytopathol 1:397–418
Ophel KM, Bird AF, Kerr A (1993) Association of bacteriophage particles with toxin production by *Clavibacter toxicus*, the causal agent of annual ryegrass toxicity. Phytopathology 83:676–681
Osawa S, Furuse K, Watanabe I (1981) Distribution of ribonucleic acid coliphages in animals. Appl Environ Microbiol 45:164–168
Randhawa PS, Civerolo EL (1986) Interaction of *Xanthomonas campestris* pv *pruni* with pruniphage and epiphytic bacteria on detached peach leaves. Phytopathology 76:549–553
Reichelderfer K, Barry JW (1995) Introduction. In: Hal FR, Barry JW (eds) Biorational pest control agents: formulation and delivery. American Chemical Society, Washington, DC, pp 28–35
Saccardi A, Gambin E, Zaccardelli M, Barone G, Mazzucchi U (1993) *Xanthomonas campestris* pv *pruni* control trials with phage treatments on peaches in the orchard. Phytopathol Mediterr 32:206–210
Schnabel EL, Fernando WGD, Meyer MP, Jones AL, Jackson LE (1999) Bacteriophage of *Erwinia amylovora* and their potential for biocontrol. Acta Hortic 489:649–654
Stein B, Ramallo J, Foguet L, Morandini M (2005) Chemical control of citrus canker in lemons [*Citrus limon* (L) Burm F] (Abstr). In: Second international citrus canker and Huanglongbing research workshop, Tucuman, Argentina, p 25
Summers WC (2005) Bacteriophage research: early history. In: Kutter E, Sulakvelidze A (eds) Bacteriophages: biology and applications. CRC Press, Boca Raton, pp 5–27
Svircev AM, Lehman SM, Kim WS, Barszez E, Schneider KE, Castle AJ (2006) Control of the fire blight pathogen with bacteriophages. In: Zeller W, Ullrich C (eds) Proceedings of the Ist international symposium on bio control of bacterial plant pathogens. Die Deutsche Bibliothek – CIP-Einheitsaufnahme, Berlin, Germany, pp 259–261
Tanaka H, Negishi H, Maeda H (1990) Control of tobacco bacterial wilt by an avirulent strain of *Pseudomonas solanacearum* M4S and its bacteriophage. Ann Phytopathol Soc Jpn 56:243–246
Tanji Y, Shimada T, Yoichi M, Miyanaga K, Hori K, Unno H (2004) Toward rational control of *Escherichia coli* O157:H7 by a phage cocktail. Appl Microbiol Biotechnol 64:270–274
Thayer PL, Stall RE (1961) A survey of *Xanthomonas vesicatoria* resistance to streptomycin. Proc Fla Hortic Soc 75:163–165
Thomas RC (1935) A bacteriophage in relation to Stewart's disease of corn. Phytopathology 25:371–372
Vidaver AK (1976) Prospects for control of phytopathogenic bacteria by bacteriophages and bacteriocins. Annu Rev Phytopathol 14:451–465
Vidaver AK, Schuster ML (1960) Characterization of *Xanthomonas phaseoli* bacteriophages. J Virol 4:300–308
Voloudakis AE, Reignier TM, Cooksey DA (2005) Regulation of resistance to copper in *Xanthomonas axonopodis* pv *vesicatoria*. Appl Environ Microbiol 71:782–789
Wagner PL, Waldor MK (2002) Bacteriophage control of bacterial virulence. Infect Immun 70:3985–3993
Wiggins BA, Alexander M (1985) Minimum bacterial density for bacteriophage replication: implications for significance of bacteriophages in natural ecosystems. Appl Environ Microbiol 49:19–23
Woods TL, Israel HW, Sherf AF (1981) Isolation and partial characterization of a bacteriophage of *Erwinia stewartii* from the corn flea beetle, *Chaetocnema pulicaria*. Prot Ecol 3:229–236
Wu WC (1972) Phage-induced alterations of cell disposition, phage adsorption and sensitivity, and virulence in *Xanthomonas citri*. Ann Phytopathol Soc Jpn 38:333–341
Wu WC, Lee ST, Kuo HF, Wang LY (1993) Use of phages for indentifying the citrus canker bacterium *Xanthomonas campestris* pv *citri* in Taiwan. Plant Pathol 42:389–395
Zaccardelli M, Saccardi A, Gambin E, Mazzucchi U (1992) *Xanthomonas campestris* pv *pruni* bacteriophages on peach trees and their potential use for biological control. Phytopathol Mediterr 31:133–140

Biofumigation

4

Abstract

There has been considerable interest in the use of certain crops as biological fumigants ahead of crop production to reduce the need for chemical fumigation, especially in tight rotations. These are crops that would be grown for their naturally occurring compounds that kill soil-borne pests. Plants in the mustard family, such as mustards, radishes, turnips and rapeseed, and Sorghum species (Sudan grass, Sorghum–Sudan grass hybrids) have shown the potential to serve as biological fumigants. Research has shown some promise in using these crops to reduce soil-borne pests. Plants from the mustard family produce chemicals called glucosinolates in plant tissue (roots and foliage). These glucosinolates are released from plant tissue when it is cut or chopped and then are further broken down by enzymes to form chemicals that behave like fumigants. The most common of these breakdown products are isothiocyanates. These are the same chemicals that are released from metham sodium (Vapam) and metham potassium (K-Pam), commonly used as chemical fumigants. Sorghums produce a cyanogenic glucoside compound called dhurrin that breaks down to release toxic cyanide when plant tissue is damaged.

Soil-borne pests and pathogens, including weed propagules, nematodes, insects, fungi, bacteria and certain other agents, can be limiting factors in the production of horticultural and other crops. One of the principal strategies used by the growers of high-value horticultural crops to combat these organisms is preplant soil disinfestation, using pesticides or other physical or biological methods. Soil fumigants are the most effective soil disinfestation chemicals, and methyl bromide (MB) is the most important soil fumigant chemical used by growers around the world. It is a broad-spectrum pesticide with excellent activity against most potential soil pests. Apart from controlling major soil-borne pathogens and pests, soil fumigation with MB and other fumigants frequently provides increased crop growth and yield responses.

MB was identified as a risk to the stratospheric ozone layer in 1992 and targeted for worldwide phase out in 1997 by means of the Montreal Protocol, an international treaty. Under the current terms of the agreement and of the federal Clean *Air* Act, preplant consumption of MB in the United States is scheduled to be gradually phased out by 2005 (USDA 2000). The impending

P.P. Reddy, *Recent Advances in Crop Protection*,
DOI 10.1007/978-81-322-0723-8_4, © Springer India 2013

loss of MB as a soil fumigant has stimulated intensive efforts to develop and implement suitable replacement strategies. Amongst the potential alternative control method being touted to replace methyl bromide is biofumigation that is amongst the most useful of the non-chemical disinfestation methods.

4.1 What Is Biofumigation?

Biofumigation is an agronomic technique that makes use of some plants' defensive systems. The main plant species in which this is found are the Brassicaceae (cabbage, cauliflower, kale and mustard), Capparidaceae (cleome) and Moringaceae (horseradish). In suitable conditions, the biofumigation technique is able to efficiently produce a number of important substances. In the above plant families, one of the most important enzymatic defensive systems is the myrosinase–glucosinolate system. With this system, tissues of these plants can be used as a soft, eco-compatible alternative to chemical fumigants and sterilants. In a number of countries over the past few years, several experiments have been carried out to evaluate the effectiveness of the myrosinase–glucosinolate system, in particular using the glucosinolate-containing plants as a biologically active rotation and green manure crop for controlling several soil-borne pathogens and diseases. The use of this technique is growing, and it is studied in several countries at a full-field scale (the USA, Australia, Italy, the Netherlands and South Africa), thus triggering the interest of some seed companies, with a positive effect on the 'biofumigation' seed market, which is significantly growing year after year. New potential has also been found for the dehydrated plant tissues and/or for defatted meal pellets production and use. An intense discussion amongst researchers of this topic in the various countries seems to be of fundamental importance particularly to define and develop future common strategies. The aim of this chapter is to describe current strategies to include biofumigation as part of an integrated approach to MB replacement in horticulture.

4.2 Advantages of Biofumigation

- The biofumigation achieved by disking Brassica residues into the soil can significantly suppress weeds, nematodes and soil-borne plant pathogens.
- Brassicas, including numerous mustard species, provide biomass to the soil.
- Biofumigation also improves physical and biological soil characteristics.
- Other benefits of biofumigation include improved soil texture, increased water holding capacity and improved soil microbial community structure.
- Brassica cover crops are known to reduce run-off and preserve nitrogen.
- Green manuring with *Brassica juncea* produced a delayed but remarkable increase of potentially mineralisable nitrogen.
- Green manures can also increase nutrient availability through weathering of soil mineral components. This weathering may be caused by the production of acids by microorganisms during the decomposition of the green manure.
- Biofumigation is responsible for increased infiltration rate, reduced wind erosion and reduced soil compaction.

4.3 Modes of Utilisation

Biofumigation can involve glucosinolate (GSL)-containing plants as rotation crops, or intercrops, by incorporating fresh plant material as green manure or utilising processed plant products high in GSLs such as seed meal or dried plant material treated to preserve isothiocyanates (ITC) activity.

4.3.1 Crop Rotation/Intercropping

Biofumigation by rotation crops or intercrops, where above-ground material is harvested or left to mature above ground, relies on root exudates of growing plants throughout the season, leaf washings or root and stubble residues. Both GSLs and ITCs have been detected in the rhizosphere of

Fig. 4.1 Growing, incorporation and mixing of green manures in soil of plant material using tractor-drawn implements

intact plants, and these have been implicated in the suppression of pests and pathogens in both natural and managed ecosystems. In strawberry production systems, rotations with *Brassica* vegetables such as broccoli have been shown to provide effective control for pathogens such as Verticillium wilt (*Verticillium dahliae*) (Subbarao et al. 2007). The greater reduction of *V. dahliae* microsclerotia and higher vigour and yield of strawberry following broccoli or Brussels sprout rotation crops compared with lettuce occurred at both infested and non-infested sites indicating benefits other than suppression of those diseases were also involved.

4.3.2 Processed Plant Products

The seed meal or oil cake by-products which remain after pressing rapeseed or mustard seed for oil constitute a convenient, high GSL material suitable for soil amendment for high-value horticultural crops. These products contain sufficient intact myrosinase to ensure effective hydrolysis of the GSLs upon wetting. Brassicaceous seed meals have demonstrated significant suppressive activity in a range of insect, nematode, fungi and weeds. Suppression of common scab (*Streptomyces scabies*) was much greater in using dried and ground post-harvest residues of *Brassica* vegetables (Gouws 2004). The suppression of bacterial wilt (*Ralstonia solanacearum*) in potato crops was significantly reduced (40–50%) by a range of *Brassica* amendments incorporated at 5 kg fresh material/m^2. Both *Rhizoctonia solani* and *Pratylenchus penetrans* responsible for apple replant disease were suppressed by both high and low GSL rapeseed meal (Mazzola et al. 2001).

4.3.3 Green Manuring

Incorporated biofumigant green manures or plough-downs (Fig. 4.1) can potentially combine the beneficial elements of rotation crops with a more concentrated release of biocidal GSL-hydrolysis products at the time of incorporation. Mojtahedi et al. (1993) demonstrated significant suppression of root-knot nematodes and increase in yield (17–25%) on potato by rapeseed (*Brassica napus*) compared with wheat green manures and provided evidence for the role of GSLs. Harding and Wicks (2001) found green manures of Indian mustard (*B. juncea*), canola (*B. napus*) and radish (*Raphanus sativus*) all reduced population of *V. dahliae* to a greater degree than a range of cereals but not more than a clover/ryegrass mixture. Suppression of common scab (*Streptomyces scabies*) was most effective with a mustard green manure (Larkin and Griffin 2007).

4.4 Biofumigation Crops (Fig. 4.2 and Table 4.1)

4.4.1 Brassica Plant Species

4.4.1.1 Rapeseed (Canola)

Two Brassica species are commonly grown as rapeseed, *Brassica napus* and *B. rapa*. Rapeseed that has been bred to have low concentrations of

Fig. 4.2 Brassicas used for biofumigation

both erucic acid and glucosinolates in the seed is called canola, which is a word derived from Canadian oil.

Annual or spring-type rapeseed belongs to the species *B. napus*, whereas winter-type or biennial rapeseed cultivars belong to the species *B. rapa*. Rapeseed is used as industrial oil, whilst canola is used for a wider range of products including cooking oils and biodiesel. Besides their use as an oil crop, these species are also used for forage. If pest suppression is an objective, rapeseed should be used rather than canola since the breakdown products of glucosinolates are thought to be a principal mechanism for pest control with these cover crops.

Rapeseed has been shown to have biological activity against plant parasitic nematodes as well as weeds. Due to its rapid fall growth, rapeseed captured as much as 135 kg of residual nitrogen per ha in Maryland. In Oregon, above-ground biomass accumulation reached 6,800 kg/ha, and N accumulation was 135 kg/ha.

Some winter-type cultivars are able to withstand quite low temperatures (10°F). This makes rapeseed one of the most versatile cruciferous cover crops, because it can be used either as a spring- or summer-seeded cover crop or a fall-seeded winter cover crop. Rapeseed grows 1–1.7 m tall.

4.4.1.2 Mustard

Mustard is a name that is applied to many different botanical species, including white or yellow mustard (*Sinapis alba*, sometimes referred to as *Brassica hirta*), brown or Indian mustard (*B. juncea*, sometimes erroneously referred to as canola) and black mustard (*B. nigra*) (Fig. 4.3) (Koch 1995).

The glucosinolate content of most mustards is very high compared to the true *Brassicas*. In the Salinas Valley, California, mustard biomass reached 8,500 lb/acre. Nitrogen content on high residual N vegetable ground reached 328 lb N/acre. Because mustards are sensitive to freezing, winter-killing at about 25°F, they are used either as a spring/summer crop or they winter-kill except in areas with little freeze danger. Brown and field mustards both can grow to 2 m tall.

In Washington, a wheat/mustard–potato system shows promise for reducing or eliminating the soil fumigant metham sodium. White mustard and oriental mustard both suppressed potato early dying (*Verticillium dahliae*) and resulted in tuber yields equivalent to fumigated soils whilst also improving infiltration, all at a cost savings of about $165/ha. Mustard green manures can be used to replace fumigants to improve infiltration in potato cropping systems. Mustards have also been shown to suppress growth of weeds.

4.4.1.3 Radish

The true radish or forage radish (*Raphanus sativus*) does not exist in the wild and has only been known as a cultivated species since ancient times. Cultivars developed for high forage biomass or high oilseed yield are also useful for cover crop purposes (Fig. 4.4). Common types include oilseed and forage radish.

Their rapid fall growth has the potential to capture nitrogen in large amounts (190 kg/ha in Maryland) and from deep in the soil profile. Above-ground dry biomass accumulation reached 9,090 kg/ha, and N accumulation reached 190 kg/ha in Michigan. Below-ground biomass of radishes can be as high as 4,200 kg/ha.

Oilseed radish is less affected by frost than forage radish but may be killed by heavy frost below 25°F. Radish grows about 2–3 ft tall. Radishes have been shown to alleviate soil compaction and suppress weeds.

In an Alabama study of 50 cultivars belonging to the genera *Brassica*, *Raphanus* and *Sinapis*,

Table 4.1 Brassica plant species used for biofumigation

Species	Selection	Sowing time	Biomass yield	Contents in GLs (μmol/g dm)	GLs yield (Moles/ha)	Main GLs	% N content (d m)	Biocidal activity
Brassica juncea	ISCI 20	Autumn	93.8 ± 1.1	12.9 ± 0.4	201 ± 10	Singrin	1.5	3
		Spring	46.5 ± 10.6	25.8 ± 0.4	246 ± 32.5		2.0	
Brassica juncea	ISCI 61	Autumn	133.5 ± 23	11 ± 1.1	183.6 ± 26	Singrin	1.4	2
		Spring	60 ± 14.1	17.7 ± 2.3	213 ± 35		2.0	
Brassica juncea	ISCI 99	Autumn	109 ± 16	16.9 ± 0.5	300 ± 49.9	Singrin	1.2	4
		Spring	54.5 ± 9.5	33.6 ± 1.6	333.8 ± 41		1.6	
Rapistrum rugosum	ISCI 4	Autumn	84 ± 33.2	27.6 ± 3.2	562 ± 158	Cheirolin	1.9	4
		Spring	65.3 ± 21.4	24.7 ± 1.0	160.7 ± 45		2.8	
Eruca sativa	Nemat	Autumn	87 ± 19.2	9.4 ± 0.6	186 ± 15	Erucin	1.5	1
		Spring	45.3 ± 4.2	12.9 ± 2.5	107.2 ± 8.5		2.0	
Brassica nigra	ISCI 27	Autumn	102.8 ± 25	17.1 ± 0.3	255.5 ± 8.1	Singrin	0.9	4
		Spring	46 ± 15.5	20.7 ± 0.2	204.5 ± 61		1.5	

Fig. 4.3 *Sinapis alba* and *Brassica juncea* used for biofumigation

forage and oilseed radish cultivars produced the largest amount of biomass in central and south Alabama, whereas winter-type rapeseed cultivars had the highest production in North Alabama.

4.4.1.4 Turnips

Turnips (*Brassica rapa* var. *rapa*) are used for human and animal food because of their edible root. Turnip has been shown to alleviate soil compaction. Whilst they usually do not produce as much biomass as other Brassicas, they provide many macrochannels that facilitate water infiltration. Similar to radish, turnip is unaffected by early frost but will likely be killed by temperatures below 25°F.

4.4.1.5 Processed Brassica Amendments

Another mode for achieving biofumigation effects is to use Brassica-derived isothiocyanate-rich materials such as seed meals (Fig. 4.5) or oils as soil amendments to achieve pesticidal effects (Lazzeri et al. 2004a, b). There may be a niche for such products, but they are likely to be regarded as pesticides by regulatory authorities and will therefore potentially face significant hurdles in implementation that the growing of a Brassica crop rich

Fig. 4.4 Daikon radish at full canopy closure, planted in August in Blacksburg, VA (Appalachian region), photographed approximately 60 days after planting

Fig. 4.5 Brassica seed meal used for biofumigation

in appropriate isothiocyanates and manipulating it to maximise their release would not face.

4.4.2 Non-Brassica Plant Species

4.4.2.1 Grasses

Members of the grass family (Gramineae) also produce a rich diversity of bioactive chemical compounds, including phenolics, glycosides, benzoxazinones and amino acids. Although many are primarily known for their allelopathic activity against other plants (Putnam and DeFrank 1983), they may also possess properties deleterious to a broad range of fungi, bacteria, nematodes and insects in soil. For example, residues of several Graminaceous crops of agronomic importance, including cultivars of barley (*Hordeum vulgare*), wheat (*Triticum aestivum*), triticale (*Xanthium triticosecale*) and oats (*Avena sativa*), all demonstrated significant, deleterious effects on soil-borne nematodes during their decomposition in soil. Phytotoxicity was evident in many test plants when they were subsequently established in the soil shortly after amendment with the residues (Stapleton 2006).

A similar phytotoxic effect was even more pronounced when field and greenhouse studies were conducted using sudex, a hybrid of Sorghum and Sudan grass (*Sorghum bicolor* × *S. sudanense*), as a cover crop. Severe allelopathic effects occurred on subsequently planted tomato, broccoli and lettuce transplants, unless a waiting period of at least 6 weeks was observed between the incorporation of sudex residues in soil and planting of the following crop (Summers et al. 2009).

Allelochemicals are plant-produced compounds (other than food compounds) that affect the behaviour of other organisms in the plant's environment. For example, Sudan grass (and Sorghum) contains a chemical dhurrin that degrades into hydrogen cyanide, which is a powerful nematicide (Luna 1993; Forge et al. 1995; Wider and Abawi 2000). Some cover crops have exhibited nematode-suppressive characteristics equivalent to aldicarb, a synthetic chemical pesticide (Grossman 1990).

4.4.2.2 Garlic and Onions

Yet another group of bioactive plants are those in the onion family (Alliaceae). Garlic and onion, especially, have been known for their bioactive properties since ancient times. The biocidal properties of garlic, onion and leek are attributed to sulphur volatiles produced during degradation of *Allium* tissues. The primary emitted compounds are thiosulfinates and zwiebelanes mainly converted in soil or in *Allium* products (extracts) to disulfides. The activities of these compounds were studied in vitro on soil pathogenic fungi and insects in order to measure their disinfection potential. These studies show a good potential for three disulfides: dimethyl disulfide (DMDS), dipropyl disulfide (DPDS) and diallyl disulfide (DADS) to inhibit several fungal species: *Aphanomyces euteiches, Colletotrichum coccodes, Fusarium moniliforme, Fusarium oxysporum radicis-cucumerinum, Phytophthora cinnamomi, Pythium aphanidermatum, Rhizoctonia solani, Sclerotium rolfsii* and *Sclerotinia slerotiorum*. The insecticidal activity of two disulfides was also evaluated on termites. The sensitivity varied with termite phenotypes, and DMDS was the most toxic disulfide. This study shows that disulfides exhibit greater toxicity for insects than for fungi.

Feasibility studies examined the decomposition of garlic and onion residues in moist soil, as related to the seed inactivation of four important agricultural weeds: black nightshade (*Solanum nigrum*), common purslane (*Portulaca oleracea*), London rocket (*Sisymbrium irio*) and barnyard grass (*Echinochloa crusgalli*). The inhibitory and herbicidal effects of the Alliaceous residues were generally mild or inconsistent when tested at soil temperatures of 23°C. However, at 39°C, which by itself was mildly inhibitory to weed seed germination, the activity of the decomposing residues was far more potent (Mallek et al. 2007).

The term 'biofumigation' has been extended to the use of *Allium* species producing compounds derived from sulphur amino acids. The primary active compounds produced in crushed *Allium* spp. are thiosulfinates, but they are rapidly transformed into active disulfides, particularly dimethyl disulfide (DMDS) which is also produced by Brassicaceae.

4.5 Mode of Action

4.5.1 The Glucosinolate–Myrosinase System

Glucosinolates are a class of sulphur compounds occurring as secondary plant products almost exclusively in families of the order Capparales: Brassicaceae, Capparaceae, Moringaceae, Tovariaceae and Resedaceae (Brown and Morra 1997). They are particularly common in the Brassicaceae, plants that are widely cultivated as important vegetable, condiment, oilseed and forage crops (Fenwick et al. 1983). Glucosinolates are organic anions possessing a thioglucose moiety, a sulfonated oxime and any one of a variety of aliphatic or aromatic R groups. At least 120 structurally different glucosinolates have been identified in 16 different families of angiosperms. Although glucosinolates themselves possess limited biological activity, enzymatic degradation by thioglucoside glucohydrolase or myrosinase results in the formation of a number of compounds including isothiocyanates (Fig. 4.6), nitriles, SCN^-, oxazolidinethione, epthionitriles and organic thiocyanates. The bioactive hydrolysis products of glucosinolates, particularly the isothiocyanates, can be used to control soil pests and weeds by incorporating glucosinolate-containing plant material in soil – a practice known as biofumigation.

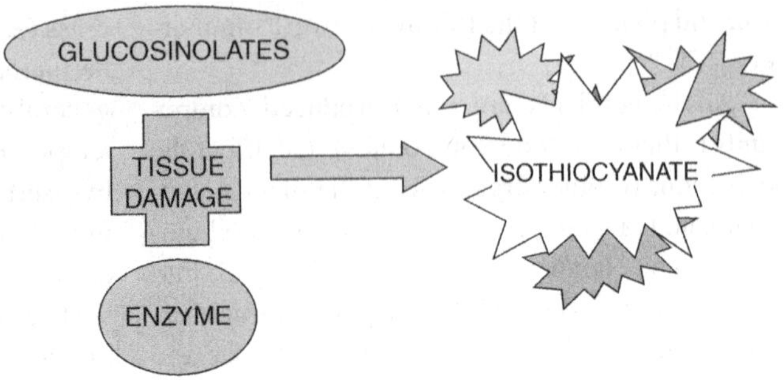

Fig. 4.6 Glucosinolate hydrolysis to release isothiocyanate

4.5.2 Non-ITC-Related Mechanisms

4.5.2.1 Trap Crops and Non-hosting

Brassicaceous green manures can also act as trap crops. Perhaps the most well-documented case involving their use for this purpose is for the control of sugar beet nematodes (*Heterodera schachtii*) in northern Europe (Muller 1999). Fodder radish (*Raphanus sativus*) and white mustard (*Sinapis alba*) cover crops have been selected and developed to be grown as green manures preceding sugar beet crops. The Brassicas are invaded by the nematodes, which develop within the roots, but have their sexual differentiation disrupted. This results in very low numbers of females in the subsequent generation, causing significant decline in the population and reducing the infection of subsequent sugar beet crops. It is a unique example related to specific nematode-resistant Brassicas rather than a general non-hosting effect.

For a green manure to be effective in disease control, it is generally desirable that it not host the pathogen in question so that a decline in population or inoculum occurs during its growth. Indeed, the fact that Brassica species are generally moderate hosts to some other important plant parasitic nematodes (e.g. *Meloidogyne* spp.) has reduced their applicability as biofumigant green manures as very careful management is required to avoid population increases on the biofumigant crop, particularly in warmer climates (Stirling and Stirling 2003). In such cases, it is possible that the suppressive effects of the incorporated tissues could still reduce the population to manageable levels, but it is generally desirable to select biofumigants that are poor hosts or to grow them at times of the year when pathogens do not build up.

4.5.3 Non-glucosinolate- or Non-isothiocyanate-Related Effects

There are several examples in the literature of significant suppression by incorporated Brassica amendments that are not associated with the glucosinolate concentration of the tissue. The most consistent of these are those related to nematode suppression. A study by Potter et al. (1998) showed clearly that the significant suppressive effects (60–95%) of six diverse Brassica species were unrelated to either the total glucosinolate content of the tissues or to the concentration of propenyl glucosinolate, the major glucosinolate found in mustard which is known to produce the biologically active propenyl isothiocyanate upon hydrolysis. Similar results have also been shown for *Meloidogyne javanica* by McLeod and Steel (1999) and in glasshouse studies of the microbial complex associated with apple replant disease. The most likely explanation for these observations is either that the incorporation of organic matter itself increased the populations of antagonistic organisms in the soil or that non-glucosinolate compounds released by Brassicas are toxic to nematodes. Indeed, many potentially biologically active, non-glucosinolate sulphur-containing compounds are released from Brassica amendments, and other products of microbial decomposition of tissues including fatty acids can also be biologically active. Irrespective of the mechanism responsible, these levels of suppression that are unrelated to glucosinolates can significantly confound the interpretation of biofumigation studies if appropriate controls are not included in the experiments.

In addition to isothiocyanates, hydrolysis of glucosinolates in incorporated biofumigants can release a range of other potentially toxic products including nitriles, epinitriles and ionic thiocyanates (Brown and Morra 1997). Although isothiocyanates are generally considered the most toxic of these hydrolysis products, they may not be released in large concentrations in all circumstances following incorporation (Bending and Lincoln 1999). Much of the weed suppression noted following incorporation of Brassicaceous seed meals is likely to arise from the ionic thiocyanate rather than the isothiocyanates. Often experimental protocols used to determine the toxicity of different amendments can favour different compounds depending on volatility (e.g. sealed containers), water solubility (irrigation/soil water content) and reactivity (soil organic matter content). To resolve the question of isothiocyanate-related suppression, it is generally desirable to correlate the level of pest suppression with measured levels of isothiocyanate released in soil, a task that is not trivial given the relatively rapid loss from soils as a result of many different processes (Brown and Morra 1997).

During plant residue decomposition, concentrations of volatile chemical compounds in the soil tend to increase with increasing temperature. This is important because toxic effects are a function of the toxicant concentration multiplied by the duration of exposure. The liberation of volatile compounds from decomposing crop residues generally occurs within a few days after their incorporation in moist soil (Fig. 4.5). Manipulation of the system via soil covers and/or heating to maximise biofumigant concentrations can be the difference between the effective and ineffective management of pests.

4.6 Disease Management

Biofumigation is the use of volatile plant chemicals for control of soil-borne plant pathogens. These products have been shown to suppress the pathogens *Botrytis cinerea, R. solani, F. oxysporum, Didymella lycopersici* and *Cladosporium fulvum* (Urbasch 1984). Although the fungicidal properties of ITCs have been identified since at least 1937 (Walker et al. 1937), only recently has the practice of using *Brassicas* for pest control been termed ‘biofumigation’ (Angus et al. 1994). The volatiles from several *Brassica* species suppressed the growth of the tomato pathogens *P. ultimum, R. solani* and *S. rolfsii* (Charron and Sams 1998, 1999; Harvey et al. 2002).

Brassicas such as oriental mustard have been associated in some reports with improved root health in a subsequent cash crop, such as potatoes, grown after a green incorporated Brassica cover crop. Brassica cover crops can be managed as a green manure – when green residues are incorporated – and this may reduce or suppress some pathogens, including *Verticillium* in potato; *Pythium, Fusarium* and *Rhizoctonia* root rots in beans; *Pythium* in lettuce; pink root in onion; *Aphanomyces, Pythium, Rhizoctonia* and *Fusarium* root rot in peas; and cavity spot and *Fusarium* in carrot (Sanders 2005).

The biocidal activity of *Brassicas* against fungal pathogens, nematodes, weeds and insects is frequently attributed to ITCs from *Brassica* tissues (Delaquis and Mazza 1995). ITCs are effective, broad-spectrum pesticides (Mithen et al. 1986; Isshiki et al. 1992), and substantial quantities of them can be produced for field application. Research has shown that black mustard (*B. nigra*) and Indian mustard (*B. juncea*) produce high levels of ITC (Tollsten and Bergström 1988) and could be utilised in a biofumigation cropping system.

Control of *V. dahliae* by *Brassica* residues has also been shown. The number of *V. dahliae* propagules in a 36-m^2 plot incorporated with 200 kg of chopped broccoli (*B. oleracea* var. *italica*) were lower than in control plots and were comparable to plots fumigated with methyl bromide + chloropicrin.

The ‘biofumigation’ properties of Brassicas as rotation, companion or green manure crops and the use of these properties in managing the important pathogens of tropical vegetable production systems have been studied.

The effectiveness of the biocides was evaluated in commercial and small-scale farmer fields in the context of integrated disease management, using strategies that included disease resistance/tolerance, non-host crop rotation and solarisation (harnessing the sun's energy to destroy some weeds and soil pathogens).

It was demonstrated that significant suppression of the organism could be achieved using Brassica green manures, and that this arose from a combination of two separate mechanisms. The first mechanism was short-term suppression (over 2–3 days) related to the release of isothiocyanates (ITCs) from the Brassica tissues. To maximise the impact of this mechanism, strategies to increase ITC release from the plant tissues and their residence time in the soil were developed into a 'best-bet' approach. This involved use of Brassica tissues which release high concentrations of toxic ITCs (e.g. mustard), incorporating around 5 kg/m^2 (5% w/w) of fresh tissue into the soil, adequate maceration of the tissues at a cellular level to release the ITCs, adequate water to facilitate hydrolysis of ITCs, rapid and thorough incorporation for complete mixing of ITCs through the soil, watering or covering to retain the volatile ITCs in the soil and targeting light-textured soils with low organic matter to reduce inactivation of ITCs. This approach could be readily adopted in highly mechanised farming systems and was very effective on light-textured soils. Field experiments in Mareeba in 2003 and 2005 showed mustard green manures reduced bacterial wilt incidence in tomatoes from 80 to 15%, with a tenfold increase in yield (from 2.5 to 20.0 t/ha). On loam and clay soils with more organic matter, there was less evidence that ITC-related suppression was occurring.

However, a second mechanism of suppression operating over a longer period was also identified. This mechanism, occurring over a period of 3–4 weeks, was related to the effects on the organism of added organic amendments. This mechanism was observed on soils with higher clay and organic matter and did not require the strict protocols listed above. Reductions of bacterial wilt incidence in potatoes (49% down to 17%) and associated yield increases (9.1–18 t/ha) in eight on-farm experiments were largely attributable to this process irrespective of the type of green manure used.

The scientists confirmed that the nitrogenous components of organic amendments could generate the same pattern of suppression as the green manures when added in pure forms to soil, and evidence suggests that it is related to the mobilisation of specific antagonistic soil bacterial populations which are yet to be identified. More work is needed to develop strategies to fully exploit this more general mechanism of suppression.

Biofumigant green manures should not host the organism of interest. Brassica species are generally considered non-hosts of bacterial wilt and although one incidence of bacterial wilt infection was observed in a mustard line from Australia in a nursery in the Philippines. Subsequent investigations in Taiwan suggest root damage is a prerequisite. However, continual vigilance to ensure biofumigants do not become susceptible to bacterial wilt is warranted.

4.6.1 Citrus Root Rot, *Phytophthora nicotianae* var. *parasitica*

Biofumigation with yellow and oriental mustard cover crops reduced citrus *Phytophthora* root rot by 90%. The beneficial fungi such as *Trichoderma* have been found at much higher levels in the mustard-treated soils. This does not occur when the soil is fumigated with chemicals. Mustard cover crops are now commonplace in the Salinas Valley, California.

4.6.2 Apple Root Rot, *Rhizoctonia solani*

Streptomyces spp. soil populations increased significantly in response to rapeseed meal (RSM) amendment applied to apple orchard soils. The vast majority of *Streptomyces* spp. recovered from the apple rhizosphere produced nitric oxide and possessed a nitric oxide synthase homologue. The transformations in the bacterial community structure are associated with the observed control of *Rhizoctonia* root rot and lesion nematode, *Pratylenchus* spp. with nitric oxide (NO) production by soil bacteria potentially having a role in the induction of plant systemic resistance (Cohen et al. 2005).

4.6.3 Stone Fruits Brown Rot, *Monilinia laxa*

Allyl isothiocyanate (AITC) from sinigrin (*Brassica carinata*) treatments was effective against brown rot, *Monilinia laxa* on stone fruits. Also butenyl isothiocyanate vapours from gluconapin (*B. rapa*) controlled the brown rot produced by *M. laxa* although with a higher initial concentration. At these concentrations, the pathogen control was obtained without any phytotoxic effects on fruits. These results have led to study the best practical large-scale application, using Brassicaceae meals as starting material to produce the active isothiocyanates for fruit biofumigation (Bernardi et al. 2004).

4.6.4 Common Scab of Potato, *Streptomyces scabies*

Brassica residue (cabbage, cauliflower, broccoli and Brussels sprouts, at 0.33%) incorporation as a disease control practice is an innovative means of managing common scab of potato (*Streptomyces scabies*) that is effective, economically feasible and non-detrimental to the environment (Gouws 2004).

4.6.5 Cauliflower Wilt, *Verticillium dahliae*

Broccoli (*Brassica oleracea* var. *italica*) is resistant to *V. dahliae* infection and does not express wilt symptoms in cauliflower. Incorporation of broccoli residues into infested soil reduces soil populations of *V. dahliae*. The overall reduction in the number of propagules after two broccoli crops was approximately 94%, in contrast to the fivefold increase in the number of propagules in infested plots without broccoli after two cauliflower crops. Disease incidence and severity were both reduced approximately by 50% ($P<0.05$) in broccoli treatments compared with no broccoli treatments. Abundance of microsclerotia of *V. dahliae* on cauliflower roots about 8 weeks after cauliflower harvest was significantly ($P<0.05$) lower in treatments with broccoli compared with treatments without broccoli. Rotating broccoli with cauliflower and incorporating broccoli residues into the soils are novel means of managing Verticillium wilt on cauliflower (Xiao et al. 1998).

4.6.6 Lettuce Drop, *Sclerotinia sclerotiorum* and *Sclerotinia minor*

Biofumigation with Fumus at 16 kg/ha gave best control of *Sclerotinia sclerotiorum* disease of lettuce (white mould). In field studies, biofumigant *Brassica* green manure crops could increase the marketable yield of a subsequent lettuce crop by reducing plant loss from *S. minor* infection. Other important benefits of green manures such as reduced soil crusting, improved infiltration, increased organic matter and the subsequent increase in beneficial microflora and microfauna are often overlooked. Soil improvement following a *Brassica* may produce longer lasting effects on disease suppression and crop health, as well as soil health.

Infection of lettuce with *S. minor* was 91 and 68% less after 'biofumigation' with yellow and oriental mustard, respectively, than after faba bean. Lettuce after yellow mustard/plastic was nearly infection-free (7.6 times lower infection score than control). Yellow mustard/plastic resulted in the largest lettuce heads. The biofumigants BQ-Mulch and Fumus significantly reduced the percentage of plants with *Sclerotinia* wilt. BQ-Mulch appeared more effective than Fumus in reducing wilted plants.

4.6.7 Tomato Southern Blight, *Sclerotium rolfsii*

Since 1999, field trials have been conducted at Knoxville Experiment Station. Significant decreases in the incidence of southern blight (*Sclerotium rolfsii*) of tomato were recorded, as well as increases in fruit yield, when integrating biofumigation into a sustainable production system. The research yielded important knowledge pertaining to the development of the appropriate production methods utilising biofumigation as a management technique. Both *Brassica* cover crops and mustard seed meal incorporations have been used. The ability to amend production soils with the spreadable meal, in conjunction with its high ITC content (3–4× leaf tissue), gives this enhanced biofumigation technique great potential. Problems associated with growing cover crops such as variable stands and weather complications are avoided when using the meal. As a fully organic product, preplant mustard meal applications could give the grower superior ability to control soil-borne pathogens.

4.6.8 Bacterial Wilt of Solanaceous Vegetables

The most serious soil-borne disease of solanaceous vegetable crops such as eggplant, tomato and potatoes in tropical environments is bacterial wilt (BW) caused by *Ralstonia solanacearum*. Biofumigation using Indian mustard (*Brassica juncea*) as green manure has been found effective in reducing the level of BW in the soil and lessening the severity of the disease in the following solanaceous crops. A commercially available Indian mustard biofumigant green manure was shown to significantly reduce bacterial wilt in a following potato crop, resulting in spectacular yield increases (from 0.3 to 22 t/ha). The fact that such results were obtained with no purposeful selection of Brassicas for high biofumigant properties indicates there may be significant scope to improve the level of suppression achieved.

On-farm trials utilising chopped radish, cabbage, broccoli, cauliflower and mixed crucifers showed that soil incorporated with crucifers had significantly lower bacterial wilt (BW) incidence (21.13–44.27%) compared with untreated potato plots (57.79%). The lowest percentage of BW incidence was obtained with mixed crucifers at 21.13%. The lowest number of infected tubers was obtained with chopped radish at 2.28 t/ha. The mixed treatment also gave the highest yield of 11.44 t/ha, but this was comparable to the yield in the cauliflower treatment (11.04 t/ha). The untreated plot had the lowest yield of 6.67 t/ha. Economic analysis based on the on-farm trials for BW utilising chopped radish, cabbage, broccoli, cauliflower and mixed crucifers showed that white potato grown in soil incorporated with crucifers gave higher yield (7.92–11.44 t/ha) and return on investment (ROI) (133–192.84%). Mixed crucifers obtained the highest marketable yield and ROI. Untreated plots had the lowest yield (6.67 t/ha) and ROI (113.97%) (Abragan et al. 2008).

The reduction and long-term elimination of bacterial wilt (*Ralstonia solanacearum*) from the soil by incorporating especially mustard or radish plants in large amounts into the soil immediately before planting tomatoes reduced the incidence of bacterial wilt by 50–70% in the Philippines.

Green manuring with 20-day-old canola crop gave significant protection against soil-borne inoculum of the fungus and significantly reduced inhibition of eyes germination, sprout killing, stem canker index, stolon canker index, black scurf disease (*Rhizoctonia solani*) index and reduction in potato yield. Canola and rapeseed rotations reduced the incidence and severity of black scurf by 54–73% relative to the oat rotation. Chinese mustard reduced black scurf by 50–85% relative to a ryegrass control (Larkin and Griffin 2007).

4.7 Nematode Management

Cultivation and incorporation of cover/rotation crops, especially Brassicaceae plants, occasionally suppress soil-borne diseases, including nematodes. Utilisation of volatile toxic compounds, such as isothiocyanates generated from the glucosinolates in such crops, for soil-borne disease control is generally termed biofumigation (Kirkegaard et al. 1993). Isothiocyanates are known to possess broad pesticidal activity against weeds, bacteria, fungi and nematodes (Matthiessen and Kirkegaard 2006) and are similar or relative to compounds found in commercial soil fumigants, such as metham sodium and dazomet, which release methyl isothiocyanate into the soil. Nematode suppression by biofumigation with Brassicaceae plants has been attempted in several regions worldwide, mainly with rapeseed (*Brassica napus*) and Indian mustard (*Brassica juncea*) (Mojtahedi et al. 1991, 1993; Ploeg and Stapleton 2001; Stirling and Stirling 2003; Rahman and Somers 2005).

The efficacies for soil-borne disease control were generally variable and inconsistent, a trend which seemed more salient with nematodes than with soil-borne fungi (Matthiessen and Kirkegaard 2006).

Before selecting a plant for biofumigation purposes, the host range of the target nematode(s) needs to be checked, because most Brassicaceae used as cover crops are hosts of some important nematodes, such as *Meloidogyne* spp. (McLeod and Steel 1999; McLeod et al. 2001), and host Brassicaceae plants may increase nematode populations instead of reducing them. Biofumigant plants that are non-hosts or poor hosts, or are nematode resistant, are strongly preferred. A study in Australia showed that *M. javanica* reproduced on *B. juncea* and *B. napus* at rates which increased the nematode population in the soil during their growth period in hot seasons; however, nematode proliferation could be prevented if these plants were grown in the winter under low temperatures (Stirling and Stirling 2003). Use of dried pellets of glucosinolate-containing plants can solve the problem of nematode population build-up in the field whilst adding a biofumigation effect (Lazzeri et al. 2004b). Again, to obtain maximum control efficacy from biofumigation, several factors must be considered in addition to host status for nematodes, including the optimum plant growth stage for soil incorporation (probably when the glucosinolate content reaches maximum levels), biomass of the plant, and adequate soil moisture for plant degradation and gas diffusion into the soil whilst preventing rapid gas escape from the soil to the atmosphere. Variations in glucosinolate profiles and concentrations exist amongst Brassicaceae crops and their developmental stages.

An effective way of enhancing the nematode-control efficacy of biofumigation in small areas is to combine the incorporation of Brassicaceae plants with soil tarping using plastic film, which may prevent rapid emission of volatile nematicidal compounds from the soil to the atmosphere, and increase the soil temperature via a soil-solarisation effect if performed in hot seasons. Elevation in soil temperatures due to soil solarisation has been shown to improve the control efficacy of amendments with Brassicaceae crops (Gamliel and Stapleton 1993; Ploeg and Stapleton 2001). Combinations of sublethal soil temperatures (30–38.8°C) and the biofumigation effect (toxic volatile compounds) may have a synergistic effect on nematode-control efficacy. Sublethal temperatures may render nematodes more sensitive to toxic compounds or to antagonistic microorganisms. Because nematode control has been obtained by soil solarisation plus biofumigation with non-Brassicaceae plants, such as pepper plant residues, this method does not appear to be specific to glucosinolate-containing plants. Other studies on the effects of fresh organic amendments and soil tarping on fungal pathogens have suggested that toxic compounds, elevated temperature, anaerobic and reducing soil conditions and biological activity are involved in the control (Gamliel and Stapleton 1993).

Nitriles are another group of toxic compounds derived from glucosinolates. Simple nitriles have been shown to be involved in direct and indirect plant defence responses upon attack of *Pieris rapae* (Lepidoptera). The role of nitriles in nematode suppression is not clear.

Different mustards (e.g. *Brassica juncea* var. *integrifolia* or *B. juncea* var. *juncea*) should be used as intercrop on nematode infested fields. As soon as mustards are flowering, they are mulched and incorporated into the soil. Whilst incorporated plant parts are decomposing in a moist soil, nematicidal compounds of this decomposing process do kill nematodes. Two weeks after incorporating plant material into the soil, a new crop can be planted or sown (it takes about 2 weeks for the plant material to decompose and stop releasing phytotoxic substances, i.e. chemicals poisonous to plants).

It is recommended to alternate the use of agricultural residues with green manure, especially from Brassicae, using 5–8 kg/m^2 of green matter, although combinations of legumes and grass can be applied. In the case of the use of green manure cultivated in the same field, fast-growing plants should be used to be incorporated at least 30 days after having been planted, to avoid the increase of pathogen populations. Planting Brassicae after biofumigation can serve as bioindicators of possible phytotoxicity, because the germination of these seeds is sensitive to phytotoxic substances. At the same time, they are very sensitive to nematodes and permit the detection of areas in the crop where biofumigation is not effective. They act like trap plants and like biofumigants when incorporated into the soil.

In Spain, successful application of biofumigation was achieved in strawberries, peppers, cucurbits, tomato, Brassicae, cut flowers, citrus and banana. Biofumigation has also been recently applied to Swiss chard and carrot crops. The most utilised biofumigants have been goat, sheep and cow manure and residues from rice, mushroom, Brassicae and gardens.

4.7.1 Citrus Nematode, *Tylenchulus semipenetrans*

Biofumigation with yellow and oriental mustard cover crops reduced citrus nematode (*Tylenchulus semipenetrans*) by 90%. The beneficial fungi such as *Trichoderma* have been found at much higher levels in the mustard-treated soils. This does not occur when the soil is fumigated with chemicals. Mustard cover crops are now common place in the Salinas Valley, California. Citrus nematode suppression was 92% greater after oriental and yellow mustard (except yellow mustard/plastic) than after cereal or legume, indicating mustard allelochemicals possibly suppressing nematodes.

In vitro assays revealed that isothiocyanates from Brassicaceae plants were nematicidal to juveniles of *M. incognita* and *T. semipenetrans* at concentrations as low as 10 mM (Lazzeri et al. 2004a, b). However, several studies have shown that nematode-control efficacy and nematode reproduction in treated soils are not correlated with glucosinolate contents in plants (Potter et al. 1998; McLeod and Steel 1999). Although glucosinolates are thought to play an important role in nematode suppression, non-glucosinolate compounds, such as other sulphur-containing compounds (Bending and Lincoln 1999), as well as biological and physiological factors, may also be involved (Mazzola et al. 2001). Although the effects of soil type on biofumigation efficacy have not been well studied, sandier soils with low organic matter content appear to allow better performance of the biofumigant.

4.7.2 Potato Columbian Root-Knot Nematode, *Meloidogyne chitwoodi*

Mojtahedi et al. (1993) reported that incorporation of *B. napus* shoots into *M. chitwoodi* infested soil reduced nematodes to very low levels in potato. The amendment protected host plant roots growing in the zone of *B. napus* incorporation from nematode infestation for up to 6 weeks. Soil incorporation rates of 4% (w/w) killed nearly all second-stage juveniles, whereas rates of 6% were required to prevent hatching of juveniles from egg masses.

4.7.3 Potato Root-Knot Nematode, *Meloidogyne incognita*

Soil application of shredded mixed crucifers showed the highest percentage reduction in *M. incognita* population of 83.6% in potato crop followed by radish (79.2%), broccoli (74.2%) and cabbage (68.4%). The use of mixed crucifers was recommended because it had the highest percentage reduction in nematode counts both on station (83.57%) and on farm (86.7%) and highest marketable yield (11.44 t/ha) and return on investment (ROI) of 192%.

4.7.4 Potato Cyst Nematode, *Globodera rostochiensis*

Black mustard, Caliente mustard, marigolds and Sudan grass have the potential to reduce potato cyst nematode (*Globodera rostochiensis*) by biofumigation. Chopped and incorporated into soil, the green manure releases chemicals that 'fumigate' the soil killing of the cysts as well as fungal spores and weed seeds.

4.7.5 Potato Root Lesion Nematode, *Pratylenchus penetrans*

On-farm research in western Idaho showed that rapeseed green manures decreased soil populations of root lesion nematodes (*Pratylenchus penetrans*) on potato to a greater extent than did Sudan grass green manures. Fall Sudan grass should be ploughed down after it is stressed (i.e. the first frost, stopping irrigation). Winter rapeseed and canola should be incorporated in very early spring (Cardwell and Ingham 1996).

4.7.6 Tomato Root-Knot Nematode, *Meloidogyne incognita*

Brassica juncea sel. ISCI 99 acted mainly as a biofumigant, whilst *Eruca sativa* cv. Nemat revealed an interesting trap crop effect on *M. incognita* infecting tomato. Also, the use of pellets derived from Brassicaceae species incorporated into the soil before the transplanting of tomato has shown a lower gall index in the roots. The oil extract of Brassicaceae species reduced the juvenile nematode population in the soil during the whole tomato cultivation cycle (Colombo et al. 2008). All the treatments produced a tomato yield significantly higher than with untreated soil.

4.7.7 Carrot Root-Knot Nematode, *Meloidogyne incognita*

Soil population density of *M incognita* was significantly lower in the carrot plots treated with green manure or hay of *B. juncea* or with green manure of *E. sativa*, either alone or combined with seed meal pellet, compared with untreated control. Nematicidal effects of biofumigation treatments did not differ from that of fenamiphos. Moreover, commercial pellet and *B. juncea* green manure and hay resulted also in carrot yield almost double the control, whereas lower increases were provided by *E. sativa* green manure, either alone or combined with the pellet.

4.7.8 Muskmelon Root-Knot Nematode, *Meloidogyne incognita*

In muskmelon crop, *E. sativa* green manure resulted in a nematicidal effect (on root-knot nematodes) statistically similar to cadusaphos and higher than *B. juncea* amendment. Muskmelon yields recorded for chemical treatment and *B. juncea* were 79.2 and 70.8 t/ha, respectively, and were three times higher than untreated control (De Mastro et al. 2008b).

4.7.9 Lettuce Root-Knot Nematode, *Meloidogyne incognita*

Highest top weight and plant height of lettuce were noted in broccoli- and cabbage-amended soil. Root weight was highest in broccoli- and cauliflower-amended soil. Significantly lower number of galls was recorded in resistant cultivar Great Lakes as compared to that of susceptible cultivar Tyrol. No egg masses were produced in cultivar Great Lakes. Significant reductions in second-stage juveniles (*M. incognita*) in the soil were found in mustard and radish amendment.

4.7.10 Aster Root-Knot Nematode, *Meloidogyne incognita*

In the aster crop, the highest suppression of *M. incognita* population and the lowest gall formation on aster roots were achieved after *E. sativa* green manure and pellet, alone or in combination, which was also significantly lower than fenamiphos. A consistent but lower reduction of nematode population was found after *B. juncea* green manure. Control plots showed significantly lower stem length and diameter, number of flowers per stem and dry matter compared to soil treated with *E. sativa* and *B. juncea* green manures or pellet (De Mastro et al. 2008a).

4.7.11 Sugar Beet Cyst Nematode, *Heterodera schachtii*

Following the major breakthrough in 1985 that oil radish variety RSO1841 significantly reduced sugar beet cyst nematode (SBCN) population, field studies were conducted with Pagletta, Nemex and R184, and more reduction in nematode population was achieved. Higher quality of sugar beet crop was generally produced on soils planted with oilseed radish or white mustard residues than with the regular management practices. In the recent greenhouse studies, it was found that two new oil radish varieties Defender and Comet significantly reduced the population of *H. schachtii* (95%). In the first experiment, mustard Concerta produced 35% more above-ground biomass than radish Colonel, and the viable cysts declined 29 and 19% in oil radish and mustard treatments, respectively. In the second experiment, radish Adagio produced significantly more above-ground biomass than mustard Metex. In the third experiment, biomass (top, root and total) production of oil radish Dacapo was significantly higher than mustard Metex. A fourth experiment indicated that maximum beet yield (t/acre) was with the Luna (37.0) planted plots followed by the Defender (36.0) plots. Recent studies in 2007 proved that amongst six varieties of green manure crops, maximum beet yield (t/acre) was from Defender (36.8), Colonel (36.1) and Arugula (36.4) planted pots. At present, Defender is the most economical variety highly suitable for the SBCN management (Hafez and Sundararaj 2004).

4.7.12 Root-Knot Nematode in Roses

Mexican marigold, also known as *Tagetes*, has been successfully used in the control of root-knot nematode in roses by a Kenyan farmer (report on ToT for alternatives to the use of methyl bromide for soil fumigation in Brazil and Kenya).

4.8 Weed Management

4.8.1 Glasshouse

A mixture of redroot pigweed, wild mustard and wild oat seeds was planted into potting media amended with 0.5 and 1.0 MT/ha canola, oriental mustard or yellow mustard seed meal, along into untreated potting media, and weed seedling emergence and biomass were recorded. Yellow mustard seed meal amendment reduced wild mustard and redroot pigweed seedling emergence by 68 and 70%, respectively, and biomass by 68 and 80%, respectively, compared to the untreated control, but canola and oriental mustard meal treatments had little effect on the broadleaf weeds. In contrast, oriental mustard meal reduced wild oat emergence and biomass most effectively. Yellow mustard meal treatments affected not only weed seedling emergence but seedling growth and mortality. Overall, oriental and yellow mustard meal treatments had the greatest phytotoxic effect on weed emergence and biomass and have potential herbicidal uses (Brown et al. 2006).

4.8.2 Rapeseed for Weed Control

'Dwarf Essex' rapeseed was planted on June 14 and was mowed, chopped and rototilled into the soil on August 30. In some plots, plastic tarps were then laid immediately on the soil in order to combine the action of solarisation and biofumigation. In each plot, seeds of milkweed, curly dock and foxtail were buried. These packets of seeds were retrieved in October and counted the number of seeds that germinated.

Biofumigation with rapeseed alone killed 95% of milkweed seeds, 68% of foxtail seeds and 51% of dock seeds. Rapeseed plots that were tarped for solarisation killed 94% of dock and 100% of milkweed. However, high temperatures under tarps apparently stimulated foxtail. Twice as many foxtail seeds germinated after solarisation as in the control.

4.8.2.1 Potato Weeds

In greenhouse trials on potato, *Brassica hirta*, *B. juncea* and *B. napus* chopped residues added to soil inhibited emergence of several small-seeded weed species from 20 to 95% and reduced growth of weed seedlings from 8 to 90%. Redroot pigweed and barnyard grass, *Echinochloa crus-galli*, germination and growth were inhibited to various degrees by eight individual isothiocyanates, with methyl and allyl isothiocyanates being the most inhibitory. Meadow foam, *Limnanthes alba*, seed meal added to soil in greenhouse studies at 1% or more by weight reduced fresh weight of redroot pigweed by greater than 90% (Boydston et al. 2004).

4.8.2.2 Onion Weeds

Biofumigation is used as an alternative for managing agricultural weeds on onion by using Ida Gold mustard, hairy vetch and winter wheat. These crops are grown for several weeks, and then tilled into the soil to release their bioactive compounds during decomposition, to help manage troublesome weeds such as morning glory and yellow and purple nutsedge in onion fields.

4.8.3 Onion and Garlic Amendments for Weed Control

Results of this study showed that exposure to 39°C soil temperature was consistently deleterious to seed survival in barnyard grass (*Echinochloa crusgalli*), common purslane (*Portulaca oleracea*), London rocket (*Sisymbrium irio*) and black nightshade (*Solanum nigrum*), as compared to 23°C. Significant effects of the onion (*Allium cepa*) and garlic (*A. sativum*) amendments on weed seed viability were common but less consistent than with the other experimental factors. No differences in weed seed viability due to soil amendment with onion versus garlic were found in the 2000 experiment and only in barnyard grass and black nightshade in the 2001 experiment. On the other hand, seed viability differences due to amendment concentration (1 vs. 3% by weight) were found in barnyard grass, black nightshade and London rocket in the 2000 experiment, but no concentration differences were found in the 2001 experiment. In experiments, barnyard grass, common purslane and London rocket seeds were less viable after longer incubation in the microcosms, whilst black nightshade was not significantly affected by exposure time.

The results indicate that the *Allium* spp. amendments, especially when combined with elevated soil temperature (solarisation), may contribute to effective weed management strategies for both conventional and organic production.

4.9 Insect Management

4.9.1 Masked Chafer Beetle

Masked chafer beetle larvae (*Cyclocephala* spp.) are a common soil-inhabiting pest of turf grasses, agronomic crops including corn and soybean, and some ornamental plants. The damage they cause is similar to damage caused by larvae of other scarabaeid pests, including the Japanese beetle (*Popillia japonica*) and European chafer (*Rhizotrogus majalis*).

AITC concentration at 0.25, 4, 8 and 24% was positively correlated to larval mortality (Fig. 4.7). The mortality of grubs in the *B. juncea* treatments was not significantly different ($P<0.05$) from those in the AITC controls, indicating that AITC is the predominant toxin from *B. juncea* responsible for masked chafer beetle larval mortality.

Tilling a winter cover crop of *B. juncea* into the soil and covering with plastic mulch (to help retain volatile compounds) are potential methods of biofumigation.

4.9.2 Termites

The termite, *Reticulitermes grassei*, is more sensitive than *R. santonensis* for each disulfide. It is also noted the greater toxicity of dimethyl disulfide (DMDS) in comparison to dipropyl disulfide (DPDS) and diallyl disulfide (DADS).

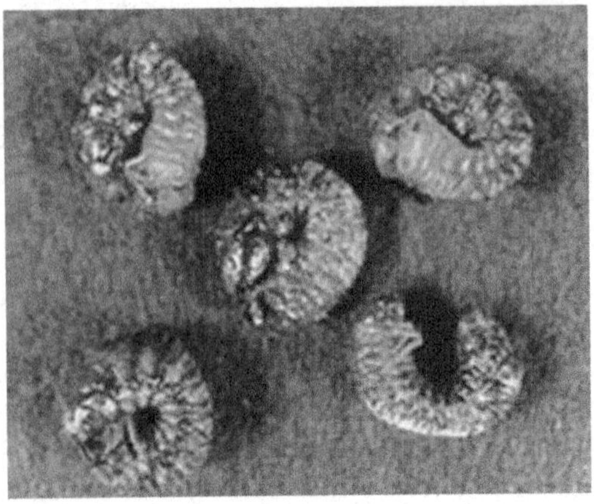

Fig. 4.7 Dead larvae from soil amended with 8% *Brassica juncea* tissue

This result is in accordance with a previous study (Auger et al. 2002), and it can be concluded that isopterans and hymenopterans are more sensitive than dipterans, lepidopterons and coleopterans (about ten times more).

4.9.3 Bruchids

Dimethyl disulfide (DMDS) and dipropyl disulfide (DPDS) lead to the mortality of Coleopteran *Bruchidus atrolineatus* (Nammour et al. 1989). Sulphur compounds also have an action on the Coleopteran *Callosobruchus maculatus* and its parasitoid *Dinarmus basalis* (Dugravot et al. 2002).

4.9.4 Wireworms

The effectiveness of four new biocidal seed meals from *Brassica carinata* sel. ISCI7, *Eruca sativa* cv. Nemat, *Barbarea verna* sel. ISCI100 and *Sinapis alba* cv. Pira applied broadcast and in furrow was evaluated using wireworms (*Agriotes brevis, A. sordidus, A. ustulatus*) put, at different times, in vials or pots filled with sandy soil and planted with winter wheat or maize. The first two biocidal meals caused a very high larval mortality and prevented wireworms from damaging crop seedlings. The best results were obtained with broadcast applications. Insecticide activity of volatiles produced by the glucosinolate enzymatic hydrolysis disappeared in 2–3 days. This makes the treatment particularly safe from an environmental point of view but opens some practical problems to set up an effective procedure to cause high wireworms mortality in open field conditions. It can be hypothesised that biocidal seed meals can give a high wireworm's control effect only if all the following conditions occur concurrently:

- Homogeneous broadcast biocidal seed meal application
- Effective and prompt incorporation into the soil
- Suitable soil temperature and humidity conditions
- Treatment with biocidal meal when there is the presence of most of the wireworms
- Population in the upper part of the soil

All the information about the biology and behaviour of different *Agriotes* species has to be carefully taken into consideration to optimise application procedure. In Italy, the most suitable conditions should be usually found in early spring and early autumn with bare, warm and humid soils, just before and after crop growing season. In order to maximise wireworm population control, at least 10–15 days with soil temperature over 18–20°C after cold or dry conditions should pass so that also the larvae in little mobile larval stages (pre-moulting, moulting and hardening stages) in the deeper soil layers can turn into the feeding stage and move upwards looking for food.

According to the biology of the different *Agriotes* species, it may be also implemented for effective population management; for example, applying biocidal seed meals in the very suitable period of presence of first instar larvae can cause a significant population reduction in next years (Furlan et al. 2004).

4.9.5 Black Vine Weevil

Black vine weevil (*Otiorhynchus sulcatus*) larvae cause significant root damage and have major economical impact in the horticultural industry. There are no effective pesticides or pest management strategies available for control of larvae in field or nursery landscape situations.

Pacific Gold (*Brassica juncea*) or Ida Gold (*Sinapis alba*) seed meal at 3, 6 and 12% quantities was evaluated against black vine weevil. Pacific Gold meal treatments resulted in 100% weevil mortality, whilst Ida Gold meal had no effect on weevil mortality. In Petri-dish studies, relatively small amounts of Pacific Gold meal were effective at killing black vine weevil larvae. Soil amended with seed meal of Pacific Gold and Ida Gold, however, resulted in phytotoxicity and stunted nursery tree growth in both glasshouse and field studies. Glasshouse phytotoxicity studies on nursery trees showed varying results. Ida Gold meal showed phytotoxicity on cedar growth, whilst Pacific Gold proved toxic to dogwood tree growth. With cherry and fir, Pacific Gold treatments resulted in higher growth compared to the control. Phytotoxicity of Pacific Gold meal also resulted in stunted rhododendron and ash trees compared to pure bark mulching. In conclusion, Pacific Gold seed meal has very high potential as a biopesticide to be used in controlling black vine weevil larvae in horticultural tree production although phytotoxic effects need to be considered (Seamons et al. 2004).

4.9.6 Aphids

Toba et al. (1977) demonstrated that wheat *Triticum aestivum* delayed the frequency and severity of aphid-transmitted non-persistent viruses in cantaloupe by controlling the aphid vector.

Hooks et al. (1998) planted zucchini (*Cucurbita pepo*) between rows containing buckwheat (*Fagopyrum esculentum*) and yellow mustard (*Sinapis alba*) and found that densities of *Aphis gossypii* and *Bemisia argentifolii* and the conveyance of aphid-transmitted diseases and squash silver leaf on zucchini were reduced in these living mulch treatments when compared with monoculture plantings. Additional research by Frank (2004) and Frank and Liburd (2005) has shown that living mulch plants containing buckwheat and white clover have higher numbers of natural enemies than synthetic mulch or bare-ground plots.

4.9.7 Citrus California Red Scale

The mortality of adult red scale, *Aonidiella aurantii*, on detached orange fruits treated with a mineral oil emulsion activated by Biofence technology was 100%, significantly greater than mineral oil alone (47.5%). With the aim of improving biodegradability and lowering environmental impact of the formulation, mineral oils were substituted by vegetable one with no statistical differences. Even this time, vegetable oil alone gave 42.5% mortality which was significantly lower than the treatment with new technology (97.5%). A clear 'dose-effect' of liquid formulation was demonstrated. Linear regression between values of mortality and doses showed $R^2=0.936$ and $F=29.01$ on mineral oil and $R^2=0.944$ and $F=33.9$ on vegetable oil (Rongai et al. 2006).

4.9.8 Fungus Gnat on House Plants

Two rates of *Brassica juncea*, 'Pacific Gold', seed meal (1.7 and 4.2 g), no meal, 4.2 g *B. napus* 'Athena' canola seed meal (control) and 10 ml/L Gnatrol (standard) were applied to four common houseplant species in 10-cm-diameter pots by spreading the meal on the surface of the potting medium. Two weeks post treatment, fungus gnat larval mortality was significantly higher with the 4.2 g *B. juncea* seed meal treatment than the other treatments. Plants showed significant phytotoxic post-treatment effects with the *B. juncea* seed meal treatments at both doses. *Brassica juncea* seed meal is toxic to fungus gnat larvae, but it may show signs of phytotoxicity in some plant species, suggesting that all target plant species should be evaluated prior to use of this product (McCaffrey et al. 2006).

4.9.9 Cowpea Pests

The extracts of garlic (*Allium sativum*) bulb and African nutmeg (*Monodora myristica*) seeds were evaluated both applied at 10% in field trials during 1997–1998 against insect pest control on cowpea. All treatments exhibited significant protection from spotted pod borer *Maruca vitrata* and pod-sucking Hemiptera damage compared with untreated control. *A. sativum* and *M. myristica* treated plots recorded 37.3% and 35.0% and 36.1% and 42.7% pod damage in 1997 and 1998, respectively, compared to 67.8% and 98.95% pod damage in the untreated control. *A. sativum* and *M. myristica* treated plots recorded 24.6% and 13.0% and 30.0% and 17.2% higher grain yields in the 1997 and 1998 seasons, respectively, compared to untreated control.

4.9.10 Potato Pests

The field study on effect of intercropping on population dynamics of major insect pests and vectors of potato (cutworm, various defoliators, epilachna beetle, aphid, whitefly and other pests as well as damage caused by the pests) revealed that all the treatments, namely, potato + recommended pesticide schedule of the area, potato + onion in alternative rows (1:1) and potato + garlic in alternative rows (1:1), were statistically superior and significant over control.

4.10 Maximising Biofumigation Potential

4.10.1 Enhancing GSL Profiles

Having established that suppression by incorporated *Brassica* tissues is associated with GSL-hydrolysis products and indications of which hydrolysis products are most toxic, there are significant opportunities to enhance the biofumigation potential of Brassicaceous green manures: first, by selecting Brassicas which produce the greatest amount of the GSL precursors most toxic to the target organisms and, second, by managing the incorporation process to maximise the exposure of the organisms to the toxic compounds at the most vulnerable stage. Strategies to increase the production of GSL-hydrolysis products by Brassicas have been summarised elsewhere (Kirkegaard and Sarwar 1998) and rely upon (1) significant variation in the type and concentration in individual GSLs amongst different species, cultivars and plant parts; (2) independent variation in biomass production; and (3) differential toxicity of different hydrolysis products to particular organisms. Together, these represent opportunities to select or develop biofumigant types which may provide up to 10^5-fold increase in biofumigation potential over varieties selected at random.

4.10.2 Improving Efficacy in Field

The potential efficacy of ITCs present in *Brassica* green manures can be considered by comparison with amounts of the synthetic soil fumigant methyl ITC (MITC) which are applied (517–1,294 nmol/g). Assuming a maximum biomass of 15 t/ha, a tissue GSL concentration of 100 μmol/g and an incorporation depth of 10 cm (soil bulk density 1.4), the potential ITC production (assuming 100% conversion) would be equivalent to 1,070 nmol/g, which is in the range of commercial MITC application. Although the efficiency of conversion from incorporated tissues can be as low as 15% (Borek et al. 1997), several ITCs have been shown to be up to 10 times more toxic than MITC. It is difficult, however, to predict the impact of the lower concentrations of ITC released over an extended time period relative to commercial fumigant applications. More information is required on the fate, persistence and efficacy of the biocidal compounds released from incorporated tissues.

The impact of the biocidal compounds released from the tissue can also be increased by matching the timing of incorporation and release of biocides to the most vulnerable stages of the pest organism's life cycle. Whilst sufficient time must be provided for the organic material to decompose to avoid both physical and potential allelopathic interferences in the following crop, delaying for too long may allow some pathogens to recover.

Further studies are in progress to improve our understanding of the accumulation of GSLs in *Brassica* plants, the fate and activity of the biocidal hydrolysis compounds released from incorporated tissue in the soil and the most effective ways of incorporating biofumigant crops into integrated pest management strategies.

4.10.3 Increasing ITC Production Using Plant Stress

Brassicas increase production of GCs in response to plant stress (e.g. nutrient stress, mechanical damage, disease). For this reason, a field trial was conducted to determine if various stress treatments applied to *B. juncea* could increase its production and release of ITCs into soil. These stresses were applied prior to the incorporation of the crop into soil (12 weeks after sowing) and included low levels of contact herbicide (bipyridyl at 5 mL/100 L water), mechanical damage (puncturing of roots and shoots) and a defence elicitor. Controls consisted of a non-stressed treatment and a fallow.

Although biofumigant treatments did not kill *Phytophthora cactorum*, they suppressed its growth by 10% compared to fallowed plots. Similarly, biofumigant treatments reduced weed biomass by an average of 30%. These results show a clear potential for biofumigation to suppress pathogen and weed populations in the field. There was no evidence that stress treatments increased the efficacy of biofumigation against *P. cactorum* or weeds, although ITC concentrations are yet to be fully analysed.

Biofumigation is based on the use of glucosinolate-containing plants for the control of soil-borne pests and diseases. Upon tissue damage, glucosinolates are hydrolyzed by endogenous enzymes (myrosinase), and a range of biologically active compounds are formed. Isothiocyanates (ITCs) are the quantitatively dominating products formed at neutral pH. Most of these compounds are volatile and only sparingly soluble in aqueous systems, and depending on the R-group structure and the presence of nucleophiles, further transformation of ITCs occurs. At lower pH and in the presence of certain molecules able to deliver two redox equivalents, the proportion of nitriles increases at the expense of ITC.

The effect of ascorbic acid and glutathione on the production of nitriles at pH 5 was investigated by micellar electrokinetic capillary chromatography (MECC). The presence of 0.25 μmol ascorbic acid increased the production of nitriles although at higher concentrations the proportion of nitriles decreased. Increasing amounts of GSH favoured the production of nitriles (40% of the total degradation products were nitriles in the presence of 2 μmol GSH). The oxidation of GSH gives the redox equivalents needed for the liberation of the sulphur from the unstable intermediate of the glucosinolate hydrolysis leading to the formation of the nitrile.

4.11 Future Outlook

It is important to note that success with biofumigant crops depends on a number of factors. The following are some suggestions to achieve the best results:

- Plant biofumigant crop varieties selected or bred for higher levels of active compounds if available.
- Produce as much biomass of the biofumigant crop as possible. This requires a good crop stand, fertility and sufficient growing time. The more biomass that is produced and that is incorporated, the more chemical is released. However, as plants mature, they will reach a point where levels of these active chemicals will decline and the plants should not be allowed to seed. There is also the practical consideration that it is difficult to do a good job of incorporation with too much biomass. With a crop like Sudan grass, this means you cannot let it get too tall.
- The plant material must be thoroughly damaged so that enzymes can convert glucosinolates into isothiocyanates or dhurrin is converted into cyanide. This means that there is need to chop the material as much as possible and work it into the soil as quickly as possible so as to not lose the active compounds to the air. A delay of several hours can cause significant reductions in biofumigant activity. The finer the chop, the more biofumigant is released.

- The material should be incorporated as thoroughly as practical to release the biofumigant chemical throughout the root zone of the area that is to be later planted to vegetables. Poor distribution of the biofumigant crop pieces in the soil will lead to reduced effectiveness.
- Sealing with water or plastic after incorporation will improve the efficacy (as with all fumigants). Soil conditions should not be overly dry or excessively wet.

References

Abragan FN, Justo VP, Minguez LT, Abalde JA, Salvani JB, Maghanoy CC Jr (2008) Evaluating biofumigation for soil-borne disease management in white potato. Third international biofumigation symposium, Canberra, Australia

Angus JF, Gardner PA, Kirkegaard JA, Desmarchelier JM (1994) Biofumigation: Isothiocyanates released from Brassica roots inhibit growth of the take-all fungus. Plant Soil 16:107–112

Auger J, Dugravot S, Naudin A, Abo-Ghalia Pierre D, Thibout E (2002) Possible use of *Allium* allelochemicals in integrated control. IOBC WPRS Bull 25:295

Bending GD, Lincoln SD (1999) Characterization of volatile sulphur-containing compounds produced during decomposition of *Brassica juncea* tissues in soil. Soil Biol Biochem 31:95–103

Bernardi R, Mari M, Leoni O, Casalini L, Cinti S, Palmieri S (2004) Biofumigation for controlling post-harvest fruit pathogens. First international biofumigation symposium, Florence, Italy

Borek V, Elberson LR, McCaffry JP, Morra MJ (1997) Toxicity of rapeseed meal and methyl isothiocyanate to larvae of the black vine weevil (Coleoptera: Curculionidae). J Econ Ent 90:109–112

Boydston RA, Al-Khatib K, Vaughn SF, Collins HP, Alva AK (2004) Weed suppression using cover crops and seed meals. First international biofumigation symposium, Florence, Italy

Brown PD, Morra MJ (1997) Control of soil-borne plant pests using glucosinolate -containing plants. Adv Agron 61:192–204

Brown J, Hamilton M, Davis J, Brown D, Seip L (2006) Herbicidal and crop phytotoxicity of Brassicaceae seed meals on strawberry transplants and established crops. Second international biofumigation symposium, Moscow, ID, USA

Cardwell D, Ingham R (1996) Management practices to suppress Columbia root-knot nematode. Pacific Northwest Sustainable Agriculture, Oct, p 6

Charron CS, Sams CE (1998) Macerated *Brassica* leaves suppress *Pythium ultimum* and *Rhizoctonia solani* mycelial growth. In: Proceedings of the annual international conference on methyl bromide alternatives and emissions reductions, Orlando, FL

Charron CS, Sams CE (1999) Inhibition of *Pythium ultimum* and *Rhizoctonia solani* by shredded leaves of *Brassica* spp. J Am Soc Hortic Sci 124:462–467

Cohen MF, Yamasaki H, Mazzola M (2005) *Brassica napus* seed meal soil amendment modifies microbial community structure, nitric oxide production and incidence of Rhizoctonia root rot. Soil Biol Biochem 37:1215–1227

Colombo A, Cataldi S, Marano G (2008) Effectiveness of biofumigation technique to control the southern root-knot nematode (*Meloidogyne incognita*) in Sicily. Third international biofumigation symposium, Canberra, Australia

De Mastro G, D'Addabbo T, Verdini L, Radicci V (2008a) Biofumigation in greenhouse for the control of root-knot nematodes. Third international biofumigation symposium, Canberra, Australia

De Mastro G, D'Addabbo T, Verdini L, Radicci V (2008b) Control of carrot root-knot and cyst nematodes by biofumigating treatments. Third international biofumigation symposium, Canberra, Australia

Delaquis PJ, Mazza G (1995) Antimicrobial properties of isothiocyanates in food preservation. Food Technol 49:73–84

Dugravot S, Sanon A, Thibout E, Huignard J (2002) Susceptibility of *Callosobruchus maculatus* (Coleoptera: Bruchidae) and its parasitoid *Dinarmus basalis* (Hymenoptera: Pteromalidae) to sulphur-containing compounds: consequences on biological control. Environ Entomol 31:550

Fenwick GR, Heaney RK, Mullin WJ (1983) Glucosinolates and their breakdown products in food and food plants. Crit Rev Food Sci Nutr 18:123–201

Forge TA, Russel EI, Diane K (1995) Winter cover crops for managing root lesion nematodes affecting small fruit crops in the Pacific Northwest. Pacific Northwest Sustainable Agriculture, p 6

Frank DL (2004) Evaluation of living and synthetic mulches on zucchini for control of homopteran pests. MS thesis, University of Florida, Gainesville, FL, USA

Frank DL, Liburd OE (2005) Effects of living and synthetic mulches on the population dynamics of whiteflies and aphids, their associated natural enemies and insect-transmitted plant diseases in zucchini. Environ Entomol 34:857–865

Furlan L, Bonetto C, Patalano G, Lazzeri L (2004) Potential of biocidal meals to control wireworm populations. First international biofumigation symposium, Florence, Italy

Gamliel A, Stapleton JJ (1993) Effect of soil amendment with chicken compost or ammonium phosphate and solarization on pathogen control, rhizosphere microorganisms and lettuce growth. Plant Dis 77:886–891

Gouws R (2004) Biofumigation as alternative control measure for common scab on seed potatoes in South Africa. First international biofumigation symposium, Florence, Italy

Grossman J (1990) New crop rotations foil root-knot nematodes. Common sense pest control, Winter, p 6

Hafez SL, Sundararaj P (2004) Biological and chemical management strategies in the sugar beet cyst nematode management. In: Proceedings of the Winter Commodity Schools – 2004, University of Idaho, Moscow, ID, USA, pp 243–248

Harding RB, Wicks TJ (2001) *In vitro* suppression of soil-borne potato pathogens by volatiles released from *Brassica* residues. In Porter IJ (ed) Proceedings of the 2nd Australasian soil-borne diseases symposium, Lorne, Australia, pp 148–149

Harvey SG, Hannahan HN, Sams CE (2002) Indian mustard and allyl isothiocyanate inhibit *Sclerotium rolfsii*. J Am Soc Hortic Sci 127:27–31

Hooks CRR, Valenzuela HR, Defrank J (1998) Incidence of pest and arthropod natural enemies in zucchini grown in living mulches. Agri Ecosys Environ 69:217–231

Isshiki K, Tokuoka K, Mori R, Chiba S (1992) Preliminary examination of allyl isothiocyanate vapour for food preservation. Biosci Biotech Biochem 56:1476–1477

Kirkegaard JA, Sarwar M (1998) Biofumigation potential of brassicas. I. Variation in glucosinolate profiles of diverse field-grown brassicas. Plant Soil 201:71–89

Kirkegaard JA, Gardner PA, Desmarchelier JM, Angus JF (1993) Biofumigation using Brassica species to control pests and diseases in horticulture and agriculture. In: Wrattenand N, Mailer RJ (eds) Proceedings of 9th Australian research assembly on Brassicas. Agricultural Research Institute, Wagga, pp 77–82

Koch DW (1995) Brassica utilization in sugar beet rotations for biological control of cyst nematode. SARE project report LW91-022, Western Region SARE, Logan, Utah

Larkin RP, Griffin TS (2007) Control of soil-borne potato diseases using Brassica green manures. Crop Prot 26:1067–1077

Lazzeri L, Leoni O, Bernardi R, Malaguti L, Cinti S (2004a) Plants, techniques and products for optimizing biofumigation in full field. Agroindustria 3:181–188

Lazzeri L, Leoni O, Bernardi R, Malaguti L, Cinti S (2004b) Plants, techniques and products for optimizing biofumigation in the full field. Agroindustria 3:281–287

Luna J (1993) Crop rotation and cover crops suppress nematodes in potatoes. Pacific Northwest Sustainable Agriculture, pp 4–5

Mallek SB, Prather TS, Stapleton JJ (2007) Interaction effects of *Allium* spp residues, concentrations and soil temperature on seed germination of four weedy plant species. Appl Soil Ecol 37:9

Matthiessen JN, Kirkegaard JA (2006) Biofumigation and enhanced biodegradation: opportunity and challenges in soil-borne pest and disease management. Crit Rev Plant Sci 25:235–265

Mazzola M, Granatstein DM, Elfving DC, Mullinis K (2001) The suppression of specific apple root pathogens by *Brassica napus* seed meal amendment regardless of glucosinolate content. Phytopathology 91:673–679

McCaffrey JP, Morra MJ, Main ME (2006) *Brassica juncea* seed meal for fungus gnats control associated with house plants with consideration of phytotoxicity to specific plant species. Second international biofumigation symposium, Moscow, ID, USA

McLeod RW, Steel CC (1999) Effects of Brassica-leaf green manures and crops on activity and reproduction of *Meloidogyne javanica*. Nematology 1:613–624

McLeod RW, Kirkegaard JA, Steel CC (2001) Invasion, development, growth and egg –laying by *Meloidogyne javanica* in Brassicaecae crops. Nematology 3:463–472

Mithen RF, Lewis BG, Fenwick GR (1986) *In vitro* activity of glucosinolates and their products against *Leptosphaeria maculans*. Trans Br Mycol Soc 87:433–440

Mojtahedi H, Santo GS, Hang AN, Wilson JH (1991) Suppression of root-knot nematode population with selected rapeseed cultivars as green manure. J Nematol 23:170

Mojtahedi H, Santo GS, Wilson JH, Hang AN (1993) Managing *Meloidogyne chitwoodi* on potato with rapeseed as green manure. Plant Dis 77:42–46

Muller J (1999) The economic importance of *Heterodera schachtii* in Europe. Helminthologia 36:205–213

Nammour D, Auger J, Huignard J (1989) Mise en évidence de l'effet insecticide de composes soufrés (disulfures et trisulfures) sur *Bruchidus atrolineatus* (Pic) (Coléoptère: Bruchidae). Insect Sci Appl 10:49

Ploeg AT, Stapleton JJ (2001) Glasshouse studies on the effects of time, temperature and amendment of soil with broccoli plant residues on the infestation of melon plants by *Meloidogyne incognita* and *M javanica*. Nematology 3:855–861

Potter MJ, Davies K, Rathjen AJ (1998) Suppressive impact of glucosinolates in *Brassica* vegetative tissues on root lesion nematode *Pratylenchus neglectus*. J Chem Ecol 24:67–80

Putnam AR, DeFrank J (1983) Use of phytotoxic plant residues for selective weed control. Crop Prot 2:81

Rahman L, Somers T (2005) Suppression of root-knot nematode (*Meloidogyne javanica*) after incorporation of Indian mustard cv Nemfix as green manure and seed meal in vineyards. Austr Plant Pathol 34:77–83

Rongai D, Lazzeri L, Cerato C, Palmieri S, Patalano G (2006) A liquid formulation as biopesticide for the control of California red scale (*Aonidiella aurantii* Maskell). Second international biofumigation symposium, Moscow, ID, USA

Sanders D (2005) Growers guidelines. American Vegetable Grower, October 2005. University of California SAREP Online Cover Crop Database, Mustards

Seamons C, Brown J, McCaffrey J, Lloyd J (2004) Using *Brassicaceae* seed meal as a soil fumigant in horticultural situations. First international biofumigation symposium, Florence, Italy

Stapleton JJ (2006) Biocidal and allelopathic properties of graminaceous crop residue amendments as influenced by soil temperature. Proc Calif Conf Biol Control 5:179–181

Stirling GR, Stirling AM (2003) The potential of Brassica green manure crops for controlling root- knot nematode

(*Meloidogyne javanica*) on horticultural crops in a subtropical environment. Austr J Exp Agric 43: 623–630
Subbarao KV, Kabir Z, Martin FN, Koike ST (2007) Management of soilborne diseases in strawberry using vegetable rotations. Plant Dis 91:964–972
Summers CG, Mitchell JP, Prather TS, Stapleton JJ (2009) Sudex cover crops can kill and stunt subsequent tomato, lettuce and broccoli transplants through allelopathy. Calif Agric 63:40
Toba HH, Kishaba AN, Bohn GW, Hield H (1977) Protecting muskmelon against aphid-borne viruses. Phytopathology 67:1418–1423
Tollsten L, Bergström G (1988) Headspace volatiles of whole plants and macerated plant parts of *Brassica* and *Sinapis*. Phytochemistry 27:4013–4018
Urbasch I (1984) Production of C6-wound gases by plants and the effect on some phytopathogenic fungi. Z Naturforsch 39c:1003–1007
USDA (2000) Economic implications of the methyl bromide phase out. US Department of Agriculture, Agriculture Information Bulletin #756, pp 77–105
Walker JC, Morel S, Foster HH (1937) Toxicity of mustard oils and related sulphur compounds to certain fungi. Am J Bot 24:536–541
Wider TL, Abawi GS (2000) Mechanism of suppression of *Meloidogyne hapla* and its damage by green manure of Sudan grass. Plant Dis 84:562–568
Xiao CL, Subbarao KV, Schulbach KF, Koike ST (1998) Effects of crop rotation and irrigation on *Verticillium dahliae* microsclerotia in soil and wilt in cauliflower. Phytopathology 88:1046–1055

Biotechnological Approaches 5

Abstract

Biotechnology offers many opportunities for agriculture and provides the means to address many of the constraints placed to productivity. It uses the conceptual framework and technical approaches of molecular biology and plant cell culture systems to develop commercial processes and products. With the rapid development of biotechnology, agriculture has moved from a resource-based to a science-based industry, with plant breeding being dramatically augmented by the introduction of recombinant DNA technology based on knowledge of gene structure and function. The concept of utilising a transgenic approach to host plant resistance was realised in the mid-1990s with the commercial introduction of transgenic maize, potato and cotton plants expressing genes encoding the insecticidal δ-endotoxin from *Bacillus thuringiensis*. Similarly, the role of herbicides in agriculture entered a new era with the introduction of glyphosate-resistant soybeans in 1995. Currently, the commercial area planted to transgenic crops is in excess of 90 million hectares with approximately 77% expressing herbicide tolerance, 15% expressing insect-resistance genes and approximately 8% expressing both traits. Despite the increasing disquiet over the growing of such crops in Europe and Africa (at least by the media and certain NGOs) in recent years, the latest figures available demonstrate that the market is increasing, with an 11% increase between 2004 and 2005.

The human population is ever increasing, with conservative estimates predicting that the population will rise to approximately ten billion by 2050. Thus, the major challenges facing the world are to feed and provide shelter for a world population that is increasing at an exponential rate. Agriculture must play a major role in achieving these goals both by providing ever-increasing food yields and an ever-increasing supply of natural products required by industry. An immediate priority for agriculture is to achieve maximum production of food and other products in a manner that is environmentally sustainable and cost-effective, against a backdrop of climate change which not only is predicted to result in the loss of agricultural land as a consequence of rising sea levels but is also likely to have a major impact on the dynamics of pest populations.

Historically, pathogen infection and insect infestation of staple crops have led to food

P.P. Reddy, *Recent Advances in Crop Protection*,
DOI 10.1007/978-81-322-0723-8_5, © Springer India 2013

shortages and considerable economic losses. Resistant varieties have, therefore, been developed by plant breeders for a number of years to reduce such losses, but pathogens are eventually able to overcome this resistance. Many pesticides have been developed to combat crop losses, with the consequence that plant disease control has become heavily dependent on these compounds. Yet the use of pesticides has also resulted in significant costs to public health and the environment. Plant biotechnology (gene isolation and plant transformation techniques), together with conventional breeding programmes, could make significant contributions to sustainable agriculture. In this regard, there has been intensive research in agricultural biotechnology aimed at crop protection. Strategy to improve crop production through genetic engineering involves protecting crops from pests. Insects, viruses, bacteria, fungi, nematodes and weeds can all impair agricultural productivity. This chapter will describe the main defence-related mechanisms that plants display to cope with pathogen infection. Subsequently, the current status of the genes identified for resistance against virus, fungi, insects and nematodes, with an emphasis on their role in resistance to pathogens in transgenic plants, shall be discussed.

5.1 Role of Transgenic Crops in Agriculture

Applications of transgenic technology in agriculture have clearly defined benefits, not least in providing greater sustainability in terms of improved levels of crop protection resulting in higher yields and reduced pesticide application. However, a major challenge facing this new industry is in the identification of suitable genes for transfer that will confer the desired agronomic traits. In terms of insect resistance, several different classes of bacterial-, plant- and animal-derived proteins have been shown to be insecticidal towards a range of economically important insect pests from different orders, with the midgut being the prime target (Gatehouse and Gatehouse 1999). Of these, the *Bt* toxins are the most commercially relevant, since, to date, *Bt*-expressing crops are the only insect-resistant transgenic crops to have been commercialised.

5.2 Genes in Defence Against Diseases

5.2.1 Fungal Resistance

Fungal diseases have been one of the main causes of crop losses for years. Control of fungal diseases has traditionally involved three strategies: husbandry techniques, such as crop rotation and confinement of contaminated soil and plant material, breeding of resistant crop cultivars and application of fungicides. Although conventional plant breeding has made a significant impact by improving crop resistance to many important diseases, it still requires extended periods of time to develop a new variety and there may not be any natural sources of resistance to major diseases available to the breeder (Oerke 1994). In modern agriculture, farmers employ a variety of fungicides, but their use is being restricted because of their high costs and growing concerns about the degradation of the environment. Furthermore, the excessive use of fungicides frequently results in the development of resistant fungal strains. The advent of advanced molecular techniques for plant breeding has allowed the development of crops resistant to a number of fungal pathogens (Table 5.1). There have been mainly two approaches to generate broad-spectrum, fungal-resistant crops: one relies on the overexpression of genes encoding antifungal proteins (AFP) and the other aims to enhance pathogen perception, by manipulation of certain pathogen and plant *R* genes.

5.2.1.1 Antifungal Proteins

Many of the genes induced by the plant disease-resistance responses described encode proteins with antifungal activity, which probably have an important role against fungal infection. The AFP strategy involves the constitutive expression

Table 5.1 Fungal resistance in transgenic plants

Plant	Transgene(s)	Pathogen	Reference
Alfalfa	Peanut resveratrol synthase	*Phoma medicaginis*	Hipskind and Paiva (2000)
Brassica napus	Bean chitinase	*Rhizoctonia solani*	Broglie et al. (1991)
	Tomato/tobacco chitinase	*Cylindrosporium conc.*	Grison et al. (1996)
		Sclerotinia sclerotiorum	Grison et al. (1996)
		Phoma lingam	Grison et al. (1996)
Carrot	Tobacco chitinase + β-1, 3-glucanase	*Alternaria dauci*	Melchers and Stuiver (2000)
	Tobacco AP24	*Alternaria radicina*	Melchers and Stuiver (2000)
		Cercospora carotae	Melchers and Stuiver (2000)
		Erysiphe heraclei	Melchers and Stuiver (2000)
Potato	AP24	*Phytophthora infestans*	Liu et al. (1994)
	Glucose oxidase	*Phytophthora infestans*	Wu et al. (1995)
		Verticillium dahliae	Wu et al. (1995)
	Osmotin	*Phytophthora infestans*	Li et al. (1999)
	Tobacco class II catalase	*Phytophthora infestans*	Yu et al. (1999)
	Aly AFP	*Verticillium* sp.	Liang et al. (1998)
	Soybean β-1, 3-glucanase	*Phytophthora infestans*	Borkowska et al. (1998)
	PR-1a, SAR 8.2	*Peronospora tabacina*	Alexander et al. (1993)
Tobacco		*Phytophthora parasitica*	Alexander et al. (1993)
		Pythium sp.	
	Class III chitinase	*Phytophthora parasitica*	Alexander et al. (1993)
	Class I chitinase	*Rhizoctonia solani*	Alexander et al. (1993)
	Bean chitinase	*Rhizoctonia solani*	Broglie et al. (1991)
	Barley RIP	*Rhizoctonia solani*	Logemann et al. (1992)
	Serratia marcescens chitinase	*Rhizoctonia solani*	Logemann et al. (1992)
	Barley chitinase + β-1, 3-glucanase	*Rhizoctonia solani*	Jach et al. (1995)
	Barley chitinase + RIP	*Rhizoctonia solani*	Jach et al. (1995)
	Sarcotoxin IA	*Rhizoctonia solani*	Mitsuhara et al. (2000)
		Pythium aphanidermatum	Mitsuhara et al. (2000)
	Pseudomonas syringae hrmA	*Phytophthora parasitica*	Shen et al. (2000)
	Oxalate decarboxylase	*Sclerotinia sclerotiorum*	Kesarwani et al. (2000)
	Radish AFP	*Alternaria longipes*	Terras et al. (1995)
Tomato	Tomato chitinase + β-1, 3-glucanase	*Fusarium oxysporum*	Jongedijk et al. (1995)
Rice	Rice chitinase	*Rhizoctonia solani*	Lin et al. (1995)
Wheat	Aly AFP	*Fusarium* sp.	Liang et al. (1998)

in transgenic plants of genes encoding proteins with a fungitoxic or fungistatic capacity (Table 5.2). This is an extension of the paradigm that has worked so well for insect control genes based on insecticidal proteins from *Bacillus thuringiensis*.

Plant defence mechanisms are usually accompanied by the expression of a large set of genes termed PR. At least ten families of PRPs have been identified (Dempsey et al. 1998), and the two most prominent members have been the hydrolytic enzymes chitinase and β-1, 3-glucanase, which are capable of degrading the major cell wall constituents (i.e. chitin and β-1, 3-glucan) of most filamentous fungi. Expression of either enzyme individually in a number of plants has conferred resistance against a particular pathogen, but constitutive coexpression of both enzymes confer even higher levels of resistance, suggesting a synergistic interaction between the two enzymes. Other genes coding for different PRPs have yielded comparable results.

Table 5.2 Pathogenesis-related proteins in plants

PR protein family	Enzymatic activity	Target in pathogen
PR-1	Unknown	Membrane?
PR-2	1, 3-glucanase	Cell wall glucan
PR-3	Endochitinase	Cell wall chitin
PR-4	Endochitinase	Cell wall chitin
PR-5	Osmotin-like	Membrane
PR-6	Proteinase inhibitor	Proteinase
PR-7	Proteinase	Unknown
PR-8	Endochitinase	Cell wall chitin
PR-9	Peroxidase	Plant cell wall
PR-10	RNAase?	Unknown
PR-11	Endochitinase	Cell wall chitin
Unclassified	α-Amylase	Cell wall α-glucan
	Polygalacturonase	Unknown
	Inhibitor protein (PGIP)	Polygalacturonase

Constitutive expression in tobacco of PR1a, a protein with unknown biological function, increased resistance to two oomycete pathogens (Alexander et al. 1993). However, fungal resistance is significantly enhanced when more than one gene is employed. Future studies are necessary to identify different combinations of PR proteins that might confer effective broad-spectrum protection.

In addition to PR proteins, a broad family of small, cysteine-rich AFP has been characterised (Broekaert et al. 1995, 1997). This family includes plant defensins, thionins, lipid transfer proteins (LTP) and hevein- and knottin-type peptides and has been shown to possess antifungal activity in vitro against a broad spectrum of fungal pathogens (Broekaert et al. 1997). They seem to inhibit fungal growth by permeabilisation of fungal membranes (Thevissen et al. 1999). Plant defensins have been isolated from seeds of a variety of plants and shown to be induced upon pathogen infection in an SA-independent manner (Terras et al. 1995). They have been used to confer resistance against *Alternaria longipes* in tobacco (Broekaert et al. 1995). Thionins and LTP are also induced in an SA-independent manner by infection of a variety of plant pathogens in many plant tissues (Broekaert et al. 1997). Overexpression of an endogenous thionin gene in *Arabidopsis* conferred protection against *Fusarium oxysporum* (Epple et al. 1997), whereas expression of a barley non-specific LTP in *Arabidopsis* and tobacco conferred enhanced resistance to the bacterial pathogen *Pseudomonas syringae* (Molina and Garcia-Olmedo 1997). In the only example where hevein- and knottin-type peptides were overexpressed in transgenic plants, the resultant tobacco plants were not any more resistant to infection by *Alternaria longipes* than control plants (De Bolle et al. 1993). This might be explained by the high susceptibility of the antifungal activity of hevein- and knottin-type peptides to the presence of inorganic cations (De Bolle et al. 1993).

Finally, there are several AFP that do not fall into any of the classes described above. For example, the H_2O_2-producing enzyme oxalate oxidase has been shown to accumulate in barley infected by *Erysiphe graminis* (Zhou et al. 1995). Interestingly, Wu et al. (1995) demonstrated that constitutive expression of another H_2O_2-producing enzyme, glucose oxidase, provided disease resistance to a range of plant pathogens, including *Phytophthora infestans*, *Erwinia carotovora* and *Verticillium* wilt disease.

Most of the reports described above are based exclusively on observations of the increased fungal resistance of transgenic plants tested in climate-controlled growth chambers or greenhouse facilities. The challenge now is to translate such results into a significant outcome in the field.

5.2.1.2 Plant *R* Genes

The genetically controlled induction of HR is triggered in plant–pathogen interactions only if the plant contains a disease-resistance protein (R) that recognises the correspondent avirulence (Avr) protein from the pathogen. In the absence of a functional *R* gene or avirulence gene product, no recognition occurs and disease ensues. This indicates that the factors controlling HR are quite specific and that they do not provide resistance to more than a limited number of races or pathotypes. To engineer broad-spectrum disease resistance relying on HR, two approaches have been employed, although they are still in the early stages: one is based on the transfer of an avirulence gene (i.e. the *Cladosporium fulvum avr9* gene) into a plant containing the corresponding resistance gene (i.e. the tomato *Cf-9* gene) and its subsequent expression under the control of a promoter inducible by fungal pathogens. Pathogen-induced expression of the *Avr* gene will then provoke a resistance reaction manifested by a HR. A localised HR will be induced preventing further spread of any invading pathogen, followed by a general defence response. The other approach is based on the overexpression of *R* genes.

The first approach has been tested experimentally three times, twice in tomato using the *AvrPto* gene (Melchers and Stuiver 2000) and the other in tobacco using an elicitin gene coupled to a pathogen-inducible promoter (Keller et al. 1999). In both cases, increased resistance to a broad spectrum of diseases, including both fungal and viral, was obtained. Nevertheless, the main limitation of this approach is still the limited number of *Avr* genes available.

The second approach has benefited from the cloning and analysis of over 20 *R* genes isolated from seven plant species, including both monocots and dicots (Martin 1999). Resistance genes involved in race-specific interactions often provide full disease resistance and are well known from conventional breeding programmes (Rommens and Kishore 2000). In spite of the great diversity in lifestyles and pathogenic mechanisms of disease-causing organisms, it was somewhat surprising that *R* genes were found to encode proteins with sequence similarities and conserved motifs (Martin 1999). They have been classified into five classes according to the structural characteristics of their predicted protein products: intracellular protein kinases, receptor-like protein kinases with an extracellular leucine-rich repeat (LRR) domain, intracellular LRR proteins with a nucleotide binding site (NBS) and a leucine zipper motif, intracellular NBSLRR proteins with a region with similarity to the toll and interleukin-1 receptor (TIR) proteins and LRR proteins that encode membrane-bound extracellular proteins.

Overexpression of *Pto* (an *R* gene) in tomato elicited an array of defence responses including microscopic cell death, SA accumulation and PR gene expression, and the plants showed increased resistance to several pathogenic bacteria and fungi (Tang et al. 1999). Similarly, overexpression of the *Bs2* pepper *R* gene in tomato allowed enhanced resistance against *Xanthomonas campestris* and bacterial spot disease. Isolation and analysis of plant *R* genes is an extremely active area of research, and considering the large number of genes already available and the current work to isolate more genes in many laboratories around the world, there will be substantial progress in this field in the short term.

Overexpression of signalling components that lie downstream of *R* genes is another possible strategy to increase disease resistance. In the first successful example of this approach, the *NPR1* gene was overexpressed in *Arabidopsis*, and the resultant transgenic plants exhibited significant increases in resistance to *Pseudomonas* and *Peronospora* pathogens (Cao et al. 1998). Manipulation of downstream components such as *NPR1* potentially allows activation of only specific defence pathways. This might be one way to avoid agronomic problems associated with constitutive activation of *R*-gene-mediated pathways such as HR.

5.2.1.3 Other Strategies

In addition to the strategies described above, other approaches to control fungal disease are based on the manipulation of the levels of phytoalexins which are small, broad-spectrum antimicrobial compounds whose synthesis and accumulation is frequently associated with HR. For example, increased resistance to *Botrytis cinerea* has been observed in tobacco expressing a grapevine stilbene synthase gene, which increased the level of the phytoalexin resveratrol (Hain et al. 1993). However, there are several caveats to engineering phytoalexin-mediated resistance. Many of these compounds are synthesised via complex pathways, which, in order to alter existing phytoalexin structure or content, would require the manipulation of several genes, making the whole process technically more demanding. In addition, phytoalexins are often toxic to the pathogen as well as to the plant. Thus, some type of inducible expression system may be required. A novel and promising approach to engineer broad-spectrum resistance relies on the use of antimicrobial peptides. These peptides are ubiquitous (Gabay 1994) and show a strong antimicrobial activity. Expression in potato plants of a synthetic gene encoding an N-terminus-modified, cecropin-melittin cationic peptide chimera resulted in significant resistance against several bacterial and fungal phytopathogens (Osusky et al. 2000). It is likely that this type of approach will be employed considerably in the near future to engineer a range of disease-resistant plants.

5.2.1.4 *Trichoderma harzianum* as a Biological Control Agent

Biological control of soil-borne plant pathogens is an attractive approach to control fungal diseases (Chet 1990). When effective, it is a significant method of pest control not only because it eliminates the use of fungicides but also because if the introduced biocontrol agent becomes properly established, it does not require repeated applications. *Trichoderma* spp. is amongst the most studied biocontrol agents (Chet 1990). Several species of *Trichoderma* have been isolated and found to be effective biocontrol agents of various soil-borne plant pathogenic fungi under greenhouse and field conditions (Chet and Inbar 1994). *Trichoderma harzianum* has proved to be the most effective species, and it has been shown to attack a range of economically important soil-borne plant pathogenic fungi (Chet 1990). *Trichoderma* can be added to the soil as a powder, wheat bran or a peat–wheat bran preparation. It can be sprayed or injected, painted on tree wounds and inserted in pellets in holes drilled in trees, and conidia can be applied directly to the ground or as seed coating (Chet 1990). In addition, *Trichoderma* may contribute to the overall plant defence response as it has been recently shown that *Trichoderma* application induces systemic resistance mechanisms in cucumber plants (Yedidia et al. 1999).

Trichoderma may have three modes of action as part of its antagonistic activity, antibiosis, competition and mycoparasitism. However, mycoparasitism has been suggested as the main mechanism involved in the antagonism as a biocontrol agent (Haran et al. 1996). In order to attack a fungal cell, *Trichoderma* must degrade the cell wall. Given the composition of most fungal cell walls, it has been suggested that chitinases, proteases and β-1, 3-glucanases are the main enzymes involved in the mycoparasitic process (Elad et al. 1982; Haran et al. 1996). However, it is likely that the co-ordinated action of all hydrolases produced by *Trichoderma* is required for a complete dissolution of the cell wall. Indeed, a number of *Trichoderma* isolates are able to secrete different kinds of hydrolytic enzymes into the medium when grown in the presence of cell walls of phytopathogenic fungi (Geremia et al. 1993). In addition, the production of several of these lytic enzymes by *Trichoderma* is induced during the parasitic interaction (Haran et al. 1996; Flores et al. 1997).

As of late, researchers have been trying to increase the effectiveness, stability and biocontrol capacity of *Trichoderma* spp. by altering the levels of different hydrolytic enzymes. Several improved *Trichoderma* strains have, therefore, been obtained which display an increased antifungal activity by overexpression of a proteinase (Flores et al. 1997) or a 33-kDa chitinase (Limon et al. 1999). For that reason, considering the activity and specificity of many fungal enzymes, mycoparasitic fungi may serve as excellent sources of genes for disease resistance (Lorito et al. 1998). In spite of the success achieved, treatments with *T. harzianum* have usually not

Fig. 5.1 Papaya ringspot virus resistance. *Left* – susceptible, *right* – resistant (Rainbow)

been as effective as the use of some fungicides. Treatment with *T. harzianum* has, therefore, been combined with other cultural practices to implement integrated pest management (Hall 1991).

5.2.2 Viral Resistance

There is much interest in the genetic engineering of viral disease-resistant plants, and some success has been obtained with several virus diseases, the best known of which is papaya ringspot (Fig. 5.1). Papaya cv. Rainbow is resistant to ringspot virus.

5.3 Genes in Defence Against Insect Pests

5.3.1 Plants Expressing *Bacillus thuringiensis* (*Bt*) Toxins

Insect infestation has caused heavy losses in many agricultural crops for years. Depending on the crop, it is estimated that losses range from 5 to 30% (FAOSTAT 2000). Control of insects has traditionally employed application of pesticides and to some extent biocontrol agents such as *Bacillus thuringiensis* (*Bt*). *Bt* is a soil-dwelling bacterium of major agronomic and scientific interest. Whilst the subspecies of this bacterium colonises and kills a large variety of host insects, each strain tends to be highly specific. Toxins for insects in the orders Lepidoptera (butterflies and moths), Diptera (flies and mosquitoes), Coleoptera (beetles and weevils) and Hymenoptera (wasps and bees) have been identified (de Maagd et al. 2001), but interestingly, none with activity towards Homoptera (sap suckers) have, as yet, been identified, although a few with activity against nematodes have been isolated. Further, there is little evidence of effective *Bt* toxins against many of the major storage insect pests.

These *Bt* toxins (also referred to as δ-endotoxins; *cry* proteins) exert their pathological effects by forming lytic pores in the cell membrane of the insect gut. On ingestion, they are solubilised and proteolytically cleaved in the midgut to remove the C-terminal region, thus generating an 'activated' 65–70 kDa toxin. The active toxin molecule binds via domains I and II to a specific high-affinity receptor in the insect midgut epithelial cells. Following binding, domain I is inserted into the membrane where it forms pores with other toxin molecules; this results in cell death by colloid osmotic lysis, followed by death of the insect (de Maagd et al. 2001). A number of putative receptors have been identified and include aminopeptidase N proteins (Luo et al. 1997), cadherin-like proteins (Gahan et al. 2001) and glycolipids (Denolf 1996), although Griffitts et al. (2001) suggest that a common carbohydrate motif may explain why a single toxin can bind to at least two receptors that are completely unrelated in sequence.

Transgenic plants expressing *Bt* toxins were first reported in 1987 (Vaeck et al. 1987), and following this initial study, numerous crop species have been transformed with genes encoding a range of different *cry* proteins targeted towards different pests species. Since bacterial *cry* genes (genes encoding *Bt* toxins) are rich in A/T content compared to plant genes, both the full-length and truncated versions of these *cry* genes have had to undergo considerable modification of codon usage and removal of polyadenylation sites before successful expression in plants (de Maagd et al. 1999). These studies have been extensively reviewed, and the reader is referred elsewhere (Shelton et al. 2002). Crops expressing *Bt* toxins were first commercialised in the mid-1990s, with the introduction of *Bt* potato and cotton.

The cloning of the δ-endotoxin of *Bt* has allowed the generation of transgenic plants containing the gene. For the past 10 years, genes coding for δ-endotoxins from different *Bt* subspecies, individually or in combination, have been used to protect crops against insects (Sanchis and Lereclus 1999). Currently, corn, potato and cotton plants expressing different synthetic *Bt* are commercially available; they show meaningful protection against different insects such as European corn borer, Colorado potato beetle and bollworm infestations, respectively (Dempsey et al. 1998). Recently transgenic rice containing a fusion gene derived from cry IA(b) and cry IA(c) was field tested in natural and repeated heavy manual infestation of two lepidopteron insects, leaf folder and yellow stem borer (Tu et al. 2000). The transgenic hybrid plants showed high resistance against both insect pests without reduced yield.

Currently, the commercial area planted to transgenic crops is in excess of 90 million hectares with approximately 77% expressing herbicide tolerance, 15% expressing insect-resistance genes and approximately 8% expressing both traits. Despite the increasing disquiet over the growing of such crops in Europe and Africa (at least by the media and certain NGOs) in recent years, the latest figures available demonstrate that the market is increasing, with an 11% increase between 2004 and 2005 (James 2005). To date, there are no reports of resistance in pest populations having evolved in the field to transgenic *Bt*-expressing plants (Ferry et al. 2006). However, resistance has evolved to the lower *Bt* levels found in *Bt* bacterial sprays used in organic agriculture (Tabashnik et al. 1996). The cultivation of *Bt*-expressing crops has brought some substantial gains to the farming community both in terms of increased yields and lower production costs. For example, the costs for producing *Bt* cotton in China compared to isogenic non-*Bt* cotton varieties were approximately fivefold less, representing significant savings. This saving was primarily due to reduced pesticide application. Similarly, benefits in India include a 70% reduction in insecticide applications in *Bt* cotton fields, resulting in a saving of up to US$ 30/ha in pesticide costs, with an increase of approx 85% in yield of harvested cotton (Christou et al. 2006). Furthermore, expression of *Bt* has also resulted in improved crop quality as a consequence of decreased levels of *Fusarium* infestation and fumonisin mycotoxin production; this benefit is particularly important in food crops such as maize.

5.3.2 Transgenic Plants Expressing Inhibitors of Insect Digestive Enzymes

A quite different alternative to control insect infestation has been the use of proteinase inhibitors (PIs). These compounds can inhibit various digestive enzymes (proteinases) found in the gut of many insects. The synthesis of some PI is stimulated by wounding, including insect attack, whereas others are induced by pathogen infection (Ryan 1990). The wound-inducible serine PIs from tomato have been studied the most extensively. They were divided into two groups based on sequence and molecular weight

(PI 1=8 kDa; PI I=12 kDa) (Ryan 1990). Several components in the induction pathway leading to PI synthesis have been identified, including systemin, various intermediates of the octadecanoid pathway and jasmonic acid (Ryan and Pearce 1998). Constitutive expression of different types of PI, including serine (Hilder and Boulter 1999) and cysteine (Irie et al. 1996), in transgenic plants reduces predation by inhibiting important digestive enzymes in the insect gut. Nevertheless, similarly to the *Bt* δ -endotoxin, insects have developed resistance against PI (Girard et al. 1998). For a more effective and durable resistance, it may be necessary to combine different strategies.

Strategies to exploit endogenous resistance mechanisms (Gatehouse 2002) by way of genes encoding such defensive proteins were obvious candidates for enhancing crop resistance to insect pests. Interfering with digestion, and thus affecting the nutritional status of the insect, is a strategy widely employed by plants for defence and has been extensively investigated as a means of producing insect-resistant crops (Gatehouse et al. 2000). Numerous studies since the 1970s have confirmed the insecticidal properties of a broad range of protease inhibitors from both plant and animal sources (Gatehouse et al. 2000). Proof of concept for exploiting such molecules for crop protection was first demonstrated with expression of a serine protease inhibitor from cowpea (CpTI), which was shown to significantly reduce insect growth and survival (Hilder et al. 1987). These studies were subsequently extended to include a greater range of target pests (Xu et al. 1996) and a broader range of inhibitors and plant species, including economically important crop species (De Leo et al. 2001).

Since many economically important coleopteran pests predominantly utilise cysteine proteases for protein digestion, inhibitors for this class of enzyme (cystatins) have been investigated as a means for controlling pests from this order. Oryzacystatin, a cysteine protease inhibitor isolated from rice seeds, is effective towards both coleopteran insects and nematodes when expressed in transgenic plants (Pannetier et al. 1997). Similarly, the cysteine/aspartic protease inhibitor equistatin, from sea anemone, is also toxic to several economically important coleopteran pests, including the Colorado potato beetle (Outchkourov et al. 2003). More recent studies have included the stacking of different families of inhibitors to increase the spectrum of activity (Abdeen et al. 2005).

Many protease inhibitors have been identified from many plant species, which are active against certain insect species both in vivo assays against insect gut proteases and in vitro artificial diet bioassays. Protease inhibitors act on the enzymes responsible for the protein catabolism in insect midgut. Proteases in insect include serine, cysteine, aspartic and metalloprotease that catalyse the release of amino acids from dietary protein and so provide the nutrients crucial for normal growth and development. Transgenic tobacco plants with cowpea trypsin inhibitor genes (CpT1) are resistant to larvae of *Heliothis virescens*.

α-Amylase inhibitors present in seeds are active against insects which are highly specific. α-Amylase inhibitor acts on larvae of two seed-feeding beetles: the cowpea weevil, *Callosobruchus maculatus*, and azuki bean weevil, *C. chinensis*. Transgenic pea plants with α-amylase inhibitor gene have been found to produce 1.2% of α-amylase inhibitor protein in seeds and confer resistance to both cowpea and azuki bean weevils.

A major limitation, however, to this strategy for control of insect pests arises from the ability of some lepidopteron and coleopteran species to respond and adapt to ingestion of protease inhibitors by either overexpressing native gut proteases or producing novel proteases that are insensitive to inhibition (Jongsma and Bolter 1997). Thus, detailed knowledge about the enzyme–inhibitor interactions, both at the molecular and biochemical levels, together with detailed knowledge on the response of insects to exposure to such proteins is essential to effectively exploit this strategy. The concept of inhibiting protein digestion as a means of controlling insect pests has been extended to inhibition of carbohydrate digestion. For example, inhibitors of α-amylase have been expressed in transgenic plants and shown to confer resistance to bruchid beetles (Marsaro et al. 2005).

5.3.3 Transgenic Plants Expressing Lectins

Lectins, found throughout the plant and animal kingdoms, form a large and diverse group of proteins identified by a common property of specific binding to carbohydrate residues, either as free sugars or, more commonly, as part of oligo- or polysaccharides. Many physiological roles that have been attributed to plant lectins include defence against pests and pathogens (Peumans and Vandamme 1995). Lectins (carbohydrate-binding proteins) found in seeds are involved in defence against the attack of bacteria, fungi and insects. Lectin binds to midgut epithelial cells of insects and leads to disruption of critical cellular functions. Transgenic sweet potato with lectin genes is resistant to various insects.

Although some lectins are toxic to mammals and are thus not suitable candidates for transfer to crops for enhanced levels of protection, this is by no means universal. Many lectins are not toxic to mammals yet are effective against insects from several different orders (Gatehouse et al. 1995), including homopteran pests such as hoppers and aphids (Foissac et al. 2000). This finding has generated significant interest, not least since no *Bts* effective against this pest order have been identified to date. One such lectin is the snowdrop lectin (*Galanthus nivalis* agglutinin, GNA). Both constitutive and phloem-specific (Rss1 promoter) expressions of GNA in rice are effective means of significantly reducing survival of rice brown plant hopper (*Nilaparvata lugens*) and green leafhopper (*Nephotettix virescens*), both serious economic pests of rice (Tinjuangjun et al. 2000). GNA has been expressed in combination with other genes encoding insecticidal proteins, including the *cry* genes (Maqbool et al. 2001). When a linear transgene construct lacking vector backbone sequences was used to generate transgenic rice plants, the subsequent levels of transgene expression were two- to fourfold higher than plants transformed with whole plasmids (Tinjuangjun et al. 2000). Although lectins such as GNA and ConA are not as effective against aphids as they are against hoppers, they nonetheless have significant effects on aphid fecundity when expressed in potato (Gatehouse and Gatehouse 1999) and wheat (Stoger et al. 1999).

The precise mode of action of lectins in insects is not fully understood although binding to gut epithelial cells appears to be a prerequisite for toxicity. In the case of rice brown plant hopper, GNA not only binds to the luminal surface of the midgut epithelial cells but also accumulates in the fat bodies, ovarioles and throughout the haemolymph, suggesting that the lectin is able to cross the midgut epithelial barrier and pass into the insect's circulatory system, resulting in a systemic toxic effect (Powell et al. 1998). One of the receptors for GNA in brown plant hopper gut is a subunit of ferritin, indicating that GNA may be interfering with metal homeostasis within the insect (Du et al. 2000).

As with protease inhibitors, the levels of protection conferred by expression of lectins in transgenic plants are generally not high enough to be considered commercially viable. However, the absence of genes with proven high insecticidal activity against homopteran pests may well mean that transgenic crops with partial resistance may still find acceptance in agriculture, especially if expressed with other genes that confer partial resistance, or if introduced into partially resistant genetic backgrounds.

5.3.4 Transgenic Plants Expressing Novel Insecticides

Generating insecticidal transgenic crops harbouring genes from nonconventional sources is an extremely active area, with amongst others, foreign genes from plants (e.g. enzymes inhibitors and novel lectins) (Tinjuangjun et al. 2000; Rahbe et al. 2003), and animal sources including insects (e.g. biotin-binding proteins) (Burgess et al. 2002), neurohormones (Fitches et al. 2002), venoms and enzyme inhibitors (Christeller et al. 2002) being a major focus.

The development of second-generation transgenic plants with greater durable resistance might result from the expression of multiple insecticidal genes such as the Vip (vegetative insecticidal proteins) produced by *Bacillus thu-*

ringiensis during its vegetative growth. The benefit of such an approach is a broader insect target range than conventional *Bt* proteins and the proposed expectation to control current *Bt*-resistant pests due to the low levels of homology between the domains of the two protein classes (de Maagd et al. 2003).

With *Bt* toxins as the classical reference, toxins from other insect pathogens provide a potential repository of novel insecticidal compounds. *Photorhabdus* spp. are bacterial symbionts of entomopathogenic nematodes which are lethal to a wide range of insects (Chattopadhyay et al. 2004). *Photorhabdus* toxin expression in *Arabidopsis* caused significant insect mortality (Liu et al. 2003).

5.3.5 Transgenic Plants Expressing Fusion Proteins

In spite of these successes, it is worth mentioning that *Bt* δ-endotoxins have not been effective against all insects and, most importantly, that insects have developed resistance against different δ-endotoxins (Tabashnik et al. 2000). For that reason, alternative insecticidal proteins are being actively pursued. Several such proteins have been identified, including Vip3A and cholesterol oxidases (Dempsey et al. 1998). Various endogenous proteins, which are synthesised in response to insect attack, could potentially be used to engineer pest-resistant plants. One such protein is systemin, the first plant polypeptide hormone discovered. Systemin is phloem mobile and is an essential component of the wound-inducible systemic signal transduction system leading to the transcriptional activation of the defensive genes (Ryan 2000). Systemin is processed from a larger prohormone protein, called prosystemin, by proteolytic cleavages, and it is being suggested that overexpression of prosystemin in transgenic plants may confer protection against insect invasion (Schaller 1999).

The concept of 'gene stacking' has recently been extended to the development and use of fusion proteins. Such proteins not only provide a means of increasing durability but also provide a 'vehicle' for more effective targeting of insecticidal molecules, including peptides. It thus offers an alternative/complementary strategy to address potential limitations in conventional transgenic insect pest control. For example, recognition of binding sites in the insect gut is an important factor determining the toxicity of *Bt*. Enhancing toxin-binding capabilities should thus extend host range and delay resistance. *Bt* is believed to bind primarily to aminopeptidase N or cadherin membrane proteins, whilst the generation of a fusion protein with the non-toxic B chain of ricin (RB) was shown to extend the binding of *Bt* to include specific glycoproteins. Transgenic plants expressing the *Bt*-fused RB demonstrated that the addition of the RB-binding domain provided a wider repertoire of receptor sites within target species and significantly enhanced the levels of toxicity of *Bt*. For example, survival of the armyworm, *Spodoptera littoralis*, a species of insect not sensitive to *Bt*, was reduced by approximately 90% when feeding on transgenic maize expressing the fusion, compared to plants expressing either *Bt Cry*1Ac alone or the RB-binding domain (Mehlo et al. 2005). Expression of the fusion protein resulted in the insect becoming sensitive to *Bt*.

Not only do fusion proteins have potential for use in transgenic crops but also to improve the efficacy of biopesticide-based sprays. Neuropeptides potentially offer a high degree of biological activity and thus provide an attractive alternative pest management strategy. There are major drawbacks to their use, particularly as topical sprays. They are unlikely to be rapidly absorbed through the insect cuticle to their site of action and are prone to proteolysis and rapid degradation in the environment. Should they survive the application process and are then taken up by the insect, they are then unlikely to survive the conditions of the insect gut or be delivered to the correct targets within the insect. The discovery that snowdrop lectin (GNA) remains stable and active within the insect gut after ingestion and that it is able to cross the gut epithelium provides an opportunity for its use as a 'carrier molecule' to deliver other peptides to the circulatory system of target insect species. This strategy effectively delivered the insect neuropeptide hormone, allatostatin, to the haemolymph of the tomato moth, *Lacanobia oleracea*. Subsequent expression of the fusion protein in potato further provided proof of concept for the efficacy of fusion proteins, as a means of delivery. The results demonstrated significant reduction in mean larval weight when compared to the controls. GNA can be used to deliver insecticidal peptides isolated from the venom of the spider *Segestria florentina* (SFI1) to the haemolymph of *L. oleracea* (Fitches et al. 2004). Neither the GNA nor the SFI1 moieties alone were acutely toxic; the SFI1/GNA fusion was insecticidal to first-stage larvae, causing 100% mortality after 6 days. This spider venom neurotoxin is believed to irreversibly block the presynaptic neuromuscular junctures. Such venom toxins show high degrees of specificity and thus lend themselves to environmentally benign pest management strategies.

5.3.6 Genetically Altered Bacterium

Another strategy of potential importance for insect control involves a genetically altered bacterium. The organism – a strain of corn-root-colonising bacteria called *Pseudomonas fluorescens* – has been genetically changed so it produces an endotoxin that is a potent insecticide for certain pests, including black cutworm. The gene to produce the toxin was transferred from another bacterium, *Bacillus thuringiensis*, which itself has been marketed as a biological insecticide for more than 20 years. The recombinant bacterium can be freeze-dried and coated directly on seeds before planting, or it can be sprayed onto the fields. Tests indicate that the non-recombinant parental *P. fluorescens* strain remains viable for only 8–14 weeks in the field; then it dissipates and appears to have no long-term effects. Although the current recombinant strain affects a small range of insects, the company developing it intends it to be a prototype for products that could be marketed within the next few years. Successful work at another company has focused on transfer of the toxin gene into plants themselves, which makes them self-protecting against certain insects, notably the tobacco hornworm. In a similar approach, a search is under way for genes controlling resistance or toxins against nematodes.

5.3.7 Benefits of Transgenic Insect-Resistant Crops

- Promotes greater sustainability of natural resources by reducing use of energy and chemicals (more target use of pesticides and reduction in use of fossil fuels)
- Reduction in land/water contamination through reduced pesticide usage
- Preserving natural habitats for biodiversity (more efficient use of land)
- Reduced impact on non-target organisms, including beneficial insects
- Enhancing safety of food crops by reducing mycotoxin contamination
- Increased yield

5.3.8 Impact of Insect-Resistant Transgenic Crops on Natural Enemies

Assessing the environmental consequences of transgenic crop species is an important precursor to their becoming adopted in agriculture. The expression of transgenes that confer enhanced levels of resistance to insect pests is of particular significance as it is aimed at manipulating the biology of organisms in a different trophic level to that of the plant. Recent research has identified potential risks to beneficial non-target arthropods via bitrophic interactions, involving the plant and an herbivorous insect, and tritrophic interactions, those involving the plant, the pest insect and its natural enemy, particularly in relation to arthropod biodiversity (Obrist et al. 2006).

There are two major routes for insecticidal transgene products to impact on exposed natural enemies (predators and parasitoids) at higher trophic levels through the tritrophic interaction: (1) through direct exposure to the product as it accumulates in the pest and (2) through indirect effects on the growth and development of the pest that influence subsequent growth and developmental processes in the parasitoid or predator. The distinction between these two mechanisms is of considerable environmental significance, but discriminating between them is not straightforward (Down et al. 2001). Several lepidopteran species tested contained *Bt* toxin after consumption of transgenic tissue and therefore provided a potential route of secondary *Bt* exposure (Head et al. 2001). The level of toxin was uniformly low, but it was still biologically active. Concerns over the use of plant-derived insecticidal proteins, such as protease inhibitors and lectins, are perhaps greater as they usually do not cause rapid and complete mortality of the target insect pest. There is no doubt that they do have a significant effect on insect survival, but their major contribution in crop protection is to reduce the build-up of pest populations on plants. Thus, they will be readily available for subsequent parasitism and predation, although many predators, such as carabids, do scavenge dead prey.

Despite this opportunity for exposure, most studies to date have demonstrated that although the predator/parasitoid is exposed to the transgene product, which in many cases can be detected in the natural enemy, it has little effect. For example, exposure of the parasitoid *Eulophus pennicornis* to GNA via parasitism of *Lacanobia oleracea* larvae reared on GNA-expressing plants failed to cause any deleterious effects, and in some instances, parasitoid performance was actually improved (Bell et al. 2001). Furthermore, GNA had no deleterious effects on the parasitoid *Meterous gyrator* (Wakefield et al. 2006). Conversely, cowpea trypsin inhibitor was deleterious at the third trophic level, but these effects were considered to be indirect, as a result of poor performance of the pest larvae on CpTI-expressing potato plants. There is also little evidence that predators are much affected. The exposure of the predatory stink bug *Podisus maculiventris* to pest larvae (*L. oleracea*) reared on either GNA-expressing or CpTI-expressing potato plants had no significant effects on nymphal survival or weight (Bell et al. 2003). Those insects reared on GNA did show a significant lengthening of preadult development. GNA had no deleterious effects on two-spot ladybird *Adalia bipunctata* when fed GNA-dosed aphids from artificial diets or aphids colonising GNA-expressing potato plants (Down et al. 2003). Interestingly, the cysteine protease inhibitor OC-1 has no effect on *Harmonia axyridis* predating diamondback moth larvae (DBM, *Plutella xylostella*) reared on OC-1-expressing oilseed rape plants, despite these predators relying predominantly on cysteine proteases for proteolytic digestion. In the early stages of development, the predators performed better on the DBM fed with the transgenic oilseed rape than the controls. The predators were able to modulate enzyme activity in response to dietary protease inhibitors (Ferry et al. 2003). Carabid beetles could circumvent the inhibitory effects of the serine protease inhibitor MTI-2 expressed in oilseed rape and delivered through the prey by modulation of their digestive proteases profile (Ferry et al. 2005). In this study, expression of MTI-2 was selected to target serine proteases, since carabids rely predominantly on this class of protease for protein digestion.

5.4 Genes in Defence Against Nematode Pests

Plant parasitic nematodes are obligate parasites that cause billions of dollars in losses annually to the world's farmers (Williamson 1998). They have been divided into ectoparasites and endoparasites. The sedentary endoparasites of the family Heteroderidae cause the most economic damage and include two groups: the cyst nematodes, which include the groups *Heterodera* and *Globodera*, and the root-knot nematodes (genus *Meloidogyne*). Root-knot nematodes infect a broad range of plant species, whereas cyst nematodes have a narrower host range. With increasing restrictions on the use of chemical pesticides, the use of host resistance for nematode control has grown in importance.

Although an important constituent of current nematode management strategies is the incorporation of natural resistance, one must be aware of the fact that there may not be appropriate resistance loci available for many crops. In addition, it is not a given fact that a particular gene will function effectively in heterologous hosts. Attempts to transfer *Mi*-mediated resistance into tobacco have not so far been successful (Williamson 1998). Furthermore, nematodes can eventually develop virulence, which may limit the effectiveness of this approach. Clearly, novel strategies to control nematode infestation are required combining existing with new approaches.

Biotechnology has a role to play in incorporation of resistance against nematodes and biological control of plant nematodes. A number of genes that mediate nematode resistance have now been or soon will be cloned from a variety of plant species. Nematode-resistance genes are present in several crop species and are an important component of many breeding programmes including those for tomato, potato, soybeans and cereals (Trudgill 1991). Several resistance genes have been mapped to chromosomal locations or linkage groups, and some of them have been cloned. The first nematode-resistance gene to be cloned was *Hs1pro-1*, a gene from a wild relative of sugar beet conferring resistance to *Heterodera schachtii* (Cai et al. 1997). The cDNA, under the control of the CaMV 35S promoter, was able to confer nematode resistance to sugar beets transformed with *Agrobacterium rhizogenes* in an in vitro assay (Cai et al. 1997). The *Mi* gene from tomato conferred resistance against a root-knot nematode and an aphid in transgenic potato (Rossi et al. 1998). The gene *Mi* is a true *R* gene, characterised by the presence of NBS and LRR domains. Recently, *Gpa2*, a gene that confers resistance against some isolates of the potato cyst nematode *Globodera pallida*, was identified (Van Der Vossen et al. 2000). This gene shares extensive homology with the Rx1 gene that confers resistance to potato virus X suggesting a similarity in function (Van Der Vossen et al. 2000).

Embryo rescue technique and protoplast fusion or somatic hybridisation have been used in nematology to overcome interspecific plant breeding problems with a view to incorporate nematode resistance.

5.4.1 Embryo Rescue Technique

Embryo rescue technique involves hybridisation between economic plant species and wild relatives followed by excision and culturing of hybrid zygotic embryo on nutrient media. One of the classical examples of successful application of embryo rescue technique in plant nematology is the synthesis of tomato cultivars resistant to root-knot nematodes. Cultivated tomato species, *Lycopersicon esculentum*, is highly susceptible to attack by *Meloidogyne* spp., but the related wild species *L. peruvianum* has a high degree of resistance against the nematodes. A cross between these two species yielded a hybrid in which case the endosperm does not develop and the embryo aborts. This problem was solved by carefully isolating and culturing the hybrid embryo on tissue culture media (Smith 1944). This technique thus provides a mechanism for transfer of nematode-resistance gene from *L. peruvianum* into cultivated species of tomato, *L esculentum*.

5.4.2 Protoplast Fusion/Somatic Hybridisation

In somatic hybridisation, plant tissues from two parental sources are digested with a mixture of pectinase and cellulase in an osmotically stable solution to produce protoplasts. Protoplasts are naked cells surrounded by a plasma membrane and have the capacity to fuse spontaneously or are induced to fuse by way of chemicals (polyethylene glycol, dextrin and polyvinyl alcohol) or electrofusion. These fused protoplasts are capable of cell wall regeneration. After cell division, a single nucleus may be formed containing the chromosomes of one of the parent cell or, more frequently, selective elimination of chromosomes of one of the parental species during cell division and tissue formation.

This technique of protoplast fusion has been used as a means of transferring root-knot nematode-resistance factors from *Solanum sisymbriifolium* into eggplant, *S. melongena.* S1 and S2 progenies of *S. sisymbriifolium* plant regenerated from callus not only showed resistance to *M. javanica* but also retained resistance to

M. incognita (Fassuliotis and Bhatt 1982). The nematode did not develop beyond second-stage larvae. Transfer of nematode resistance by protoplast fusion between sexually incompatible plant species holds promise in future.

5.4.3 Recombinant DNA Technology

Application of genetic engineering made possible the synthesis of transgenic plants. Transgenic plant is one in which foreign DNA gene (a transgenic) is stably incorporated in the genome early in embryonic development and so is present in both somatic and germ cells, is expressed in one or more tissues and is inherited to offspring in a Mendelian fashion. The aim of gene transferring technology is to improve a top variety in one additional character without disturbing the rest of the genome. For gene transfer, three methods are available: vector-mediated gene transfer (*Agrobacterium tumefaciens*); microinjection by capillaries, laser micro beams or micro projectiles; and direct transfer.

This technique of recombinant DNA technology has been used in tomato to transfer *Mi* gene which shows resistance to *M. arenaria, M. incognita* and *M. javanica,* but not to *M. hapla* (Gilbert and Mc Guire 1956).

A lot of information is being generated with respect to identification of resistance gene loci against various nematode pests in different crops of importance. Particular interest has been shown in the chromosomal location of resistance to *Globodera rostochiensis* in potato (Pireda et al. 1993), *Heterodera schachtii* in beet (Jung et al. 1992) and root-knot nematode, *Meloidogyne* spp. (Klein-Lankhorst et al. 1991; Messeguer et al. 1991). The identification and cloning of such genes would facilitate their transfer and expression in transgenic crops. In case of potato, a major dominant locus *Gro* 1 conferring resistance to some pathotypes of *G. rostochiensis* was mapped by restriction fragment length polymorphism (RFLP) to potato linkage group IX; the nearest neighbouring RFLP markers are CP 51 and TG 20(a) (Barone et al. 1990). Other groups are developing high-resolution RFLP maps around the *Mi* genes in tomato. The *Mi* genes originating from wild tomato species *L. peruvianum* confer resistance to all major species of *Meloidogyne.* The *Mi* gene is located at position 44 on chromosome 6 and is closely linked to the leaf calyx markers of *yv* (yellow virescent) and acid phosphatase-1 (Aps-1). The work of Messeguer et al. (1991) led to the production of a more detailed consensus linkage map of chromosome 6. Four co-dominant RFLP markers (TG 297, CD 14, Tom 25 and GP 79) can be used as markers for selection of the *Mi* gene. The extent of *L. peruvianum* DNA segment in different *Mi* gene-containing cultivars was also determined. Three groups of cultivars (ABC) with different amounts of *L. peruvianum* DNA were found. Only in the cultivars of group A, there is linkage between *Mi* and Aps-1. For cultivars of group B, the marker Gp 79 can be used, and new markers are needed for nematode resistance in group C cultivars.

Many patents have already been made in the area of transgenic plants resistant to nematodes. A transgenic potato with a novel nucleic acid molecule capable of increasing the resistance of a plant to nematode infection comprising a nucleotide sequence encoding a polypeptide with a collagenase (EC-3.4.24.3) activity or its precursor has been patented by Stratford and Shields (1995). A lectin from garlic has been patented to confer resistance against wide array of nematode pests (Geoghegan et al. 1995). Similarly, protease from *Paecilomyces lilacinus* (patented information) can be used to construct nematode-resistance transgenic plants (Den-Belder et al. 1994). A transgenic plant with nematode resistance may be prepared by introduction of an RNAase gene under control of a promoter which induces RNA production in response to nematode attack of root tissue, resulting in prevention of formation of nematode feeding sites. This strategy has been patented by Monsanto, USA.

5.5 Long-Term Impact of Genetically Modified Plants

There are three main concerns about the long-term impact of engineering genes for disease resistance in transgenic plants: (1) the resistance of the plants to pathogen attack once they are grown in large scale, (2) the development of resistance by the pathogen and (3) the possible phenotypic alterations of the plant. The application of biotechnology in agriculture has had great success in the generation of commercially useful insect- and virus-resistant crops (Dunwell 2000). However, the first commercially available *Bt*-expressing cotton crop (grown in 1996) showed mixed success (Dempsey et al. 1998). Clearly, detailed and sufficiently extensive studies about the large-scale agronomic performance of each new variety grown in different conditions are necessary. However, until the new variety is grown on a large scale in the appropriate place, one may not determine precisely the actual resistance to pathogen attack.

Resistant plants can impose a selective pressure that results in development of resistance in the pathogen, which is not an uncommon event (Tabashnik et al. 2000). There exists an antagonistic co-evolution between plants and pathogens that is constantly selecting for genotypes that can overcome the other's defences (Stahl and Bishop 2000). Clearly, the introduction of a single transgene, therefore, may not be sufficient to achieve durable and broad-spectrum disease resistance. A combination of transgenic strategies will be needed to ensure durability of resistance. By combining several methods for pest control, one may reduce the probability that any of these methods will soon become obsolete as a result of adaptation by the pest or disease-causing agent. The increasing availability of resistance genes is allowing the generation of transgenic plants with resistance to various pests. However, since constitutive expression of certain *R* and *Avr* genes may have deleterious effects on the plant (Hone'e et al. 1995), a judicious choice of the genes to be transferred, combined with detailed molecular studies, will be necessary to achieve durable resistance together with an optimal field performance (Rommens and Kishore 2000). Furthermore, multiple *R* genes may compete for the signalling components during a mixed infection, thereby interfering with the activation of defences (Bent 1996). There still remains a series of questions that need to be addressed concerning the response to pathogens of the engineered plants. For instance, what is the likelihood that plants engineered for disease resistance will provide an environment for the development of a novel pathogen that exhibits an increased host range or is resistant to currently available control methods? How durable will the engineered resistance prove to be once crops are grown on a large scale? Will the different *R* genes, cloned and those to be cloned, prove to be functional in heterologous plant species?

5.6 Future Trends

The field of plant resistance control is undergoing a very active and exciting period, during which major breakthroughs are being made. Increased availability of cloned *R* genes will permit their testing in different plant backgrounds. In contrast with the success in the production of insect- and virus-resistant crops, the production of fungal-resistant crops with commercially useful levels of resistance has not been achieved. However, it is likely that commercial introduction of fungal-resistant crops can be expected within 4–8 years. In the same way that plant breeders are continually developing new varieties that contain the most effective combination of existing characteristics, there is a similar trend with transgenic crops. Many laboratories are experimenting with 'pyramiding of genes', which consists of the introduction of multiple genes conferring different characters. A good example of this is a potato line containing seven transgenes that will confer resistance to insects, fungi and virus and will alter other phenotypic characteristics (Dunwell 2000).

Overexpression of signalling components that lie downstream of *R* genes is an interesting approach that is currently being tested. This may allow activation of only certain defence pathways and may avoid agronomic problems associated with constitutive activation of some *R*-mediated pathways. The use of antimicrobial peptides to engineer broad-spectrum resistance is a promising and powerful approach that will be used considerably in the near future.

The increased use of inducible promoters is another current trend to manipulate specific pathways or to express *R* genes in a much more controlled manner (Shen et al. 2000). Finally, in addition to the progress being made on the plant side of the equation, an understanding of the genetic make-up of pathogens and the critical genes involved in the pathogenesis process are expected to open new avenues in crop protection.

5.7 Conclusions

Time has demonstrated that biotechnology can provide very clear benefits to agriculture, not least with the increasing contribution it can make towards sustainability. Indeed, globally there has been a steadily increasing market for genetically modified (enhanced) crops, particularly for production of cotton and animal feeds, with the acreage now in excess of 90 million hectares. The fears voiced over the environmental impact of the technology, particularly in terms of deleterious consequences for biodiversity, including effects on natural enemies such as predators and parasitoids have not been realised. There is no room for complacency, and it is essential that all novel technologies are thoroughly investigated before their release. It is important that these investigations are carried out in comparison with their conventional counterparts to ensure a meaningful evaluation. Thus, in agricultural terms, biotechnology must be evaluated in comparison with current conventional practices, for example, chemical and biopesticide application for pest control. It is not suggested that biotechnology is necessarily used as a stand-alone technology, but rather that it is used as a component of integrated pest management.

References

Abdeen A, Virgos A, Olivella E, Villanueva J, Aviles X, Gabarra R, Prat S (2005) Multiple insect resistance in transgenic tomato plants over-expressing two families of plant proteinase inhibitors. Plant Mol Biol 57(2):189–202

Alexander D, Goodman RM, Gut-Rella M, Glascock C, Weyman K, Friedrich L, Maddox D, Ahl-Goy P, Luntz T, Ward E, Ryals J (1993) Increased tolerance to two oomycete pathogens in transgenic tobacco expressing pathogen-related protein 1a. Proc Natl Acad Sci USA 90:7327–7331

Barone A, Ritter E, Schanchtschabel U, Debner T, Salamini I, Gebhardt C (1990) Localization by restriction fragment length polymorphism mapping of a major dominant gene conferring resistance to the potato cyst nematode, *Globodera rostochiensis*. Mol Genet 224:177–182

Bell HA, Fitches EC, Marris GC, Bell J, Edwards JP, Gatehouse JA, Gatehouse AMR (2001) Transgenic GNA expressing potato plants augment the beneficial biocontrol of *Lacanobia oleracea* (Lepidoptera: Noctuidae) by the parasitoid *Eulophus pennicornis* (Hymenoptera: eulophidae). Transgenic Res 10:35–42

Bell HA, Down RE, Fitches EC, Edwards JP, Gatehouse AMR (2003) Impact of genetically modified potato expressing plant-derived insect resistance genes on the predatory bug *Podisus maculiventris* (Heteroptera: Pentatomidae). Biocontrol Sci Technol 13:729–741

Bent AF (1996) Plant disease resistance genes: function meets structure. Plant Cell 8:1757–1771

Borkowska M, Krzymowska M, Talarczyk A, Awan MF, Yakovleva L, Kleczkowski K, Wielgat B (1998) Transgenic potato plants expressing soybean beta-1, 3-endoglucanase gene exhibit an increased resistance to *Phytophthora infestans*. Z Naturforsch 53:1012–1016

Broekaert WF, Terras FR, Cammue BP, Osborn RW (1995) Plant defensins: novel antimicrobial peptides as components of the host defense system. Plant Physiol 108:1353–1358

Broekaert WF, Cammue BP, De Bolle MFC, Thevissen K, Desamblanx GW, Osborn RW (1997) Antimicrobial peptides from plants. Crit Rev Plant Sci 16:297–323

Broglie K, Chet I, Holiday M, Cressman R, Biddle P, Knowlton S, Mauvis CJ, Broglie R (1991) Transgenic plants with enhanced resistance to the fungal pathogen *Rhizoctonia solani*. Science 254:1194–1197

Burgess EPJ, Malone LA, Christeller JT, Lester MT, Murray C, Philip BA, Phung MM, Tregidga EL (2002) Avidin expressed in transgenic tobacco leaves confers resistance to two noctuid pests, *Helicoverpa armigera* and *Spodoptera litura*. Transgenic Res 11:185–198

Cai D, Kleine M, Kifle S, Harloff HJ, Sandal NN, Marcker KA, Kleinlankhorst RM, Salentijn EMJ, Lange W, Stiekema WJ, Wyss U, Grundler FMW, Jung C (1997) Positional cloning of a gene for nematode resistance in sugar beet. Science 275:832–834

Cao H, Li X, Dong X (1998) Generation of broad-spectrum disease resistance by overexpression of an essential regulatory gene in systemic acquired resistance. Proc Natl Acad Sci USA 95:6531–6536

Chattopadhyay A, Bhatnagar NB, Bhatnagar R (2004) Bacterial insecticidal toxins. Crit Rev Microbiol 30:33–54

Chet I (1990) Biological control of soil-borne plant pathogens with fungal antagonists in combination with soil treatments. In: Hornby D (ed) Biological control of soil-borne plant pathogens. CAB International, Wallingford, pp 15–25

Chet I, Inbar J (1994) Biological control of fungal pathogens. Appl Biochem Biotechnol 48:37–43

Christeller JT, Burgess EPJ, Mett V, Gatehouse HS, Markwick NP, Murray C, Malone LA, Wright MA, Philip BA, Watt D, Gatehouse LN, Lovei GL, Shannon AL, Phung MM, Watson LM, Laing WA (2002) The expression of a mammalian proteinase inhibitor, bovine spleen trypsin inhibitor in tobacco and its effects on *Helicoverpa armigera* larvae. Transgenic Res 11:161–173

Christou P, Capell T, Kohli A, Gatehouse JA, Gatehouse AMR (2006) Recent developments and future prospects in insect pest control in transgenic crops. Trends Plant Sci 11:302–308

De Bolle MF, David KM, Rees SB, Vanderleyden J, Cammue BP, Broekaert WF (1993) Cloning and characterization of a cDNA encoding an antimicrobial chitin-binding protein from amaranth, *Amaranthus caudatus*. Plant Mol Biol 22:1187–1190

De Leo F, Bonade-Bottino M, Ceci LR, Gallerani R, Jouanin L (2001) Effects of a mustard trypsin inhibitor expressed in different plants on three lepidopteron pests. J Biochem Mol Biol 31:593–602

de Maagd RA, Bosch D, Stiekema W (1999) *Bacillus thuringiensis* toxin-mediated insect resistance in plants. Trends Plant Sci 4:9–13

de Maagd RA, Bravo A, Crickmore N (2001) How *Bacillus thuringiensis* has evolved specific toxins to colonize the insect world. Trends Genet 17:193–199

de Maagd RA, Bravo A, Berry C, Crickmore N, Schnepf HE (2003) Structure, diversity, and evolution of protein toxins from spore-forming entomopathogenic bacteria. Annu Rev Genet 37:409–433

Dempsey DA, Silva H, Klessig DF (1998) Engineering disease and pest resistance in plants. Trends Microbiol 6:54–61

Den-Belder E, Fitters PFL, Waalnijk C (1994) PN: EP 623672; 09.11.94, Res Inst Plant Prot

Denolf P (1996) Isolation, cloning and characterization of *Bacillus thuringiensis* delta endotoxin receptors in Lepidoptera. PhD thesis, University of Gent

Down RE, Ford L, Bedford SJ, Gatehouse LN, Newell C, Gatehouse JA, Gatehouse AMR (2001) Influence of plant development and environment on transgene expression in potato and consequences for insect resistance. Transgenic Res 10:223–236

Down RE, Ford L, Woodhouse SD, Davison GM, Majerus MEN, Gatehouse JA, Gatehouse AMR (2003) Tritrophic interactions between transgenic potato expressing snowdrop lectin (GNA), an aphid pest [peach-potato aphid; *Myzus persicae* (Sulz)] and a beneficial predator (2-spot ladybird; *Adalia bipunctata* L). Transgenic Res 12:229–241

Du JP, Foissac X, Carss A, Gatehouse AMR, Gatehouse JA (2000) Ferritin acts as the most abundant binding protein for snowdrop lectin in the midgut of rice brown plant hoppers (*Nilaparvata lugens*). J Biochem Mol Biol 30:297–305

Dunwell JM (2000) Transgenic approaches to crop improvement. J Exp Bot 51:487–496

Elad Y, Chet I, Henis Y (1982) Degradation of plant pathogenic fungi by *Trichoderma harzianum*. Can J Microbiol 28:719–725

Epple P, Apel K, Bohlmann H (1997) Overexpression of an endogenous thionin enhances resistance of Arabidopsis against *Fusarium oxysporum*. Plant Cell 9:509–520

FAOSTAT (2000) http://apps.fao.org/cgi-bin/nph-db.pl?subset=agriculture

Fassuliotis G, Bhatt DP (1982) Potential of tissue culture for breeding root-knot nematode resistance into vegetables. J Nematol 14:10–14

Ferry N, Raemaekers RJM, Majerus MEN, Jouanin L, Port G, Gatehouse JA, Gatehouse AMR (2003) Impact of oilseed rape expressing the insecticidal cysteine protease inhibitor oryzacystatin on the beneficial predator *Harmonia axyridis* (multicoloured Asian ladybeetle). Mol Ecol 12:493–504

Ferry N, Jouanin L, Ceci LR, Mulligan A, Emami K, Gatehouse JA, Gatehouse AMR (2005) Impact of oilseed rape expressing the insecticidal serine protease inhibitor, mustard trypsin inhibitor-2 on the beneficial predator *Pterostichus madidus*. Mol Ecol 14:337–349

Ferry N, Edwards M, Gatehouse J, Capell T, Christou P, Gatehouse A (2006) Transgenic plants for insect pest control: a forward looking scientific perspective. Transgenic Res 15:13–19

Fitches E, Audsley N, Gatehouse JA, Edwards JP (2002) Fusion proteins containing neuropeptides as novel insect control agents: snowdrop lectin delivers fused allatostatin to insect haemolymph following oral ingestion. J Biochem Mol Biol 32:1653–1661

Fitches E, Edwards MG, Mee C, Grishin E, Gatehouse AMR, Edwards JP, Gatehouse JA (2004) Fusion proteins containing insect-specific toxins as pest control agents: snowdrop lectin delivers fused insecticidal spider venom toxin to insect haemolymph following oral ingestion. J Insect Physiol 50:61–71

Flores A, Chet I, Herrera-Estrella A (1997) Improved biocontrol activity of *Trichoderma harzianum* by overexpression of the proteinase encoding gene *prb1*. Curr Genet 31:30–37

Foissac X, Loc NT, Christou P, Gatehouse AMR, Gatehouse JA (2000) Resistance to green leafhopper (*Nephotettix virescens*) and brown plant hopper (*Nilaparvata lugens*) in transgenic rice expressing snowdrop lectin (*Galanthus nivalis* agglutinin; GNA). J Insect Physiol 46:73–583

Gabay JE (1994) Ubiquitous natural antibiotics. Science 264:373–374

Gahan LJ, Gould F, Heckel DG (2001) Identification of a gene associated with bit resistance in *Heliothis virescens*. Science 293:857–860

Gatehouse JA (2002) Plant resistance towards insect herbivores: a dynamic interaction. New Phytol 156: 145–169

Gatehouse J, Gatehouse A (1999) In: Reichcigl J, Reichcigl N (eds) Biological and biotechnological control of insect pests. CRC Press, Boca Raton, FL, pp 211–241

Gatehouse K, Powell W, Peumans EV, Damme Gatehouse J (1995) In: Pusztai A, Bardocz S (eds) Lectins biomedical perspectives. Taylor and Francis, London, pp 35–57

Gatehouse J, Gatehouse A, Bown D (2000) In: Michaud D (ed) Recombinant protease inhibitors in plants. Landes Bioscience, Austin, pp 9–26

Geoghegan I, Robertson W, Birch N, Gatehouse AMR (1995) PN: WO 9526634: 12.10.95, Axis- Genet, Cambridge, UK

Geremia RA, Goldman GH, Jacobs D, Ardiles W, Vila SB, Van Montagu M, Herrera-Estrella A (1993) Molecular characterization of the proteinase-encoding gene, *prb1*, related to mycoparasitism by *Trichoderma harzianum*. Mol Microbiol 8:603–613

Gilbert JC, Mc Guire DC (1956) Inheritance of resistance to severe root-knot from *Meloidogyne incognita* in commercial type tomatoes. Proc Ann Hortic Sci 68:437–442

Girard C, Le Metayer M, Bonade-Bottino M, Pham-Delegue MH, Jouanin L (1998) High level of resistance to proteinase inhibitors may be conferred by proteolytic cleavage in beetle larvae. Insect Biochem Mol Biol 28:229–237

Griffitts JS, Whitacre JL, Stevens DE, Aroian RV (2001) Bt toxin resistance from loss of a putative carbohydrate-modifying enzyme. Science 293(5531):860–864

Grison R, Besset-Grezes B, Schneider M, Lucante N, Olsen L, Leguay JJ, Toppan A (1996) Field tolerance to fungal pathogens of *Brassica napus* constitutively expressing a chimeric chitinase gene. Nature Biotechnol 14:643–646

Hain R, Reif HJ, Krause E, Langebartels R, Kindl H, Vornam B, Wiese W, Schmelzer E, Schreier Stocker RH et al (1993) Disease resistance results from foreign phytoalexin expression in a novel plant. Nature 361:153–156

Hall FR (1991) Pesticide application technology and integrated pest management. In: Pimentel D (ed) Handbook of pest management in agriculture. CRC Press, Boca Raton, pp 135–163

Haran S, Schickler H, Chet I (1996) Molecular mechanisms of lytic enzymes involved in the biocontrol activity of *Trichoderma harzianum*. Microbiol 142: 2321–2331

Head G, Brown CR, Groth ME, Duan JJ (2001) Cry1Ab Protein levels in phytophagous insects feeding on transgenic corn: implications for secondary exposure risk assessment. Entomol Exp Appl 99:37–45

Hilder VA, Boulter D (1999) Genetic engineering of crop plants for insect resistance – a critical review. Crop Prot 18:177–191

Hilder VA, Gatehouse AMR, Sheerman SE, Barker RF, Boulter DA (1987) A novel mechanism of insect resistance engineered into tobacco. Nature 330: 160–163

Hipskind JD, Paiva NL (2000) Constitutive accumulation of a resveratrolglucoside in transgenic alfalfa increases resistance to *Phoma medicaginis*. Mol Plant Microbe Interact 13:551–562

Hone'e EG, Melchers LS, Vleeshouwers VGAA, Vab Roekel JSC, De Wit PJGM (1995) Production of the Avr9 elicitor from the fungal pathogen *Cladosporium fulvum* in transgenic tobacco and tomato plants. Plant Mol Biol 29:909–920

Irie K, Hosoyama H, Takeuchi T, Iwabuchi K, Watanabe H, Abe M, Abe K, Arai S (1996) Transgenic rice established to express corn cystatin exhibits strong inhibitory activity against insect gut proteinases. Plant Mol Biol 30:149–157

Jach G, Gornhardt B, Mundy J, Logemann J, Pinsdorf E, Leah R, Schell J, Maas C (1995) Enhanced quantitative resistance against fungal disease by combinatorial expression of different barley antifungal proteins in transgenic tobacco. Plant J 8:97–109

James C (2005) Global status of commercialized biotech/GM crops. ISAAA Briefs No. 34. ISAAA, Ithaca, New York, pp 46

Jongedijk E, Tigelaar H, Van Roekel JSC, Bres-Vloemans SA, Dekker I, Van Den Elzen PJM, Cornelissen BJC, Melchers LS (1995) Synergistic activity of chitinases and β–1, 3 glucanases enhances fungal resistance in transgenic tomato plants. Euphytica 85:173–180

Jongsma MA, Bolter C (1997) The adaptation of insects to plant protease inhibitors. J Insect Physiol 43(10): 885–895

Jung C, Clauussen U, Harsthemke B, Her F, Herman RG (1992) A DNA library from an individual *Beta patellaris* chromosome conferring nematode resistance obtained by micro dissection of meotic metaphase chromosome. Plant Mol Biol 20:503–511

Keller H, Pamboukdijian N, Pochet M, Poupet A, Delon R, Verrier JL, Roby D, Ricci P (1999) Pathogen-induced elicitin production in transgenic tobacco generates a hypersensitive response and nonspecific disease resistance. Plant Cell 11:223–235

Kesarwani M, Azam M, Natarajan K, Mehta A, Datta A (2000) Oxalate decarboxylase from *Collybia velutipes*. Molecular cloning and its overexpression to confer resistance to fungal infection in transgenic tobacco and tomato. J Biol Chem 275:7230–7238

Klein-Lankhorst R, Reitveld P, Machiels B, Verlek R, Weide R, Gebhardt C, Koorneef M, Zabel P (1991) RFLP markers linked to the root-knot nematode resistance in gene *Mi* in tomato. Theor Appl Genet 81:661–667

Li R, Wu N, Fan Y, Song B (1999) Transgenic potato plants expressing osmotin gene inhibits fungal development in inoculated leaves. Chin J Biotechnol 15:71–75

Liang J, Wu Y, Rosenberger C, Hakimi S, Castro S, Berg J (1998) AFP genes confer disease resistance to transgenic potato and wheat plants (Abstract no. L-49). In: 5th international workshop on pathogenesis-related proteins in plants; signalling pathways and biological activities, Aussois, France

Limon MC, Pintor-Toro JA, Benitez T (1999) Increased antifungal activity of *Trichoderma harzianum* transformants that overexpress a 33 kDa chitinase. Phytopathology 89:254–261

Lin W, Anuratha CS, Datta K, Potrykus I, Muthukrishnan S, Datta SK (1995) Genetic engineering of rice for resistance to sheath blight. Bio/Technol 13:686–691

Liu D, Raghothama KG, Hasegawa PM, Bressan RA (1994) Osmotin overexpression in potato delays development of disease symptoms. Proc Natl Acad Sci USA 91:1888–1892

Liu D, Burton S, Glancy T, Li ZS, Hampton R, Meade T, Merlo DJ (2003) Insect resistance conferred by 283 kDa *Photorhabdus luminescens* protein TcdA in *Arabidopsis thaliana*. Nat Biotechnol 21:1222–1228

Logemann J, Jach G, Tommerup H, Mundy J, Schell J (1992) Expression of a barley ribosome inactivating protein leads to increased fungal protection in transgenic tobacco plants. Bio/Technol 10:305–308

Lorito M, Woo SL, Garcia I, Colucci G, Harman GE, Pintor-Toro JA, Filippone E, Muccifora S, Lawrence CB, Zoina A, Tuzun S, Scala F, Fernandez IG (1998) Genes from mycoparasitic fungi as a source for improving plant resistance to fungal pathogens. Proc Natl Acad Sci USA 95:7860–7865

Luo K, Sangadala S, Masson L, Mazza A, Brousseau R, Adang MJ (1997) The *Heliothis virescens* 170 kDa aminopeptidase functions as "receptor A" by mediating specific *Bacillus thuringiensis* Cry1A delta-endotoxin binding and pore formation. J Biochem Mol Biol 27:735–743

Maqbool SB, Riazuddin S, Loc NT, Gatehouse AMR, Gatehouse JA, Christou P (2001) Expression of multiple insecticidal genes confers broad resistance against a range of different rice pests. Mol Breed 7:85–93

Marsaro AL, Lazzari SMN, Figueira ELZ, Hirooka EY (2005) Arnylase inhibitors in corn hybrids as a resistance factor to *Sitophilus zeamais* (Coleoptera: Curculionidae). Neotrop Entomol 34:443–450

Martin GB (1999) Functional analysis of plant disease resistance genes and their downstream effectors. Curr Opin Plant Biol 2:273–279

Mehlo L, Gahakwa D, Nghia PT, Loc NT, Capell T, Gatehouse JA, Gatehouse AMR, Christou P (2005) An alternative strategy for sustainable pest resistance in genetically enhanced crops. Proc Natl Acad Sci USA 102:7812–7816

Melchers LS, Stuiver MH (2000) Novel genes for disease-resistance breeding. Curr Opin Plant Biol 3:147–152

Messeguer R, Grand M, de Vicente MC, Young ND, Tanksley ST (1991) High resolution RFLP map around root-knot nematode resistance gene (*Mi*) in tomato. Theor Appl Genet 82:529–536

Mitsuhara I, Matsufuru H, Ohshima M, Kaku H, Nakajima Y, Murai N, Natori S, Ohashi Y (2000) Induced expression of sarcotoxin IA enhanced host resistance against both bacterial and fungal pathogens in transgenic tobacco. Mol Plant Microbe Interact 13:860–868

Molina L, Garcia-Olmedo F (1997) Enhanced tolerance to bacterial pathogens caused by the transgenic expression of barley lipid transfer protein LTP2. Plant J 12:669–675

Obrist LB, Dutton A, Romeis J, Bigler F (2006) Biological activity of Cry1Ab toxin expressed by Bt maize following ingestion by herbivorous arthropods and exposure of the predator *Chrysoperla carnea*. Biocontrol 51:31–48

Oerke EC (1994) Crop protection and crop production. Elsevier, Amsterdam

Osusky M, Zhou G, Osuska L, Hancock RE, Kay WW, Misra S (2000) Transgenic plants expressing cationic peptide chimeras exhibit broad spectrum resistance to phytopathogens. Nat Biotechnol 18:1162–1166

Outchkourov NS, Rogelj B, Strukelj B, Jongsma MA (2003) Expression of sea anemone equistatin in potato. Effects of plant proteases on heterologous protein production. Plant Physiol 133:379–390

Pannetier C, Giband M, Couzi P, LeTan V, Mazier M, Tourneur J, Hau B (1997) Introduction of new traits into cotton through genetic engineering: Insect resistance as example. J Biochem Mol Biol 96:163–166

Peumans WJ, Vandamme EJM (1995) Lectins as plant defense proteins. Plant Physiol 109:347–352

Pireda O, Bonierbale MW, Plaisted RL, Brodie BB, Tanksley SD (1993) Identification of RFLP markers linked to the H1 gene conferring resistance to the potato cyst nematode (*Globodera rostochiensis*). Genome 36:152–156

Powell KS, Spence J, Bharathi M, Gatehouse JA, Gatehouse AMR (1998) Immunohistochemical and developmental studies to elucidate the mechanism of action of the snowdrop lectin on the rice brown plant hopper, *Nilaparvata lugens* (Stal). J Insect Physiol 44:529–539

Rahbe Y, Deraison C, Bonade-Bottino M, Girard C, Nardon C, Jouanin L (2003) Effects of the cysteine protease inhibitor oryzacystatin (OC-I) on different aphids and reduced performance of *Myzus persicae* on OC-I expressing transgenic oilseed rape. Plant Sci 164:441–450

Rommens CM, Kishore GM (2000) Exploiting the full potential of disease resistance genes for agricultural use. Curr Opin in Biotechnol 11:120–125

Rossi M, Goggin FL, Milligan SB, Kaloshian I, Ullman DE, Williamson VM (1998) The nematode resistance gene Mi of tomato confers resistance against the potato aphid. Proc Natl Acad Sci USA 95:9750–9754

Ryan CA (1990) Protease inhibitors in plants: genes for improving defenses against insects and pathogens. Annu Rev Phytopathol 28:425–449

Ryan CA (2000) The systemin signaling pathway: differential activation of plant defensive genes. Biochim Biophys Acta 7:112–121

Ryan CA, Pearce G (1998) Systemin: a polypeptide signal for plant defensive genes. Annu Rev Cell Dev Biol 14:1–17

Sanchis V, Lereclus D (1999) *Bacillus thuringiensis*: a biotechnology model. J Soc Biol 193:523–530

Schaller A (1999) Oligopeptide signalling and the action of systemin. Plant Mol Biol 40:763–769

Shelton AM, Zhao JZ, Roush RT (2002) Economic, ecological, food safety, and social consequences of the deployment of Bt transgenic plants. Annu Rev Entomol 47:845–881

Shen S, Li Q, He SY, Barker KR, Li D, Hunt AG (2000) Conversion of compatible plant-pathogen interactions into incompatible interactions by expression of the

Pseudomonas syringae pv syringae 61 hrmA gene in transgenic tobacco plants. Plant J 23:205–213
Smith PG (1944) Embryo culture of a tomato species hybrid. Proc Am Soc Hortic Sci 44:413–416
Stahl EA, Bishop JG (2000) Plant-pathogen arm races at the molecular level. Curr Opin Plant Biol 3:299–304
Stoger E, Williams S, Christou P, Down RE, Gatehouse JA (1999) Expression of the insecticidal lectin from snowdrop (*Galanthus nivalis* agglutinin; GNA) in transgenic wheat plants: effects on predation by the grain aphid *Sitobion avenae*. Mol Breed 5:65–73
Stratford R, Shields R (1995) WO 9530017; 09.11.95, Unilever, Rotterdam, The Netherlands
Tabashnik BE, Groeters FR, Finson N, Liu YB, Johnson MW, Heckel DG, Luo K, Adang MJ (1996) In: Brown T (ed) Molecular genetics and evolution of pesticide resistance. Oxford University Press, New York, pp 130–140
Tabashnik BE, Roush RT, Earle ED, Shelton AM (2000) Resistance to Bt toxins. Science 287:42
Tang X, Xie M, Kim YJ, Zhou J, Klessig DF, Martin GB (1999) Overexpression of *Pto* activates defense responses and confers broad resistance. Plant Cell 11:15–29
Terras FR, Eggermont K, Kovaleva V, Raikhel NV, Osborn RW, Kester A, Rees SB, Torrekens S, Van Leuven F, Vanderleyden J et al (1995) Small cysteine-rich antifungal proteins from radish: their role in host defense. Plant Cell 7:573–588
Thevissen K, Terras FR, Broekaert WF (1999) Permeabilization of fungal membranes by plant defensins inhibits fungal growth. Appl Environ Microbiol 65:5451–5458
Tinjuangjun P, Loc NT, Gatehouse AMR, Gatehouse JA, Christou P (2000) Enhanced insect resistance in Thai rice varieties generated by particle bombardment. Mol Breed 6:391–399
Trudgill DL (1991) Resistance to and tolerance of plant parasitic nematodes in plants. Annu Rev Phytopathol 29:167–193
Tu J, Zhang G, Datta K, Xu C, He Y, Zhang Q, Khush GS, Datta SK (2000) Field performance of transgenic elite commercial hybrid rice expressing *Bacillus thuringiensis* -endotoxin. Nat Biotechnol 18:1101–1104
Vaeck M, Reynaerts A, Hofte H, Jansens S, Debeuckeleer M, Dean C, Zabeau M, Vanmontagu M, Leemans J (1987) Transgenic plants protected from insect attack. Nature 328:33–37
Van Der Vossen EA, Van Der Voort JN, Kanyuka K, Bendahmane A, Sandbrink H, Baulcombe DC, Bakker J, Stiekema WJ, Kleinlankhorst RM (2000) Homologues of a single resistance-gene cluster in potato confer resistance to distinct pathogens: a virus and a nematode. Plant J 23:567–576
Wakefield ME, Bell HA, Fitches EC, Edwards JP, Gatehouse AMR (2006) Effects of *Galanthus nivalis* agglutinin (GNA) expressed in tomato leaves on larvae of the tomato moth *Lacanobia oleracea* (Lepidoptera: Noctuidae) and the effect of GNA on the development of the endoparasitoid *Meteorus gyrator* (Hymenoptera: Braconidae). Bull Entomol Res 96:43–52
Williamson VM (1998) Root-knot nematode resistance genes in tomato and their potential for future use. Annu Rev Phytopathol 36:277–293
Wu G, Shortt BJ, Lawrence E, Levine EB, Fitzsimmons C, Shaw DM (1995) Disease resistance conferred by expression of a gene encoding H2O2-generating glucose oxidase in transgenic potato plants. Plant Cell 7:1357–1368
Xu DP, Xue QZ, McElroy D, Mawal Y, Hilder VA, Wu R (1996) Constitutive expression of a cowpea trypsin inhibitor gene, CpTi, in transgenic rice plants confers resistance to two major rice insect pests. Mol Breed 2:167–173
Yedidia I, Benhamou N, Chet I (1999) Induction of defense responses in cucumber plants (*Cucumis sativus* L) by the biocontrol agent *Trichoderma harzianum*. Appl Environ Microbiol 65:1061–1070
Yu D, Xie Z, Chen C, Fan B, Chen Z (1999) Expression of tobacco class II catalase gene activates the endogenous homologous gene and is associated with disease resistance in transgenic potato plants. Plant Mol Biol 39:477–488
Zhou F, Collinge DB, Thordal-Christensen H (1995) Germin-like oxalate oxidase, a H2O2 producing enzyme, accumulates in barley attacked by the powdery mildew fungus. Plant J 8:139–145

Bio-priming of Seeds

6

Abstract

Bio-priming is a new technique of seed treatment that integrates biological (inoculation of seed with beneficial organism to protect seed) and physiological aspects (seed hydration) of disease control. It is recently used as an alternative method for controlling many seed- and soil-borne pathogens. It is an ecological approach using selected fungal antagonists against the soil- and seed-borne pathogens. Biological seed treatments may provide an alternative to chemical control. Seed priming, osmo-priming and solid matrix priming were used commercially in many horticultural crops, as a tool to increase speed and uniformity of germination and improve final stand. However, if seeds are infected or contaminated with pathogens, fungal growth can be enhanced during priming, thus resulting in undesirable effects on plants. Therefore, seed priming alone or in combination with low dosage of fungicides and/or biocontrol agents has been used to improve the rate and uniformity emergence of seed and reduce damping-off disease.

6.1 What Is Bio-priming or Biological Seed Treatment?

Biological seed treatments for control of seed and seedling diseases offer the grower an alternative to chemical fungicides. Whilst biological seed treatments can be highly effective, it must be recognised that they differ from chemical seed treatments by their utilisation of living microorganisms. Storage and application conditions are more critical than with chemical seed protectants, and differential reaction to hosts and environmental conditions may cause biological seed treatments to have a narrower spectrum of use than chemicals. Conversely, some biocontrol agents applied as seed dressers are capable of colonising the rhizosphere, potentially providing benefits to the plant beyond the seedling emergence stage (Nancy et al. 1997).

Seed treatment with biocontrol agents along with priming agents may serve as an important means of managing many of the soil- and seed-borne diseases, the process often known as 'bio-priming'. The bio-priming seed treatment developed for control of *Pythium* seed rot of sh2 sweet corn combines microbial inoculation with pre-plant seed hydration. Bio-priming involves coating seed with a bacterial biocontrol agent such as *Pseudomonas aureofaciens* AB254 and hydrating for 20 h under warm (23°C) conditions in moist vermiculite or on moist germination blotters in a self-sealing plastic bag. The seeds

P.P. Reddy, *Recent Advances in Crop Protection*,
DOI 10.1007/978-81-322-0723-8_6, © Springer India 2013

are removed before radical emergence. The bacterial biocontrol agent may multiply substantially on seed during bio-priming (Callan et al. 1990).

Bio-priming process had potential advantages over simply coating seed with *P. aureofaciens* AB254. Seed priming often results in more rapid and uniform seedling emergence and may be useful under adverse soil conditions. Sweet corn seedling emergence in *Pythium*-infested soil was increased by AB254 at a range of soil temperatures, but emergence at 10 °C was slightly higher from bio-primed seeds than from seeds coated with the bacterium (Mathre et al. 1994).

Induction of defence-related enzymes by biocontrol agents and chemicals, osmotic priming of seeds using polyethylene glycol (PEG) solution is known to improve the rate and uniformity of seed germination in several vegetable crops (Smith and Cobb 1991). The observed improvements were attributed to priming-induced quantitative changes in biochemical content of the sweet corn seeds (Sung and Chang 1993). Venkata Ratnam et al. (2001) investigated the effect of seed treatment with inducer chemicals like salicylic acid and bion on the systemic resistance of sunflower to *Alternaria helianthi*. Seed treatment with salicylic acid and bion increased the phenolic compounds synthesis in sunflower leaves.

A successful antagonist should colonise rhizosphere during seed germination (Weller 1983). Priming with PGPR increases germination and improves seedling establishment. It initiates the physiological process of germination but prevents the emergence of plumule and radical. Initiation of physiological process helps in the establishment and proliferation of PGPR on the spermosphere (Taylor and Harman 1990). Bio-priming of seeds with bacterial antagonists increases the population load of antagonist to a tune of tenfold on the seeds, thus protecting rhizosphere from the ingress of plant pathogens (Callan et al. 1990).

6.2 Procedure of Seed Bio-priming

- Pre-soak the seeds in water for 12 h.
- Mix the formulated product of bioagent (*Trichoderma harzianum* and/or *Pseudomonas fluorescens*) with the pre-soaked seeds at the rate of 10 g per kg seed.
- Put the treated seeds as a heap.
- Cover the heap with a moist jute sack to maintain high humidity.
- Incubate the seeds under high humidity for about 48 h at approximately 25–32 °C.
- Bioagent adhered to the seed grows on the seed surface under moist condition to form a protective layer all around the seed coat.
- Sow the seeds in nursery bed.
- The seeds thus bio-primed with the bioagent provide protection against seed- and soil-borne plant pathogens, improving germination and seedling growth (Fig. 6.1).

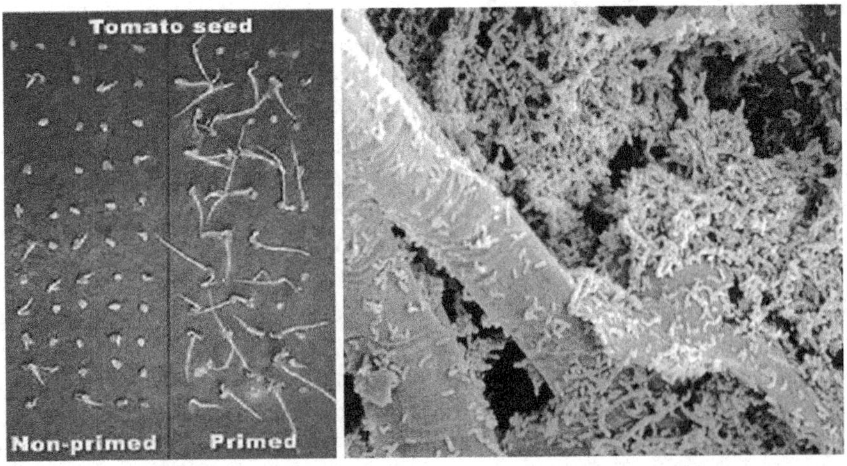

Fig. 6.1 *Left* – bio-priming of tomato seeds. *Right* – *Pseudomonas aureofaciens* AB 254 on tomato seed hairs surface

6.3 Disease Management Using Bio-priming

6.3.1 Carrot Damping Off, *Alternaria dauci, A. radicina*

Bio-priming with *Clonostachys rosea* controlled pre- and post-emergence death caused by seed-borne pathogens *Alternaria dauci* and *A. radicina* as effectively as the fungicide iprodione (Jensen et al. 2004). On highly infected seed (29% *A. radicina* and 11% *A. dauci*), bio-priming with *C. rosea* isolate IK726 reduced the incidence of *A. radicina* to <2.3% and that of *A. dauci* to <4.8%, whilst the level of both pathogens was <0.5% on bio-primed seed with a low initial infection rate.

In sand stand establishment tests, bio-priming with *C. rosea* isolate IK726 resulted in a seedling stand that was better than that of both nonprimed and hydroprimed seeds. *C. rosea* IK726 multiplied fivefold to eightfold, and carrot seeds were covered with a fine web of sporulating mycelium of *C. rosea*. The positive effect of bio-priming on healthy seedling stand remained after 5 months of storage at 4 °C, and IK726 survived at high numbers on these seeds.

6.3.2 Cowpea Root Rot, *Fusarium solani, Macrophomina phaseolina* and *Rhizoctonia solani*

On cowpea, bio-primed seed treatment (with *Trichoderma harzianum*) reduced root rot incidence (caused by *Fusarium solani, Macrophomina phaseolina* and *Rhizoctonia solani*) by 64.0 and 56.3% at pre-emergence stage and by 68.0, 60.1 and 57.1%, 64.0% at post-emergence stage after 40 and 60 days of sowing during 2004 and 2005 seasons, respectively. Therefore, fresh pod yield was increased by 44.0 and 36.1% compared with 19.5 and 11.2% in the case of Rizolex-T treatment during the same seasons, respectively. It could be noted that using of soil amendment with agricultural wastes formulated with biocontrol agents and/or bio-priming seed treatments to control soil-borne plant pathogens as a substitute of chemical fungicides is possible without any risk to human, animal and the environment (El-Mohamedy et al. 2006).

6.3.3 Faba Bean Root Rot

Under greenhouse conditions, all the tested fresh and 2 months stored, bio-primed faba bean seeds (with *Trichoderma viride*, *T. harzianum*, *Bacillus subtilis* and *Pseudomonas fluorescens*) showed a highly significant effect causing complete reduction of root rot incidence at both pre- and post-emergence stages of plant growth compared with the control treatment. The use of bio-primed seeds might be considered as a safe, cheap and easily applied biocontrol method against these soil-borne plant pathogens (El-Mougy and Abdel-Kader 2008).

6.3.4 Pea Root Rot, *Fusarium solani, Rhizoctonia solani*

In greenhouse trails, seed bio-priming enhanced effectiveness of *T. harzianum, B. subtilis* and *P. fluorescens* in the control of root rot pathogens (*Fusarium solani* and *Rhizoctonia solani*), as the highest percentages of disease reduction were recorded with bio-priming seed treatments (El-Mohamedy and Abd El-Baky 2008).

Under field conditions, seed bio-priming treatments strongly reduced pea root rot disease during two seasons. Bio-priming caused disease reduction at pre- and post- emergence stages by 72.7–84.5%, 72.2–82.9% and 67.6–80.0% after 15, 45 and 60 days after sowing, respectively. Bio-priming and seed coating with *T. harzianum* or *B. subtilis* were the most effective treatments in stimulating vegetative growth of pea plants. These treatments also caused significant increase in early and total green pod yield. Moreover, the highest values for quality parameters of pea pods, that is, pod length, pod diameter, number of seeds/pod as well as chemical contents of pod, that is, TSS, total carbohydrate and protein were recorded during two seasons (El-Mohamedy and Abd El-Baky 2008).

The present study suggested that seed bio-priming with biocontrol agents is a novel technology to replace fungicide seed treatments used in conventional agriculture to protect against seed- and soil-borne plant pathogens.

6.3.5 Soybean Damping Off

Bio-priming with *Pseudomonas aeruginosa* was the most effective treatment for controlling pre- and post-emergence damping off, with reductions in disease incidence ranging from 48.6 to 51.9% and 65.0 to 97.2%, respectively. Moreover, *P. aeruginosa* resulted in enhancement of seed germination and healthy seedling stand ranging from 32.4 to 60.0% and 56.0 to 73.9%, respectively. Bio-priming with *T. harzianum* reduced pre- and post-emergence damping off by 42.8–46.8% and 35.0–85.1%, respectively. However, *P. aeruginosa* was generally comparable to *T. harzianum* and the fungicide Benlate. Bio-priming with *P. aeruginosa* or *T. harzianum* offered an effective biological seed treatment system and an alternative to the fungicide Benlate for control of damping off of soybean caused by *C. truncatum*. Thus, bio-priming can be exploited by seed companies and organic farmers in the sustainable agriculture, which would be more economical and environmental friendly (Begum et al. 2010).

6.3.6 Coconut Leaf Rot Disease

Biocontrol agents, namely, *Bacillus subtilis*, *Pseudomonas fluorescens*, both individually and in combination, and *Trichoderma viride*, were found to be useful for bio-priming of coconut seed nuts and developing seedlings with favourable growth. More research is being undertaken at the Central Plantation Crops Research Institute, Kasargod, Kerala, India, on the role of these bio-agents in coconut seedlings in the context of reaction against pathogens of coconut leaf rot disease.

6.3.7 Sweet Corn Seed Decay, *Pythium ultimum*

Bio-priming is a combination of seed hydration and inoculation of seed with beneficial bacteria (*Pseudomonas aureofaciens*, strain AB254) to protect sweet corn from *Pythium ultimum* seed decay. Data from this study in 1995 showed earlier germination under clear plastic mulch compared to bare ground controls. Harvests were 6–9 days earlier from treatments planted under clear plastic mulch compared to controls.

Colony-forming units (CFUs) of *P. aureofaciens* strain AB254 per seed achieved with spring 1996 bio-priming ranged from 160 million (for *se* Tuxedo) to 8.5 billion (for *sh2* HMX 3364). These CFU levels are considered more than adequate for control of *P. ultimum* under most conditions, based on prior reports.

6.3.8 Maize Ear Rot, *Fusarium verticillioides*

The pure culture of *Trichoderma harzianum* was more effective in reducing the *Fusarium verticillioides* (responsible for ear rot disease of maize) and fumonisin incidence followed by talc formulation than the carbendazim-treated and untreated control. Formulations of *T. harzianum* were effective at reducing the *F. verticillioides* and fumonisin infection and also increasing the seed germination, vigour index, field emergence, yield and thousand seed weight in comparison with the control.

6.3.9 Pearl Millet Downy Mildew, *Sclerospora graminicola*

Bio-priming pearl millet seeds with *Pseudomonas fluorescens* isolate UOM SAR 14 resulted in improved growth of the plants and also induction of resistance against downy mildew disease caused by the fungus *Sclerospora graminicola*. Treatment due to *P. fluorescens* resulted in enhancement of germination, seedling vigour, plant height, leaf area, tillering capacity, seed weight (measured for 1000 seeds) and yield. The time required for flowering was advanced by 5 days. Isolate UOM SAR 14 registered higher levels of vegetative and reproductive growth; most importantly, there was a 22% increase in grain yield. The isolate also effectively induced resistance against downy mildew disease both under greenhouse and field conditions. The isolate offered protection ranging from 71 to 75%. However, the level of disease control was less than that achieved by the systemic fungicide apron. Further studies showed that the resistance induced was systemic, required a minimum of 3 days to build up and was sustained throughout the plant's life (Raj et al. 2004).

6.3.10 Rice Seed Rot (*Pythium* sp.), Damping Off (*Rhizoctonia solani*) and Brown Leaf Spot (*Helminthosporium oryzae*)

Rice seed bio-priming with *Trichoderma harzianum/T. viride/T. virens/Pseudomonas fluorescens* or mixed formulation of *T. harzianum* and *P. fluorescens* at 5–10 g/kg of seed (Fig. 6.2) gave effective control of various seed, soil and seedling diseases, especially against *Rhizoctonia solani*, *Pythium* seed rot, damping off and *Helminthosporium oryzae*.

6.3.11 Sunflower Leaf Blight, *Alternaria helianthi*

Bio-priming the sunflower seeds with *Pseudomonas fluorescens* (0.8%) gave effective control of Alternaria blight (Rao et al. 2009).

Seed treatment with *P. fluorescens* in 'jelly' as priming agent along with foliar spray of hexaconazole recorded an average yield of 1.592 t/ha and head diameter of 23.5 cm which was found on par with seed treatment with carbendazim + iprodione in water.

Studies on the effect of bio-priming of sunflower seeds with *P. fluorescens* in different priming agents, on the activity of defence-related enzymes like peroxidase, polyphenol oxidase and catalase, revealed that in peroxidase, untreated seeds and direct seed treatment with *P. fluorescens*, exhibited only one band, and seed treatment with *P. fluorescens* in 'jelly' and vermiculite showed two bands indicating increased activity of peroxidase in these seeds. Seeds treated with *P. fluorescens* with neem and apparently healthy seeds also exhibited two bands.

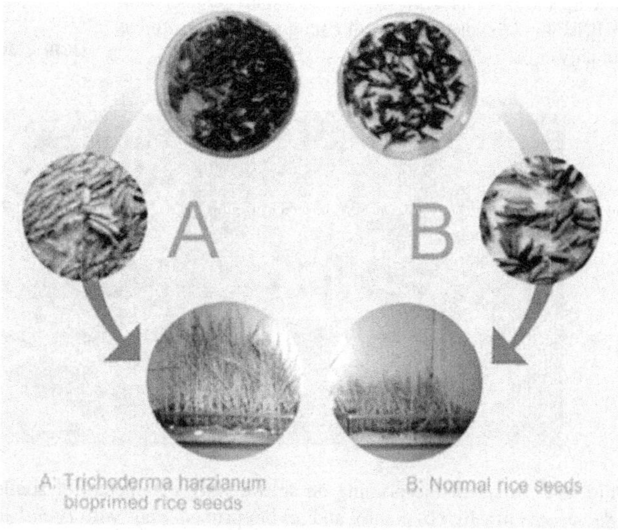

Fig. 6.2 Rice seed bio-priming with *Trichoderma harzianum* strain PBAT-43

In polyphenol oxidase, banding pattern remained same, though the intensity of bands varied over control and apparently healthy seeds. Only one band was observed in untreated seeds (control). In catalase, untreated seeds (control) and seeds treated with *P. fluorescens* + neem oil exhibited similar banding pattern with one band, and seeds treated with *P. fluorescens* in 'jelly' and *P. fluorescens* in vermiculite exhibited two bands, indicating increased activity of catalase in these treated seeds. Increased activity of peroxidase and catalase enzymes in treated seeds may result in the activation of defence genes, which in turn may lead to increased production of antimicrobial compounds like phenolics such as lignins which may be toxic to the seed-borne pathogens, and this may contribute to disease resistance (Figs. 6.3, 6.4, and 6.5).

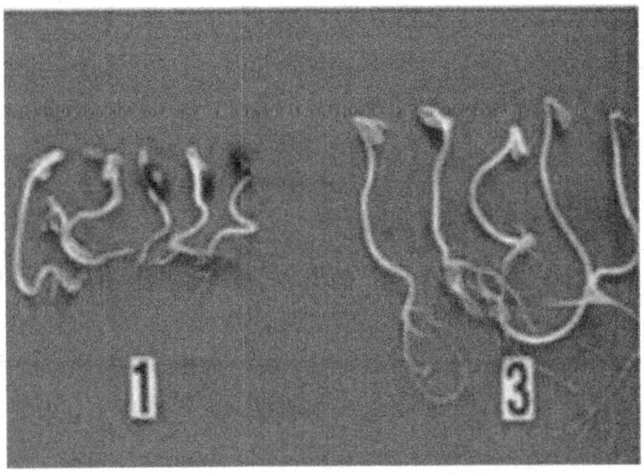

Fig. 6.3 Effect of seed treatment with chemical, bio-agents and priming agents. *1* Infected untreated seeds of KBSH 44. *3* Seeds treated with *Pseudomonas fluorescens* in jelly

6.3.12 Rape Oil Seed Blackleg, *Leptosphaeria maculans*

Seed bio-priming treatment of oilseed rape (*Brassica napus*) with the rhizobacteria *Serratia plymuthica* (strain HRO-C48) [seeds were soaked in bacterial suspensions (bio-priming) to obtain log 10^{6-7} CFU $seed^{-1}$] gave a mean blackleg disease (*Leptosphaeria maculans*) reduction of 71.6% followed by 54.0% for *Pseudomonas chlororaphis* (strain MA 342). The combined treatment was not superior to the treatment with *S. plymuthica* alone. The reduction of the disease caused by *S. plymuthica* was independent of the cultivar's susceptibility, whereas the control effect recorded with *P. chlororaphis* increased with decreasing cultivar resistance to blackleg disease (Ruba et al. 2011).

6.3.13 Sesamum Charcoal Rot, *Macrophomina phaseolina*

The sesame charcoal rot disease (*Macrophomina phaseolina*) was successfully controlled under greenhouse conditions by bio-priming sesame seeds with *Trichoderma harzianum* which produced the highest percentages of healthy mature plants (96.7%) followed by *Chaetomium bostrycoides* (90.0%), *Trichoderma* sp. (83.3%) and *T. hamatum* (80.0%).

T. harzianum and *C. bostrycoides* were significantly effective in suppressing disease incidence at seedling stage (pre- and post-emergence)

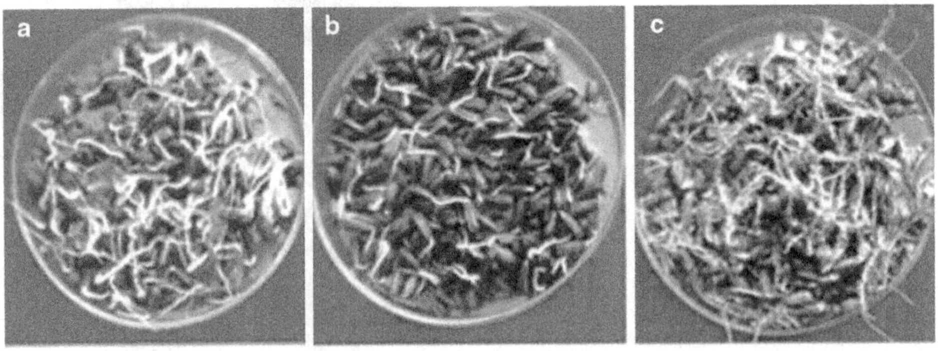

Fig. 6.4 Effect of bio-priming on seed quality parameters of sunflower. (**a**) Bio-primed seeds with *Pseudomonas fluorescens* in jelly, (**b**) control and (**c**) bio-primed seeds with *Pseudomonas fluorescens* in vermiculite

Fig. 6.5 Effect of bio-priming on sunflower plant growth

and maturity stage (charcoal rot) and consequently increased per cent healthy plants at the maturity stage. These results could be attributed to the in vitro antagonistic action of the tested microorganisms against *M. phaseolina*. Also, the antagonistic microorganism(s) could suppress the activity of a plant pathogen through the enzymatic digestion of the pathogen cell walls (Elad et al. 1983) and/or production of inhibitory volatile substances (Sankar and Sharma 2001).

6.4 Conclusions

Bio-priming seed treatments can provide a high level of protection against root rot diseases of crop plants. This protection was generally equal or superior to the control provided with fungicide seed treatment. So, it could be suggested that bio-priming (combined treatments between seed priming and seed coating with biocontrol agents) may be safely used commercially as substitute for traditional fungicide seed treatments for controlling seed- and soil-borne plant pathogens. Besides, bio-priming also improves seed germination, seedling establishment and vegetative growth. This has explored up new dimension of biological control for preventive as well as remedial for seed-borne infection by bioagents. Thus, bio-priming can be exploited by seed companies and organic farmers in the sustainable agriculture, which would be more economical and environmental-friendly.

References

Begum MM, Sariah M, Puteh AB, Zainal Abidin MA, Rahman MA, Siddiqui Y (2010) Field performance of bio-primed seeds to suppress *Colletotrichum truncatum* causing damping-off and seedling stand of soybean. Biol Control 53:18–23

Callan NW, Mathre DE, Miller JB (1990) Bio-priming seed treatment for biological control of *Pythium ultimum* pre-emergence damping-off in *sh2* sweet corn. Plant Dis 74:368–372

Elad Y, Barak R, Chet I, Henis Y (1983) Ultrastructural studies of the interaction between *Trichoderma* spp and plant pathogenic fungi. Phytopathologosche Z 107(2):168–175

El-Mohamedy RSR, Abd El-Baky MMH (2008) Evaluation of different types of seed treatment on control of root rot disease, improvement of growth and yield, quality of pea plant in Nobaria Province. Res J Agric Biol Sci 4:611–622

El-Mohamedy RSR, Abd Alla MA, Badiaa RI (2006) Soil amendment and seed bio-priming treatments as alternative fungicides for controlling root rot diseases on cowpea plants in Nobaria Province. Res J Agri Biol Sci 2:391–398

El-Mougy NS, Abdel-Kader MM (2008) Long-term activity of bio-priming seed treatment for biological control of faba bean root rot pathogens. Aust Plant Pathol 37:464–471

Jensen B, Knudsen IM, Madsen M, Jensen DF (2004) Biopriming of infected carrot seed with an antagonist, *Clonostachys rosea*, selected for control of seed-borne *Alternaria* spp. Phytopathology 94:551–560

Mathre DE, Callan NW, Schwend A (1994) Factors influencing the control of *Pythium ultimum*-induced seed decay by seed treatment with *Pseudomonas aureofaciens* AB254. Crop Prot 13:301–307

Nancy W, Don Mathre E, James B, Charles S (1997) Biological seed treatments: factors involved in efficacy. Hortic Sci 32:179–183

Raj N, Shetty N, Shetty H (2004) Seed bio-priming with *Pseudomonas fluorescens* isolates enhances growth of pearl millet plants and induces resistance against downy mildew. Int J Pest Mang 50:41–48

Rao MSL, Kulkarni S, Lingaraju S, Nadaf HL (2009) Bio-priming of seeds: a potential tool in the integrated management of Alternaria blight of sunflower. Helia 32(50):107–114

Ruba A, Mazen S, Ralf-Udo E (2011) Effect of seed priming with *Serratia plymuthica* and *Pseudomonas chlororaphis* to control *Leptosphaeria maculans* in different oilseed rape cultivars. Eur J Plant Pathol 130:287–295

Sankar P, Sharma RC (2001) Management of charcoal rot of maize with *Trichoderma viride*. Indian Phytopathol 54:390–391

Smith PT, Cobb BC (1991) Physiological and enzymatic activity of pepper seeds during priming. Physiol Plant 83:433–439

Sung FJ, Chang YH (1993) Biochemical activities associated with priming of sweet corn seeds to improve vigour. Seed Sci Technol 21:97–105

Taylor AG, Harman GE (1990) Concept and technologies of selected seed treatments. Annu Rev Phytopathol 28:321–339

Venkata Ratnam S, Narayan Reddy P, Chandrasekhara Rao K, Krishnam Raju S, Hemantha Kumar J (2001) Effect of salicylic acid and bion seed treatment in induction of systemic resistance against sunflower leaf blight. Indian J Plant Prot 29:79–81

Weller DM (1983) Colonization of wheat roots by a fluorescent pseudomonad suppressive to take all. Phytopathology 73:1548–1553

7 Disguising the Leaf Surface

Abstract

The leaf surface provides the first barrier that fungi must overcome in order to gain access to the leaf, but it also provides chemical and physical cues that are necessary for the development of infection structures for many fungal pathogens. Film-forming polymers can coat the leaf surface, acting not just as an extra barrier to infection but also disguising the cues necessary for germling development. Kaolin particle films can envelop the leaf in a hydrophobic particle film barrier that prevents spores or water from directly contacting the leaf surface and, as a result, can suppress infection. Adhesion of fungal spores to the leaf surface, which is important to keep spores on the leaf surface and for appropriate development of the fungus on the leaf surface, can be inhibited, leading to reduced infection and lesion development. Polymer and particle films have been shown to provide disease control in the field, whilst research on agents that inhibit spore adhesion on leaf surfaces is still in its infancy. There is an urgent need for research on the practicality of using these novel methods under field conditions and on ways of integrating them into current crop protection programmes.

The importance of leaf surface features in the early development and establishment of foliar pathogens suggests that interference with leaf topography will disrupt pathogen development and lead to reduction in infection. Research on a variety of agents which coat the leaf surface has shown that interference with leaf topography can indeed lead to reduction in pathogen infection. Polymer and particle films have been shown to provide disease control in the field, whilst research on agents that inhibit spore adhesion on leaf surfaces is still in its infancy. There is an urgent need for research on the practicality of using these novel methods under field conditions and on ways of integrating them into current crop protection programmes.

Various approaches that have been used to coat or disguise leaf surfaces in order to control foliar pathogens are as follows:

- Controlling disease using film-forming polymers
- Particle films as agents for control of plant diseases
- Disrupting spore adhesion to the leaf surface

P.P. Reddy, *Recent Advances in Crop Protection*,
DOI 10.1007/978-81-322-0723-8_7, © Springer India 2013

7.1 Controlling Disease Using Film-Forming Polymers

Film-forming polymers are widely used in agriculture and horticulture as antitranspirants and as spray adjuvants. Their main use as adjuvants is as filming agents to reduce weathering and pesticide efficacy and as stickers or spreaders to improve distribution and adherence of agrochemicals. Film-forming polymers used as antitranspirants include waxes, silicons and a variety of plastic polymers. When used in this role, they form a film over the stomata, increasing resistance to water vapour loss, and are used particularly on seedlings and transplants to decrease water stress and to improve water use efficiency in arid conditions. Film-forming polymers can coat the leaf surface, not just acting as an extra barrier to infection but also disguising the cues necessary for germling development.

Glenn et al. (1999) proposed a new concept of disease control using hydrophobic kaolin particle films in which a hydrophobic layer of particles kept water physically separated from the plant surface and demonstrated the concept in suppressing fire blight and apple scab. Neinhuis and Barthlott (1997) suggested that in native plant populations, water-repellent and self-cleaning plant surfaces evolved to remove surface contamination such as pathogens, water films that reduce CO_2 diffusion, soil dust particles that increase leaf temperature and salt deposits that are phytotoxic. Puterka et al. (2000) demonstrated that Fabraea leaf spot (*Fabraea maculata*) of pear was suppressed by both hydrophobic and hydrophilic kaolin particles, presumably through both a physical interference in the infection process and a lack of adherence of inoculum to the plant surface. In addition, there was an increase in pear fruit yield that was thought to result from the reflective nature of the particles that reduced plant temperature (Glenn et al. 1999, 2001a, b). Potassium silica application was effective in controlling powdery mildew due, in part, to mechanical abrasion of the hyphae (Bowen et al. 1992; Menzies et al. 1992). Whitewash reduced the incidence of insect-transmitted viruses in potato, pepper (Marco 1993) and rutabaga (Lowery et al. 1990) and suppressed the transmission of citrus stubborn disease (Gumpf et al. 1981) and papaya decline (Franck and Bar-Joseph 1992). Whitewash treatments reduced insect-transmitted disease by repelling the vector and suppressing feeding, apparently due to the reflective nature of the whitewash. A combination of aluminium, silica and titanium dioxide was effective in controlling downy and powdery mildew of grapes (Mendgen et al. 1992) through mechanisms that may include the following: (1) direct action on the hyphae, (2) interference with recognition of the plant surface and (3) stimulation of the plant's physiological defences. Antitranspirant films reduced the incidence of powdery mildew on wheat and barley (Ziv 1983; Ziv and Frederiksen 1987), general cucurbit diseases (Ziv and Zitter 1992) and powdery mildew of zinnia (Kamp 1985). Antitranspirant films form an artificial barrier on the plant surface that interferes with the infection process.

Glenn et al. (1999) demonstrated that fire blight could be suppressed by the application of a hydrophobic dust to the flowers of apple and pear. The development of liquid formulations of hydrophobic and hydrophilic particle film materials (Puterka et al. 2000) greatly improved the practicality of particle film materials. Greenhouse and field studies demonstrated that pre-emptive application of hydrophobic and hydrophilic particles reduced the incidence of fire blight (*Erwinia amylovora*) in blossoms and injured shoots of apple (Fig. 7.1) (Glenn et al. 2001a, b).

Hydrophobic and hydrophilic particles suppressed sooty blotch (*Gloeodes pomigena*) and flyspeck (*Schizothyrium pomi* anamorph *Zygophiala jamaicensis*) of apple to levels equivalent to conventional fungicides (Glenn et al. 2001a, b). Hydrophilic particles did reduce powdery mildew (*Podosphaera leucotricha* anamorph *Oidium farinosum*) russetting on apple fruit, but leaf infection was extensive (Fig. 7.2) (Glenn et al. 2001a, b).

Film-forming polymers (Bond, Designer, Nu-Film P, Spray Gard, Moisturin, Companion PCT12) inhibited germination of conidia and subsequent formation of appressoria and reduced leaf scab severity using a detached leaf bioassay. Regardless of treatment, there were no obvious trends in the percentage of conidia with one to four appressoria 5 days after inoculation. The synthetic fungicide penconazole resulted in the greatest levels of germination inhibition, appressorium development and least leaf scab severity. Under field conditions, scab severity on leaves and fruit of apple cv. 'Golden Delicious' treated with a film-forming polymer (Bond, Spray Gard, Moisturin) was less than on untreated controls. However, greatest protection in both field trials was provided by the synthetic fungicide penconazole. Higher chlorophyll fluorescence Fv/Fm emissions in polymer- and penconazole-treated trees indicated less damage to the leaf photosynthetic system as a result of fungal invasion. In addition, higher SPAD values as measures of leaf chlorophyll content were recorded in polymer- and penconazole-treated trees. Application of a film-forming polymer or penconazole resulted in a higher apple yield per tree at harvest in both the 2005 and 2006 field trials compared to untreated controls. Results suggest that application of an appropriate film-forming polymer may provide a useful addition to existing methods of apple scab management (Percival and Boyle 2009).

Fig. 7.1 Apple cv. 'Gala' shoots treated and then inoculated with *Erwinia amylovora*. Untreated control (*left*) and treated with Surround WP crop protectant (*right*)

During the field trials on the effect of antitranspirants on the water balance of sugar beet, Gale and Poljakoff-Mayber (1962) found that the incidence of powdery mildew was reduced on treated plots.

The downy mildew (*Pseudoperonospora cubensis*) of cucumber was reduced by foliar spraying of antitranspirant film kaolin, Nu-Film, Bio-Film, Folicote and polyacrylamide Anti-Stress 550. In pot experiments, antitranspirants were proved to be effective in reducing disease incidence, severity and pathogen sporulation when pre- or post-inoculated. Amongst these compounds, kaolin and Nu-Film (1.0%) were more effective in reducing spore germination counts, infected area and downy mildew lesion number. Scanning electron microscopic examination showed that kaolin antitranspirant inhibited spores germination and made the sporangia become collapsed and lose its turgidity when applied either pre- or post-inoculation. Under protected cultivation and natural infection conditions, all antitranspirants showed a remarkable

Fig. 7.2 Powdery mildew russet on 'Jonathan' apple. Left to right: untreated control, Surround WP crop protectant with Surround WP removed on right half and conventional fungicide programme

effectiveness on the reduction of disease severity when applied twice, 45 and 75 DAS, in cucumber leaves. Furthermore, all tested antitranspirants significantly increased the cucumber plant height and yield. Kaolin strongly protected cucumber against downy mildew and increased yield. Conclusively, antitranspirant film can be used as effective treatment for the control of downy mildew disease in cucumber plants under plastic houses (Haggag 2002).

In pot experiments, antitranspirants were proved to be effective in reducing disease incidence, severity and pathogen sporulation when pre- or post-inoculated. Amongst these compounds, kaolin and Nu-Film (1.0%) were more effective in reducing spore germination counts, infected area and downy mildew lesion number. Scanning electron microscopic examination showed that kaolin antitranspirant inhibited spores germination and made the sporangia become collapsed and lose its turgidity when applied either pre- or post-inoculation (Haggag 2002).

In pot experiments, Elad et al. (1989) found that Vapor Gard and Wilt Pruf reduced powdery mildew on cucumber by up to 82 and 55%, respectively, whilst under cover, Safe Pack and the fungicide fenarimol reduced infection by up to 67 and 96%, respectively. Cucumber powdery mildew was controlled most effectively by a mixture of Safe Pack and fenarimol. Film-forming polymers reduced germination of powdery mildew conidia when used at concentrations as low as 0.5%.

Hsich and Huang (1999) demonstrated that film-forming polyelectrolytes – FO4240SH, FO4490SH and FO4550SH – at 400 ppm reduced the disease severity of lily leaf blight caused by *Botrytis elliptica* in leaf-disc tests. Both FO4240SH and FO4490SH also suppressed sporulation of the pathogen on leaf discs. In greenhouse tests, the number and size of lesions on leaves of Lilium oriental hybrid cv. 'Star Gazer' were markedly reduced by FO4490SH and FO4550SH. Field trials showed that the effectiveness of FO4490SH was similar or better than that of procymidone on the reduction of lily leaf blight disease severity. The polymers had no harmful effects on the lily plants. The cationic polyelectrolytes FO4240SH, FO4490SH and FO4550SH reduced the percentage of conidial germination, inhibited germ-tube growth and also suppressed the esterase production by germ tubes of *B. elliptica*. All the above evidence indicates that the disease control achieved with polyelectrolytes is due, at least in part, to the reduction of esterase secretion by *B. elliptica*.

JMS Stylet-Oil suppressed powdery mildew on winter squash and musk melon and increased yield under field conditions. This product was also effective in controlling powdery mildew on grape clusters and leaves (Wilcox and Riegel 1997) and powdery mildew on bell pepper (Smith 1996).

Sutherland and Walters (2002) found that all three different film-forming polymers, Ethokem (ethoxylated tallow amine), Bond (synthetic latex) and Vapor Gard (1-p-menthene), provided significant control of powdery mildew infection on barley under controlled conditions. Bond and Vapor Gard altered germling development on the leaf surface, reducing germination of conidia as well as the subsequent formation of appressoria and haustoria. Thus, Bond used as a 3% spray reduced mildew infection by 63%, in comparison with the commercial fungicide cyproconazole, which reduced mildew infection by 84% (Sutherland and Walters 2002). Sutherland and Walters (2001) found that Ethokem and Bond completely inhibited linear growth of *Pyricularia oryzae* when incorporated into the growth medium at 1 and 2% and strongly inhibited linear growth of *Pyrenophora avenae* at the same concentrations. Less of an effect on linear growth of these fungi was obtained with Vapor Gard, although all three polymers decreased cell lengths and led to gross changes in hyphal morphology, including swollen, shortened cells, granulation of cytoplasm, increased branching and collapsed empty cells (Sutherland and Walters 2001).

Two mechanisms have been proposed to account for the effects of film-forming polymers in reducing fungal infection on leaf surfaces. First, it has been suggested that leaf surfaces coated with polymers are hydrophobic, leading to low water potential at infection sites. Second, coated leaf surfaces may be impenetrable due to thickness, hardiness or resistance to enzymatic attack of the film-forming polymer.

Zekaria-Oren and Eyal (1991) found that polymers applied prior to inoculation of *Puccinia recondita* f. sp. *tritici* on wheat had a greater effect on rust infection than compounds applied post-inoculation. They found that increasing the concentration of the film-forming polymers led to a progressive reduction in infection intensity, suggesting that efficacy was related to thickness and uniformity of the coat on leaf surface. These workers also found that surfaces coated with film-forming polymers interfered with fungal penetration of the leaf. They observed that both the orientation of the germinating uredospore towards the stomata and the formation of appressoria were altered on coated surfaces.

Sprayable polymers also offer a feasible and cost-effective alternative to plastic tarps for soil heating. The plastic-based polymers are sprayed on the soil surface in the desired quantity and form a membrane film, which can maintain its integrity in soil and elevate soil temperatures. Nevertheless, the formed membrane is porous and allows overhead irrigation. Stapleton and Gamliel (1993) achieved effective soil heating and a reduction in the viability of *Pythium* propagules. In Israel, a sprayable polymer product, 'Ecotex', was developed together with the technology to apply it economically on soils for various purposes (Skutelsky et al. 2000). Soil coating with this technology with a black polymer formulation resulted in a membrane film that could raise soil temperatures close to solarisation levels. Soil heating with sprayable mulch is faster than that with plastic film, but the soil also cools down to lower temperatures at night. Overall, soil temperatures under sprayable mulch are lower than those obtained under plastic film. The thickness of the sprayed coat is critical to obtaining effective heating. Soil heating using sprayable mulches was effective in controlling Verticillium wilt and scab in potato (Gamliel et al. 2001), at a level matching that achieved by solarisation using plastic films.

7.2 Particle Films as Agents for Control of Plant Diseases

Hydrophobic particle films are based on the inert material kaolin (aluminosilicate) which is treated with a water-repelling agent. Kaolin particles can be made with varying degrees of hydrophobicity by coating them with waterproofing agents such as chrome complexes, stearic acid and organic zirconate. Kaolin particle films can envelop the leaf in a hydrophobic particle film barrier that prevents spores or water from directly contacting the leaf surface and, as a result, can suppress infection. By dusting fruit trees with hydrophobic kaolin particles, Glenn et al. (1999) obtained control of arthropod pests and fungal and bacterial pathogens. However, hydrophilic kaolin particles can also provide plant disease control. Puterka et al. (2000) found that hydrophilic particle films controlled Fabraea leaf spot on pear, whilst the hydrophilic kaolin-based product Surround WP was shown to control *Zygophiala jamaicensis* and *Gloeodes pomigena* on apple fruits and *Phoma* sp. on apple leaves. Hydrophobic kaolin particle films controlled *Monilinia fructicola* on peach (Lalancette et al. 2005).

Other beneficial effects of applying kaolin particle films have been observed. Liquid formulations of both hydrophobic and hydrophilic particle films were shown to double yields of pear in field trials, whilst delayed fruit maturation, increased fruit size, increased fruit number and increased fruit yield were obtained following the use of kaolin particle films (Lalancette et al. 2005).

7.2.1 What Is Kaolin and How Does It Work?

Surround WP presents a unique form of pest control: a non-toxic particle film that places a barrier between the pest and its host plant. The active ingredient is kaolin clay, an edible mineral long used as an anti-caking agent in processed foods and in such products as toothpaste and kaopectate. There appears to be no mammalian toxicity or any danger to the environment posed by the use of kaolin in pest control.

Surround WP is sprayed on as a liquid, which evaporates, leaving a protective powdery film on the surfaces of leaves, stems and fruit. Conventional spray equipment can be used, and full coverage is important. The film works to deter insects in several ways. Tiny particles of the clay attach to the insects when they contact the tree, agitating and repelling them. Even if particles do not attach to their bodies, the insects find the coated plant or fruit unsuitable for feeding and egg laying. In addition, the highly reflective white coating makes the tree less recognisable as a host.

The standard Surround WP spray programme for plum curculio and first-generation codling moth starts at first petal fall and continues for 6–8 weekly sprays or until the infestation is over. Discontinuing sprays at this point will leave little or no residue at harvest because of rain and wind attrition. If a full season programme is used to suppress later season threats such as apple maggot, growers will need to use a scrubber/washer to remove any dust remaining on the fruit for fresh market sales. Although this residue is not considered harmful, it might be considered unsightly by consumers. However, the dust residue is not a problem for processing fruit.

Trial applications of the spray showed that where plum curculio (PC) damage was 20.30% in unsprayed checks, the treatments receiving the particle film had only 5.1% damage. Dr. Puterka is careful to say that his trials indicate suppression of PC damage rather than complete control. For the organic grower looking to achieve an economic level of control, the distinction is probably not relevant. What the researcher terms suppression in these USDA trials is very close to control, far closer than any other organically suitable option. For the non-organic grower, kaolin alone will not achieve quite as high a level of control as is ensured by the organophosphates. However, Surround WP is comparable to the OP's in that it is a broad-spectrum tool effective against most of the major pests of apple. As growers gain experience in timing applications of this new tool, efficacy data will very likely improve.

7.2.2 Horticultural Benefits

Although at first glance the film may appear to block light, Surround WP actually increases net photosynthesis and can provide secondary benefits to the trees. For overall health, according to Dr. Glenn, Surround WP keeps the tree cool so that photosynthesis can continue longer into the afternoon on hot days, after untreated trees have already shut down because of heat stress. In a 2-year study, 'Empire', when sprayed during the first six to eight weeks after petal fall, had increased yields and increased red colour. Growers have reported similar results with 'Stayman' and 'Gala'. An MSU study reported increased return bloom where Surround WP had been used the previous season. Growers in hot areas benefit from a marked reduction in sunburn damage, often 50% or greater.

7.2.3 A Systems Approach

Surround WP will be most effective when used within a well-managed agro-ecological system combining the most appropriate cultural and chemical methods for the specific orchard situation, pest complex and local climate. Such a system will integrate soil building, habitat for beneficial organisms and well-tuned nutrient and water management. Soil building and nutrient/water management are two sides of the same coin and could be considered preventive pest management. A healthy soil high in organic matter will have better water and nutrient holding capacity. Plants receiving too much or too little of either water or nutrients, particularly nitrogen, are more attractive and more susceptible to damage by insects and diseases. Good water management, through water-stress monitoring, conserves valuable (and expensive) soil nutrients, reduces contaminated run-off and conserves water. Providing habitat for beneficial organisms is like hiring millions of helpers whose sole aim in life is to eat pests. Nitrogen-fixing cover crops can do double duty as habitat for beneficial organisms if managed correctly.

Fig. 7.3 Glassy-winged sharpshooter, *Homalodisca coagulata*

Fig. 7.4 Kaolin particle film barrier as it appears on grapevine. Adequate coverage of all leaf and fruit surfaces is crucial

7.2.4 Management of Glassy-Winged Sharpshooter in Grapes

The glassy-winged sharpshooter (GWSS) (*Homalodisca coagulata*) (Fig. 7.3) can be a vector for Pierce's disease (a xylem-clogging bacterial disease caused by *Xylella fastidiosa*), an incurable bacterial disease that will generally kill a grape vine within 2 years of infection. The GWSS is a strong flier, able to travel distances of a quarter mile or more. Kaolin clay particle film (Surround® WP) can be sprayed to place a barrier between the pest and its host plant.

Surround WP has proved to be a very effective management tool for most species of leafhoppers, including the sharpshooters. In grapes, the combined effect of 'disguising' vines and interfering with feeding and probing behaviour should make transmission of the Pierce's disease bacterium less likely on Surround WP-treated surfaces (Fig. 7.4). Grapes treated with Surround WP show improved vigour, and studies have shown positive effects on juice flavour. Surround WP has proved effective in suppressing the following grape pests: several types of leafhoppers, including glassy-winged, red-headed and blue green sharpshooters and grape leafhopper, leaf rollers, Japanese beetles, thrips, grape leaf skeletoniser and June beetle.

7.2.5 Case Studies

7.2.5.1 North-Eastern US Grower Report: Eric Rice, Middletown, Maryland

One of the first orchardists to use the kaolin spray, Eric Rice, is confident that the product will help him fare better in his pack out next year. Rice, whose farm is certified organic, hopes to boost the percentage of select grade fruit from 50 to 70% of his apple crop. He expresses optimism about Surround WP's effectiveness against insects like the plum curculio, codling moth, leaf rollers, mites and aphids. It does not bother beneficials, he says, adding that the ladybugs and other predators continued to thrive in the rich ground cover of clover and grass.

Trials at the Rice orchard have shown over 90% control of the big ones – codling moth (Fig. 7.5), plum curculio (Figs. 7.6 and 7.7) and apple maggot (Fig. 7.8) – whilst Surround WP has also had a positive effect on fungal diseases like sooty blotch, flyspeck and fire blight. Rice cautions that it is not a panacea. It has had no effect on apple scab, a disease that often poses a bigger problem to growers than insect pests (initial research with kaolin focused on disease suppression, but the results were inconsistent).

Fig. 7.5 Codling moth

Rice says that Surround WP is far more useful than any other organic options available on the market. The only disadvantage he cites is the necessity of washing the clay off the fruit after harvest. Referring to the uniformly white appearance of the trees after spraying, Rice says, it looks like Christmas. People who drive by stop to inquire if something is wrong.

7.2.5.2 South-Eastern US Grower Report: Guy Ames, Fayetteville, Arkansas

I have used kaolin clay for two seasons now. Last year, I was slow in getting the first sprays on, and late May through early June was an

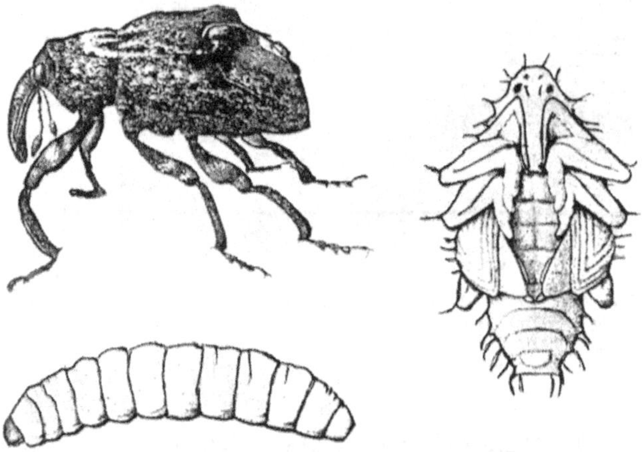

Fig. 7.6 The plum curculio

Fig. 7.7 The plum curculio: adult female on plum, showing the circular feeding punctures and the crescent-shaped egg-laying punctures

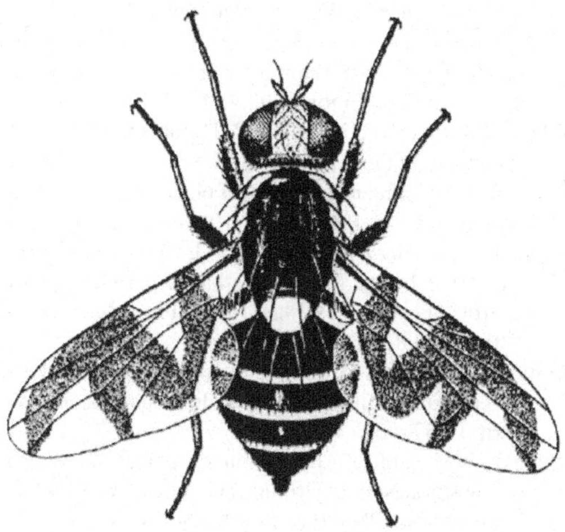

Fig. 7.8 Apple maggot

exceedingly wet period during which I had competing interests. In short, I can not claim to have given Surround a valid test in 1999.

However, in 2000 I was more attentive to keep Surround on the trees in the early part of the season. Compared to last year, I appeared to be getting good to excellent control of tarnished plant bugs and stink bugs, two of the pests that I was targeting with Surround. Other observations: It did not wash off in rain as readily as I had feared and did not seem to clog up or corrode any sprayer parts; I like being able to see my spray coverage; too early to see if it has helped to control oriental fruit moth as I hope it will.

Ames adds that Surround WP may prove key to making organic apple production economically viable in the Eastern USA by providing the long-sought organic control for plum curculio.

Another Arkansas grower reports that it is a real challenge to keep the trees covered. This is his first season using Surround WP. One technical limitation he has encountered is with his sprayer: It has been impossible to achieve full coverage of the tops of the trees using his present equipment. Full coverage is critical to achieving good pest control.

7.2.6 Economics

It seems likely that kaolin clay, when used in a well-managed IPM programme, will be an appropriate and effective reduced-risk replacement for synthetic insect pesticides. To be viable in commercial production, however, it must be *cost-effective* as well. At this time, Surround WP seems to be roughly comparable in relative cost to the commonly used OP insecticides whilst providing comparable broad-spectrum coverage. Of course, there are variables that make this a tentative statement: Relative pest pressure and rainfall, for example, influence how many sprays of Surround WP must be made in a season. Over the next few years, we will gather data on the cost of using kaolin and report our findings.

7.2.7 Compatibility

Surround WP is generally not affected by most other insecticides, miticides and fungicides. However, the user should test tank mixes before use. When mixing with other products, make up a small batch and observe slurry and film characteristics. Curdling, precipitation, lack of film formation or changes in viscosity are signs of incompatibility. Do not tank mix with sulphur or Bordeaux mixture fungicides.

Note 'Raw' kaolin clay is not the same as Surround WP. We have heard of one grower who bought a train car load of generic kaolin clay and killed most of his apple trees!

Surround WP is, at this point in time, the *only* kaolin product suitable and registered for horticultural use. The kaolin in Surround WP is processed to a very fine particulate size and combined with a sticker–spreader. Other forms of kaolin clay are phytotoxic and should be used under no circumstances.

7.2.8 Summary

Surround WP is an environmentally benign, worker-friendly and cost-effective tool that can provide many pest management and horticultural benefits. It is most effective when used in an agricultural ecosystem managed for healthy soils with well-balanced nutrient and water budgets.

Providing habitat for beneficial organisms (in the form of cover crops and unmown wild flower strips) complements the pest control benefits of kaolin clay and builds soil quality as well.

7.3 Disrupting Spore Adhesion to Leaf Surface

Adhesion of fungal spores to the leaf surface, which is important to keep spores on the leaf surface and for appropriate development of the fungus on the leaf surface, can be inhibited, leading to reduced infection and lesion development. The importance of spore adhesion to successful infection of plant surfaces suggests that disruption of this process could be useful in disease control. Stanley et al. (2002) showed that zosteric acid not only inhibited attachment of spores of *Colletotrichum lindemuthianum* but also inhibited formation of appressoria, leading to a failure to infect leaves. In fact, on intact plants, zosteric acid delayed lesion development on bean leaves following inoculation with *C. lindemuthianum*. The reduced adhesion could lower inoculum potential because either non-attached spores could be more easily detached from leaf surfaces by wind or rain splash or the spores quickly lose viability.

References

Bowen P, Menzies J, Ehret D, Samuels L, Glass ADM (1992) Soluble silicon sprays inhibit powdery mildew development in grape leaves. J Am Soc Hortic Sci 117:906–912

Elad Y, Ziv O, Aynish N, Katan J (1989) The effect of film-forming polymers on powdery mildew of cucumber. Phytoparasitica 17:279–288

Franck A, Bar-Joseph M (1992) Use of netting and whitewash spray to protect papaya plants against Nivun Haamir (NH)-dieback disease. Crop Prot 11:525–528

Gale J, Poljakoff-Mayber A (1962) Prophylactic effect of a plant transpirant. Phytopathology 52:715–717

Gamliel A, Skutelski Y, Peretz-Alon Y, Becker E (2001) Soil solarization using sprayable plastic polymers to control soil-borne pathogens in field crops. In: Proceedings of the 8th annual international research conference on methyl bromide alternatives and emission reduction, San Diego, CA, pp 10.1–10.3

Glenn DM, Puterka GJ, van der Zwet T, Byers RE, Feldhake C (1999) Hydrophobic particle films: a new paradigm for suppression of arthropod pests and plant diseases. J Econ Entomol 92:759–771

Glenn DM, Puterka GJ, Drake SR, Unruh TR, Knight AL, Baherle P, Prado E, Baugher T (2001a) Particle film application influences apple leaf physiology, fruit yield, and fruit quality. J Am Soc Hortic Sci 126:175–181

Glenn DM, van der Zwet T, Puterka G, Gundrum P, Brown E (2001b) Efficacy of kaolin-based particle films to control apple diseases. Plant Health Prog. doi:10.1094/PHP-2001-0823-01-RS

Gumpf DJ, Oldfield GN, Yokomi RK (1981) Progress in the control of citrus stubborn disease. Proc Int Soc Citric 1:457–458

Haggag WM (2002) Application of epidermal coating antitranspirants for controlling cucumber downy mildew in greenhouse. Plant Prot Bull 11:69–78

Hsich TF, Huang JW (1999) Effect of film-forming polymers on control of lily blight caused by *Botrytis elliptica*. Eur J Plant Pathol 105:501–508

Kamp M (1985) Control of *Erysiphe cichoracearum* on *Zinnia elegans*, with a polymer based antitranspirant. Hortic Sci 20:879–881

Lalancette N, Belding RD, Shearer PW, Frecon JL, Tietjen WH (2005) Evaluation of hydrophobic and hydrophilic particle films for peach crop arthropod and disease management. Pest Mange Sci 61:25–39

Lowery DT, Sears MK, Harmer CS (1990) Control of turnip mosaic virus of rutabaga with applications of oil, whitewash, and insecticides. J Econ Entomol 83:2352–2356

Marco S (1993) Incidence of non-persistently transmitted viruses in pepper sprayed with whitewash, oil, and insecticide, alone or combined. Plant Dis 77:1119–1122

Mendgen K, Schiewe A, Falconi C (1992) Biological control of plant diseases. In: Biological crop protection symposium, 24 May 1991. Pflanzenschutz-Nachrichten Bayer 45/1992,1

Menzies J, Bowen P, Ehret D, Glass ADM (1992) Foliar applications of potassium silicate reduce severity of powdery mildew on cucumber, muskmelon, and zucchini squash. J Am Soc Hortic Sci 117:902–905

Neinhuis C, Barthlott W (1997) Characterization and distribution of water-repellent, self-cleaning plant surfaces. Ann Bot 79:667–677

Percival GC, Boyle S (2009) Evaluation of film forming polymers to control apple scab [*Venturia inaequalis* (Cooke) G Wint] under laboratory and field conditions. Crop Prot 28:30–35

Puterka GJ, Glenn DM, Sekutowski DG, Unruh TR, Jones SK (2000) Progress toward liquid formulations of particle films for insect and disease control in pear. Environ Entomol 29:329–339

Skutelsky Y, Gamliel A, Kritzman G, Peretz-Alon Y, Becker E, Katan J (2000) Soil solarization using sprayable plastic polymers to control soil-borne pathogens in field crops. Phytoparasitica 28:269–270

Smith R (1996) Evaluation of materials for control of bell pepper powdery mildew. Fungicide Nematicide Tests 51:120

Stanley MS, Callow ME, Perry R, Alberte RS, Smith R, Callow JA (2002) Inhibition of fungal spore adhesion by zosteric acid as the basis for a novel, non-toxic crop protection technology. Phytopathology 92:378–383

Stapleton JJ, Gamliel A (1993) Feasibility of soil fumigation by sealing soil amended with fertilizers and crop residues containing biotoxic volatiles. Plant Prot Q 3:10–13

Sutherland F, Walters DR (2001) *In vitro* effects of film-forming polymers on the growth and morphology of *Pyrenophora avenae* and *Pyricularia oryzae*. J Phytopathol 149:621–624

Sutherland F, Walters DR (2002) Effect of film-forming polymers on infection of barley with the powdery mildew fungus, Blumeria graminis f. sp. hordei. Euro J Plant Pathol 108:385–389

Wilcox WF, Riegel DG (1997) Evaluation of fungicide programs for control of grapevine powdery mildew. Fungicide Nematicide Tests 52:86

Zekaria-Oren J, Eyal Z (1991) Effect of film-forming compounds on the development of leaf rust on wheat seedlings. Plant Dis 75:231–234

Ziv O (1983) Control of septoria leaf blotch of wheat and powdery mildew of barley with antitranspirant epidermal coating materials. Phytoparasitica 11:33–38

Ziv O, Frederiksen RA (1987) The effect of film-forming anti-transpirants on leaf rust and powdery mildew incidence on wheat. Plant Path 36:242–245

Ziv O, Zitter TA (1992) Effects of bicarbonates and film-forming polymers on cucurbit foliar diseases. Plant Dis 76:513–517

Non-pathogenic Strains

8

Abstract

Non-pathogenic (avirulent) or low virulent (hypovirulent) strains are capable of colonising infection site niches on the plants' surfaces and protecting susceptible plants against their respective pathogens. Such phenomena have been demonstrated for a considerable number of plant pathogens (*Agrobacterium* spp., *Rhizoctonia* spp., *Fusarium* spp. and *Pythium* spp.). Non-pathogenic strains of various pathogens are potential candidates for development of biocontrol preparations. Some strains are already used in agriculture. The modes of protection differ amongst the non-pathogenic strains, and one strain can protect by more than one mechanism. Competition for infection sites or for nutrients (such as carbon, iron) as well as induction of the host plant resistance have been demonstrated for several pathogens such as *Rhizoctonia* spp., *Fusarium* spp. and *Pythium* spp. Mycoparasitism was shown for *Pythium* spp. The non-pathogenic *Fusarium oxysporum* is easy to mass-produce and formulate, but application conditions for biocontrol efficacy under field conditions have still to be determined.

Fusarium oxysporum is one of the most common soil fungus in cultivated soil all over the world. It includes a large diversity of strains, all saprophytic, most parasitic (Garrett 1970; Burgess 1981). They are able to colonise to some extent plant tissues without inducing symptoms (Olivain and Alabouvette 1997), and some are pathogenic inducing root rot or tracheomycosis. Many other strains can penetrate roots but do not invade the vascular system or cause disease. The wilt-inducing strains of *F. oxysporum* cause a serious damage on many economically important agricultural crops (Benhamou et al. 1989; Hervas et al. 1998; Paul et al. 1999; Kaur 2003). *Fusarium* wilt pathogens are very host specific. Based on their host plant species and plant cultivars, there are more than 53 forms, 117 *formae speciales* and 29 varieties (Anonymous 2010). Being a soil-borne nature, the management of *Fusarium* wilt is mainly through chemical soil fumigation like methyl bromide and use of resistant cultivars. But the use of soil fumigants has been banned because of their hazardous effects to human health and environment and methyl bromide is completely phased out. The most cost-effective, environmentally safe method of control is the use of resistant cultivars, but over time, new races of the pathogen can develop which can infect

P.P. Reddy, *Recent Advances in Crop Protection*,
DOI 10.1007/978-81-322-0723-8_8, © Springer India 2013

resistant crop varieties. On the other hand, breeding for resistance is possible only if a dominant gene is known and host is monoecious. In carnation, cyclamen and flax, no dominant gene is known, and also, the palm trees are dioecious (Fravel et al. 2003). Thus, the soil-borne diseases are the target for biocontrol as conventional control by agrochemicals is problematic. With increasing awareness of deleterious effects of fungicides on the environment, growing interest in chemical-free agricultural products and time-consuming breeding programmes, the biological control of plant pathogens has achieved a considerable attention.

Fusarium oxysporum is well represented amongst the rhizosphere microflora. Whilst all strains exist saprophytically, some are well known for inducing wilt or root rots on plants, whereas others are considered as non-pathogenic. Several methods based on phenotypic and genetic traits have been developed to characterise *F. oxysporum* strains. Results showed the great diversity affecting the soil-borne populations of *F. oxysporum*. In suppressive soils, interactions between pathogenic and non-pathogenic strains result in the control of the disease. Therefore, non-pathogenic strains are developed as biocontrol agents. The non-pathogenic *F. oxysporum* strains show several modes of action contributing to their biocontrol capacity. They are able to compete for nutrients in the soil, affecting the rate of chlamydospore germination of the pathogen. They can also compete for infection sites on the root and can trigger plant defence reactions, inducing systemic resistance. These mechanisms are more or less important depending on the strain. The non-pathogenic *F. oxysporum* is easy to mass-produce and formulate, but application conditions for biocontrol efficacy under field conditions have still to be determined (Fravel et al. 2003).

Transmission of double-stranded RNA mycoviruses from hypovirulent strains to virulent strains renders the virulent strains hypovirulent. Chestnut trees infected with the chestnut blight pathogen, *Cryphonectria (Endothia) parasitica*, recovered after inoculation with transmissible hypovirulent strains. Non-pathogenic strains of various fungi are potential candidates for development of biocontrol preparations. Some strains are already used in agriculture (Sneh 1998).

Biological control of fusarial wilt diseases by non-pathogenic *Fusarium* has been reported in numerous crops in greenhouse and field trials (Mandeel and Baker 1991; Larkin et al. 1996; Larkin and Fravel 1998; Katsube and Alasaka 1997; Honda and Kawakub 1998; Minuto et al. 1997a, b; Hervas et al. 1998; Fuch et al. 1999). Non-pathogenic *F. oxysporum* isolate Fo47 which was recovered from *Fusarium* wilt suppressive soils in France has been extensively studied for the control of fusarial wilt disease of several vegetable and flower crops (Alabouvette et al. 1993, 1998).

Non-pathogenic *F. oxysporum* strains induce resistance in tomato (Bao and Lazarovits 2001), carnation (Postma and Luttikholt 1996), pea (Benhamou and Garand 2001), sweet potato (Whipps 1996) and asparagus (He et al. 2002) against pathogenic *F. oxysporum* strains. Non-pathogenic *F. oxysporum* Fo47 strain has been developed as a commercial liquid formulated product 'Fusaclean' (supplied by Nogueres, France) (Benhamou and Garand 2001). The yeast *Candida oleophila* has been developed as a commercial product 'Aspire' (Droby et al. 2002). Application of *C. oleophila* to surface wounds or to intact grapefruit elicits systemic resistance against *Penicillium digitatum*. Application of the yeast to grapefruit peel tissue increased ethylene biosynthesis, phenylalanine ammonia-lyase activity, phytoalexin accumulation and increased chitinase and ß-1, 3-glucanase protein levels (Droby et al. 2002).

However, for biological control to be implemented commercially on practical level, it is necessary to more fully understand the ecology of biocontrol agents and their interaction with host plant, pathogen and surrounding soil and rhizosphere microbial ecology (Cook 1993). Ideally, the antagonist must be ecologically fit to survive and function with the particular condition of the ecosystem. Moreover, the antagonist must be present at adequate population level and be capable of effectively interacting with the pathogen or host plant to provide acceptable disease control.

8.1 Involvement of Non-pathogenic *Fusarium* in Soil Supressiveness

Smith and Snyder (1971) and Toussoun (1975) gave the first evidence for a possible role of non-pathogenic *Fusarium* in suppressive soil and indicated that the soil suppressive to *Fusarium* wilts supported a large population of non-pathogenic *Fusarium* sp. But it was confirmed by proving Koch's postulates, that is, by demonstration that the suppressiveness disappeared after elimination of *Fusarium* by heat treatment and reappeared after reintroduction of the fungus into the heat-treated soil (Rouxel et al. 1979). Later on, several studies clearly showed that non-pathogenic *Fusarium* sp. has a potential to suppress *Fusarium* wilts in different areas of the world (Schneider 1984; Tamietti and Alabouvette 1986; Paulitz et al. 1987; Tamietti and Pramotton 1990; Larkin et al. 1993, 1996; Singh et al. 2002a, b). Although isolations of the *Fusarium* sp. from suppressive soils are an effective procedure for detecting strains able to control *Fusarium* diseases, the non-pathogenic strain of *F. oxysporum* selected by Ogawa and Komada (1984) and Postma and Rattink (1992) was obtained from stem of healthy plants. Whatever the origin, the strains of *F. oxysporum* did not have the same ability to control the *Fusarium* wilts. Therefore, screening procedures based on biotests have been developed to select the most efficient strain to control the disease. The plant pathogens controlled by non-pathogenic *Fusarium* have been listed in Tables 8.1 and 8.2.

8.1.1 Grapevine Crown Gall, *Agrobacterium vitis*

Crown gall disease of grapevine, caused by the bacterium *Agrobacterium vitis*, has become a major threat to grape production in Ontario, with no effective chemical controls available. It is particularly damaging to the *Vitis vinifera* varieties. Clean nursery stock protected against crown gall would be an invaluable contribution to the industry. It is particularly important that protected nursery stock be used when planting on sites with a previous history of crown gall. The objective of this study is to develop an effective biological control agent to 'immunise' vines against soil-borne infection by the crown gall pathogen. Recent laboratory and field testing has shown that F2/5, a non-pathogenic *A. vitis* strain from South Africa, protects grape plants when challenged with pathogenic *A. vitis* strains (Cuppels et al. 2000).

8.1.2 Tomato Wilt, *Fusarium oxysporum* f. sp. *lycopersici*

Non-pathogenic *F. oxysporum* strain Fo47 controls the incidence of Fusarium wilt. Four bioassays in which a strain of the pathogen *F. oxysporum* f. sp. *lycopersici* and Fo47 were not in direct contact were developed to evaluate whether Fo47 could induce resistance to Fusarium wilt in tomato plants. Fo47 and the pathogen were separated either physically or in time. Bioassays were carried out under hydroponic conditions (two bioassays), in potting mix or in autoclaved soil. Strain Fo47 protected tomato against Fusarium wilt in all four bioassays. Inoculation with Fo47 increased chitinase, β-1,3-glucanase and β-1,4-glucosidase activity in plants, confirming the ability of Fo47 to induce resistance in tomato. This report is the first to demonstrate that a non-pathogenic strain of *F. oxysporum* can induce resistance to Fusarium wilt in tomato plants. This result has important practical implications for biocontrol of tomato diseases under commercial conditions (Patil et al. 2011).

8.1.3 Cucumber, *Pythium ultimum*

Cucumber plants previously inoculated with non-pathogenic *Fusarium* strain Fo47 have increased resistance to *Pythium ultimum* attack. This protection is mainly associated with a strong antagonistic activity in the rhizosphere and in plants, as well as with the induction of structural and biochemical barriers that adversely affect pathogen growth and development (Benhamou et al. 2002).

Table 8.1 Diseases controlled by non-pathogenic *Fusarium oxysporum*

Disease	Causal organism	Reference
Fusarium wilt of gladiolus	*F. oxysporum* f. sp. *gladioli*	Magie (1980)
Fusarium wilt of cucumber	*F. oxysporum* f. sp. *cucumerianum*	Mandeel and Baker (1991) and Paulitz et al. (1987)
Fusarium wilt of strawberry	*F. oxysporum* f. sp. *fragariae*	Tezuka and Makino (1991)
Fusarium wilt of carnation	*F. oxysporum* f. sp. *dianthi*	Minuto et al. (1997b) and Lemanceau et al. (1993)
Fusarium wilt of sweet potato	*F. oxysporum* f. sp. *batatas*	Ogawa et al. (1996)
Fusarium wilt of spinach	*F. oxysporum* f. sp. *spinaciae*	Katsube and Alasaka (1997)
Fusarium wilt of rakkyo	*F. oxysporum* f. sp. *allii*	Honda and Kawakub (1998)
Fusarium wilt of radish	*F. oxysporum* f. sp. *raphani*	Toyota et al. (1995)
Fusarium wilt of basil	*F. oxysporum* f. sp. *bascilis*	Minuto et al. (1997a) and Larkin and Fravel (2002a, b)
Fusarium wilt of musk melon	*F. oxysporum* f. sp. *cucumerianum*	Larkin and Fravel (1999a, b)
Fusarium wilt of chickpea	*F. oxysporum* f. sp. *ciceri*	Hervas et al. (1998), Paul et al. (1999), Singh et al. (2002b), Kaur (2003) and Kaur et al. (2007b)
Fusarium wilt of tomato	*F. oxysporum* f. sp. *lycopersici*	Larkin and Fravel (1996)
Fusarium wilt of pea	*F. oxysporum* f. sp. *pisi*	Benhamou and Garand (2001)
Fusarium wilt of flax	*F. oxysporum* f. sp. *lira*	Duijff et al. (1999).
Fusarium wilt of cyclamen	*F. oxysporum* f. sp. *cyclaminis*	Minuto et al. (1997b)
Pythium damping off	*Pythium ultimum*	Benhamou et al. (2002)
Fusarium wilt of banana	*F. oxysporum* f. sp. *cubense*	Gerlach et al. (1999) and Ting et al. (2007)
Fusarium root and stem rot of cucumber	*F. oxysporum* f. sp. *radicis – cucumerianum*	Abeysinghe (2009)

Table 8.2 Non-pathogenic *Fusarium* species responsible for induction of induced resistance

Plant sp.	Non-pathogenic strain	Challenging pathogen	Reference
Cucumber	*F. oxysporum*	*F. oxysporum* f. sp. *cucumerianum*	Mandeel and Baker (1991)
Tomato	*F. oxysporum*	*F. oxysporum* f. sp. *lycopersici*	Kroon et al. (1992), Olivain et al. (1995) and Fuchs et al. (1997)
Tomato	*F. oxysporum* CS1 CS20	*F. oxysporum* f. sp. *lycopersici*	Larkin and Fravel (1999a, b)
Pea	*F. oxysporum*	*F. oxysporum* f. sp. *pisi*	Benhamou and Garand (2001)
Cucumber	*P putida A* 12 and MR	*F. oxysporum* f. sp. *cucumerianum*	Park et al. (1988)
Watermelon	Non-pathogenic *Fusarium* sp.	*F. oxysporum* f. sp. *cucumerianum*	Biles and Martyn (1989) and Larkin and Fravel (2002a, b)
Ipomea tricolour	*F. oxysporum* 1012	*F. oxysporum*	Shimizu et al. (2000)

8.1.4 Cyclamen Wilt, *Fusarium oxysporum* f. sp. *cyclaminis*

The ability of antagonistic strains of *Fusarium* spp. to control Fusarium wilt of cyclamen caused by *F. oxysporum* f. sp. *cyclaminis* was tested under glasshouse conditions over 3 years. Several antagonistic strains of *F. oxysporum* and one strain of *F. moniliforme*, applied alone or in mixtures, were able to decrease significantly ($P = 0.05$) the incidence of Fusarium wilt. Biological control was consistent especially when the antagonists were applied both by mixing a chlamydospore talc preparation in the potting substrate ($3\times10^{4-5}\times10^{5}$ cfu/ml of soil) 2 weeks before transplant and by dipping plant roots at transplant in a conidial suspension ($1\times10^{7-5}\times10^{8}$ cfu/ml). The combination of the benzimidazole fungicide, carbendazim and antagonistic *Fusarium* spp. generally increased the efficacy of control. Sodium alginate and kaolin formulations of *F. oxysporum* antagonistic strain 251/2 were not effective in reducing Fusarium wilt on cyclamen, whilst the same strain applied as chlamydospores dispersed in talc or as conidial suspension controlled the pathogen (Minuto et al. 1997b).

8.1.5 Flax Wilt, *Fusarium oxysporum*

In soil-less culture of vegetables and flowers in greenhouses, Fusarium diseases may induce severe damage. Under these growing conditions, biological control could be achieved by application of selected strains of fluorescent *Pseudomonas* or non-pathogenic *Fusarium oxysporum*. Seventy-four strains of fluorescent *Pseudomonas* were tested for their ability to reduce the incidence of Fusarium wilt of flax when applied either alone or in association with one preselected non-pathogenic strain of *Fusarium oxysporum* (Fo47). Four classes were established, based on the effect of bacteria on disease severity, on their own or in association with Fo47. Most of the strains did not modify the percentage of wilted plants. However, 10.8% of them, although having no effect on their own, significantly improved the control attributable to Fo47. One of these bacterial strains (C7) was selected for further experiments. Two trials conducted under commercial-type conditions demonstrated the effectiveness of the association of the bacterial strain C7 with the non-pathogenic *Fusarium* strain Fo47 to control Fusarium crown and root rot of tomato, even when each antagonistic microorganism was not efficient by itself. The yields were not significantly different in the protected plots in comparison with the healthy control (Lemanceau and Alabouvette 1991).

8.2 Selection of Non-pathogenic *Fusarium*

The minor differences in morphology of different species make the identification of *Fusarium* spp. from soil a difficult task, and also, the differences in cultural conditions can cause variations in *Fusarium* spp. (Doohan 1998). The non-pathogenic strains of *F. oxysporum* were first isolated from natural *Fusarium*-suppressive soils (Smith and Snyder 1971; Alabouvette 1990; Postma and Rattink 1992; Larkin et al. 1996; Edel et al. 1997). These are also found from the plant rhizosphere and rhizoplane, without inducing any symptoms (Elias et al. 1991; Olivain and Alabouvette 1999; Kaur 2003; Kaur et al. 2007a). It is occasionally difficult to distinguish *F. oxysporum* from several other species belonging to the sections Elegans and Liseola (Fravel et al. 2003). The plant pathogenic, saprophytic and biocontrol strains of *F. oxysporum* are morphologically impossible to differentiate (Fravel et al. 2003). For selecting pathogenic and non-pathogenic isolates of *F. oxysporum*, a bioassay method was suggested by Robert and Kraft (1971). These 3-week-old seedlings of test crop grown on sterilised soil were aseptically transferred into different test tubes containing the spore suspension (1×10^{4} cfu ml^{-1}) of different isolates of *F. oxysporum*. The tubes kept on a rotary shaker at 50 rpm for 24 h for development of disease. Based on this technique, Kaur (2003) isolated non-pathogenic *Fusarium* collected from rhizosphere and rhizoplane of chickpea in India and categorised isolates of *Fusarium* as non-pathogenic: No wilt symptom appeared up to 7 days of incubation; least pathogenic-wilt symptoms developed after 6–7 days and <33 % wilting occurred; moderately pathogenic-wilt symptoms developed after 4–5 days and <66 % wilting occurred; and highly pathogenic symptoms appeared after two days and plants wilted on 4th day and >67 % wilting occurred (Kaur 2003; Kaur et al. 2007a, b).

Besides this, a classification system using vegetative compatibility groups (VCG) amongst the *F. oxysporum* strains was anticipated (Puhalla 1985) based on pairing nitrate non-utilising mutants.

This system was later on modified by standardisation of numbering of VCG (Kistler et al. 1998). In this system, some *formae speciales* correspond to a single VCG, whilst others include several VCGs. Amongst non-pathogenic populations, many isolates are single-member VCGs and some are even self-incompatible (Gordon and Okarnoto 1992; Kondo et al. 1997; Steinberg et al. 1997; Fravel et al. 2003). Thus, this system was useful for the determination of VCG but cannot be used as a universal tool to identify *formae speciales* or non-pathogenic isolates (Fravel et al. 2003). These difficulties have been overcome by use of molecular tools like ITS-RFLP (Edel et al. 1995), IGS-RFLP (Appel and Gordon 1994) which were developedtosupportmorphologicalidentifications of different *Fusarium* spp.

Edel et al. (1995) developed a polymerase chain reaction (PCR)-based restriction fragment length polymorphism (RFLP) method targeting a fragment of the ribosomal rDNA that includes the internal transcribed spacer (ITS) region for the identification of *Fusarium* species. Edel et al. (2000) also developed an rDNA-targeted oligonucleotide probe and PCR assay specific for *F. oxysporum*. Mishra et al. (2003) developed a PCR-based assay for rapid identification of some *Fusarium* species. This technique is based on the ITS region of the rDNA. Nel et al. (2006) designed a PCR-based RFLP analysis of the intergenic spacer region of the ribosomal RNA operon to characterise the non-pathogens. The PCR-RFLP analysis with species-specific primers FOF1 and FORI was conducted to confirm the identity of non-pathogenic *F. oxysporum* isolates from pathogenic *F. oxysporum*.

8.3 Modes of Action of Non-pathogenic *Fusarium*

A thorough understanding of the mechanisms of action is needed to maximise consistency and efficacy of biocontrol. The mechanisms of action associated with non-pathogenic *F. oxysporum* can be divided into two broad categories: direct antagonism of the non-pathogenic strains to the pathogen and indirect antagonism mediated through the host plant.

8.3.1 Direct Antagonism

Generally speaking, mechanisms of direct microbial antagonism include parasitism, antibiosis and competition. Thus far, there is no evidence of either parasitism or antibiosis amongst strains of *F. oxysporum*, but data from many studies support the role of competition. Competition can be divided into saprophytic competition for nutrients in the soil and rhizosphere, and competition for infection sites on and in the root.

8.3.1.1 Competitive Interactions in the Soil and Rhizosphere

In the absence of any evidence of antibiosis between non-pathogenic and pathogenic strains of *F. oxysporum*, the hypothesis of trophic interactions was proposed to explain the role of non-pathogenic *F. oxysporum* in the mechanisms of soil suppressiveness. More precisely, the hypothesis of competition for carbon sources was proposed based on the fact that a single addition of glucose to a suppressive soil was sufficient to make the soil conducive (Louvet et al. 1976). Couteaudier and Alabouvette (1990) demonstrated the validity of the hypothesis of competition for carbon between strains of *F. oxysporum* by comparing the growth kinetics of a small collection of strains of *F. oxysporum* introduced into a sterilised soil amended with various amounts of glucose. Modelling of the growth curve (Couteaudier and Steinberg 1990) enabled calculation of the growth rate and the yield coefficient (i.e. the number of propagules formed per unit of glucose consumed) for each strain. Results showed a great diversity amongst the seven strains compared. The yield coefficient varied from 1×10^6 to 8×10^8 propagules formed per mg of glucose consumed. Six of these strains were then confronted with a 7th strain, the pathogenic strain *F. oxysporum* f. sp. *lini* (Foln3) resistant to benomyl. Each strain was introduced into sterilised soil in combination with the pathogenic strain Foln3 at five different inoculum ratios. By following the kinetics of growth of each strain in mixture, it was possible to calculate a 'competitiveness index' for each strain. These indices ranged from 1.3 to 3.5, indicating a large diversity in the ability of these six strains to compete in soil with the pathogenic strain *F. oxysporum* f. sp. *lini*. Lemanceau et al. (1993) have confirmed, in vitro, that carbon was the major nutrient for which a pathogenic strain of *F. oxysporum* f. sp. *dianthi* was competing in soil-less culture with the biocontrol agent Fo47. These results were confirmed by Larkin and Fravel (1999a, b) who demonstrated that isolate Fo47 significantly inhibited pathogen chlamydospore germination in soil at glucose concentrations of 0.2 mg g^{-1} soil and greater. In addition, germ tube growth also was significantly reduced in soil containing Fo47 compared with untreated soil. By contrast, the biocontrol isolate of *F. oxysporum* CS-20 had no effect on germination or germ tube development of the pathogen. Competition for nutrients has also been shown to be involved in the mode of action of other isolates of non-pathogenic *F. oxysporum* such as strain 618.12 (Postma and Rattink 1992) and strains C5 and C14 (Mandeel and Baker 1991).

8.3.1.2 Competitive Interactions on the Root Surface and in Plant Tissues

Competition also occurs on root surfaces. Mandeel and Baker (1991) postulated that the root surface had a finite number of infection sites that could be protected by increasing the inoculum density of the non-pathogenic strain. Using GUS-transformed strains, Olivain and Alabouvette (1999) clearly showed that both pathogenic and non-pathogenic strains were able to actively colonise the surface of the tomato root. Both were able to penetrate the epidermal cells and colonise the upper layers of cortical cells. The plant reacted to this fungal invasion by expressing defence reactions (wall thickenings, intracellular plugging) that were more intense in the case of the non-pathogen. Benhamou and Garand (2001) observed the same defence response in transformed pea root confronted with Fo47. As a result, these defence reactions always prevented the non-pathogen to reach the stele, although the pathogenic strain grew quickly towards the vessels, which were invaded. These observations are concordant with the hypothesis of competition between strain of *F. oxysporum* for infection sites at the root surface and for root tissue colonisation. Indeed, both strains colonised the same spots at the root surface and showed great similarities in the colonisation process of the root. These observations are also in agreement with previous results obtained by Eparvier and Alabouvette (1994) who utilised another approach to demonstrate that pathogenic and non-pathogenic strains were competing for root colonisation.

Postma and Luttikholt (1996) considered the hypothesis of direct competition between two strains of *F. oxysporum* in the vessels of the host plant. They compared the growth in the stele of carnation of a pathogenic strain of *F. oxysporum* f. sp. *dianthi* and of several non-pathogenic strains after artificial inoculation of these strains, alone or in combination into vessels of the plant. They showed that some non-pathogenic strains were able to reduce the stem colonisation by the pathogen resulting in a decrease of disease severity. Locally induced resistance or direct competition between strains within the vessels could cause this disease-suppressive effect after mixed inoculation into the stem. These observations are in agreement with the results of Ogawa and Komada (1984) who selected a non-pathogenic strain of *F. oxysporum* able to control Fusarium wilt of sweet potato when it was introduced into the stem of the plant. Taken together, these results support the existence of competition between pathogenic and non-pathogenic *F. oxysporum*, not only for infection sites at the root surface but also inside plant tissues.

The competitive ability of a non-pathogenic strain partly determines its capacity to establish in soil and in the plant rhizosphere and is probably involved in its capability to colonise the root surface. Nagao et al. (1990) demonstrated that different strains have different capacities to colonise heat-treated soil. Moreover, when flax was grown in soil fully colonised by the non-pathogenic strains, root colonisation was also drastically different. There was no correlation between the population density of the biocontrol strains in soil and their capacity to effectively colonise the roots.

Saprophytic colonisation of soil depends not only on the fungal strain but also on biotic and abiotic soil characteristics. The ecological parameters affecting the biotic component of these strains have seldom been studied. Colonisation of the root surface and root tissues probably depends not only on the fungal strain but also on the plant species and plant cultivar. The compatibility between strains of non-pathogenic *F. oxysporum* and the plant species or plant cultivar has not been thoroughly investigated. Hervas et al. (1997) studied the interactions between the genotype of chickpea, inoculum concentrations of the pathogen, and application of a strain of non-pathogenic *F. oxysporum*. The plant cultivar can also influence the *Fusarium* population in soil. Larkin et al. (1993) found that the watermelon cultivar 'Crimson Sweet' created its own suppressive soil via its root exudates, which increased populations of beneficial *F. oxysporum* whilst other watermelon cultivars did not. Clearly, many factors must be considered to maximise biocontrol efficacy (Hervas et al. 1997).

8.3.2 Indirect Antagonism: Induction of Systemic Resistance

It is well established that preinoculation of a plant with an incompatible strain of *F. oxysporum* (either a non-pathogenic strain or a pathogenic strain belonging to another *forma specialis*) results in the mitigation of symptoms when the plant is later inoculated with a compatible pathogen (Matta 1989). This phenomenon is considered as an expression of induced systemic resistance, a general plant defence response to microbial infection or various stresses. Induced systemic resistance (ISR) has been extensively studied, since it could explain the disease control provided by non-pathogenic strains of *F. oxysporum*. Biles and Martyn (1989) were the first to attribute to ISR the control of Fusarium wilt of watermelon achieved by several strains of non-pathogenic *F. oxysporum*. Many investigators have used a split root method to study ISR in Fusarium (Biles and Martyn 1989; Mandeel and Baker 1991; Kroon et al. 1992; Olivain et al. 1995; Fuchs et al. 1997; Larkin and Fravel 1999a, b). Papers reported experiments where a non-pathogenic strain applied to some roots of a host plant can delay symptom expression induced by the pathogen separately applied to other roots or directly into the stem of the plant. Since there is no direct interaction between the two microorganisms, the observed disease reduction is attributed to increased plant defence reactions in response to root colonisation by the non-pathogenic strain. Indeed, it has been clearly established that most of the non-pathogenic *F. oxysporum* actively colonise at least the upper layers of root cells as shown in Fig. 8.1 for Fo47 and CS-20. These observations are correlated with the fact that both Fo47 and CS-20 are able to induce ISR in some plant species.

When competition is the main mode of action, typically, the population of the biocontrol fungus must be at least as large, if not larger, than that of the pathogen population in order to achieve control; whereas, when ISR is the main mode of action, control can often be achieved when the pathogen population is much greater than that of the biocontrol fungus. Larkin and Fravel (1999a, b) significantly reduced wilt incidence in tomato when the pathogen population was up to 1,000 times greater than that of strain CS-20. By contrast, strain Fo47, which functions mainly through competition, is only effective when it is introduced at concentrations 10–100 times higher than the pathogen concentration.

ISR is correlated with both enzymatic changes in the plant often leading to induction of the physical barriers discussed above (Benhamou and Garand 2001). Tamietti et al. (1993) found increased activity of several plant enzymes related to plant defence reactions in tomato plants transplanted in sterilised soil infested with a strain of non-pathogenic *F. oxysporum*. Fuchs et al. (1997) attributed the biocontrol activity of the non-pathogenic strain Fo47 to induce resistance in tomato, correlated with an increased activity of chitinase, β-1,3-glucanase and β-1,4-glucosidase. Duijff et al. (1998) showed that the non-pathogenic strain Fo47 although not very effective in inducing systemic resistance in tomato induced an increase of PR proteins.

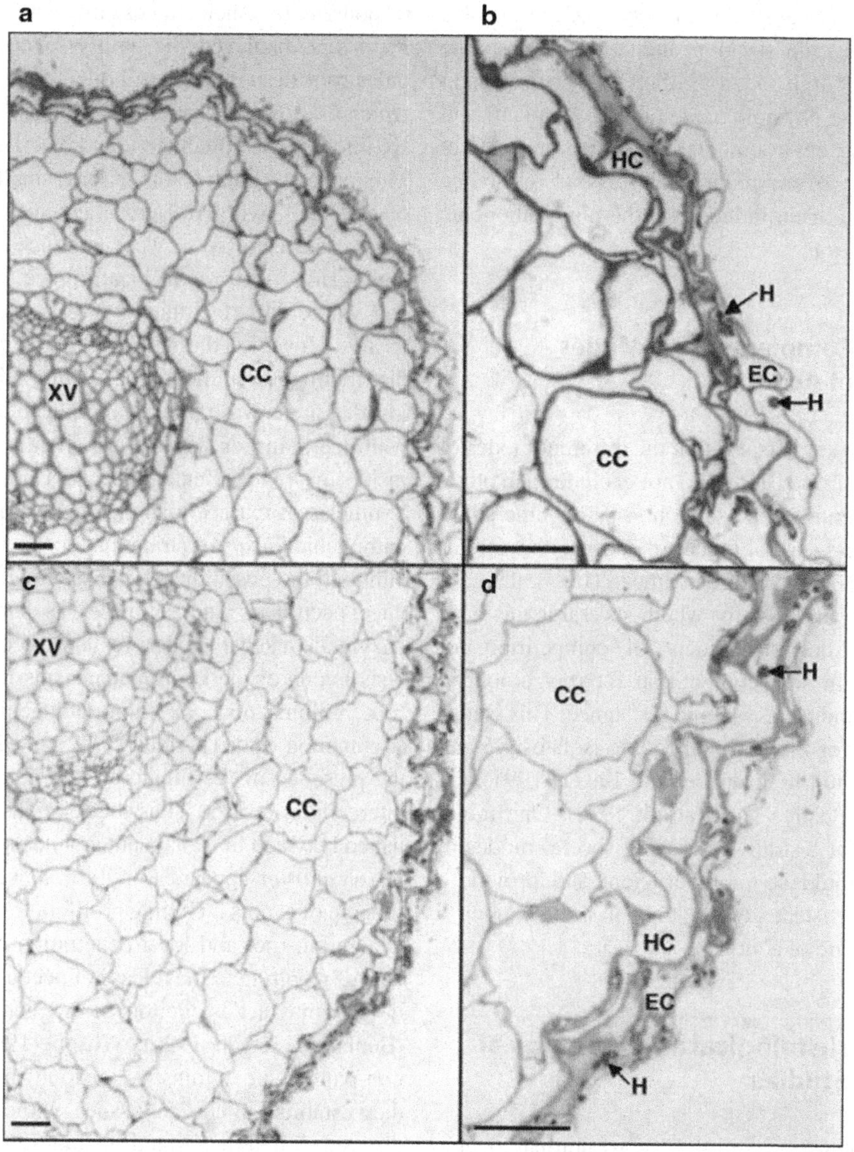

Fig. 8.1 Light micrographs of transverse semi-thin sections of flax roots stained with toluidine blue O and methylene blue O: (**a**) and (**b**) roots inoculated with *Fusarium oxysporum* CS20, (**c**) and (**d**) with *F. oxysporum* Fo47. Hyphae (*H*) of the fungi are observed at the root surface, in the epidermis (*EC*) and hypodermis (*HC*) cells. Colonised cells showed defence reactions limiting the growth of the fungi. The cortex and the stele did not appear colonised. *CC* cortical cells, *XV* xylem vessels. Magnification: (**a**) and (**c**) ×200; (**b**) and (**d**) ×600

Recorbet et al. (1998) showed an overall increased activity of constitutive glycosidase isoforms in response to infection by *F. oxysporum* f. sp. *lycopersici* that did not occur in roots colonised with non-pathogenic strains. These contrasted results all obtained with strain Fo47 applied to tomato demonstrate that the biochemical response of the plant is not clearly understood and must be accurately described, before this system can be compared to other plant–pathogen models where the cascade of biochemical events is better known.

Finally, when the main mode of action of a non-pathogenic strain is induction of systemic resistance, it is obvious that this phenomenon implies the physiological state of the plant, and fluctuating environmental conditions may affect the ability of the plant to express its resistance to the pathogen, induced by the non-pathogenic *F. oxysporum*.

8.3.3 Complementary Modes of Action

It must be emphasised that the different modes of action described above do not exclude each other. On the contrary, the same non-pathogenic strain can express several modes of action either simultaneously or at different times. This is the case for the strain Fo47 for which several teams have reported the involvement of competition for nutrients in soil, competition for root colonisation and induced systemic resistance. This is also the case for the non-pathogenic isolates C5 and C14 isolated by Mandeel and Baker (1991) and 618–12 (Postma and Rattink 1992). One might expect that a strain expressing several modes of action would be more efficient and provide a more consistent control than a strain having a single mode of action.

8.4 Histological and Cytological Studies

The cytology of root infection by non-pathogenic *Fusarium* with that associated to a pathogenic *Fusarium* strain was first studied by Benhamou and Garand (2001). They found the non-pathogenic strain grew actively at the root surface and colonised a number of epidermal and cortical cells, whereas in roots inculcated with pathogenic strain, the fungus multiplied a massive elaboration of hemispherical wall appositions and deposition of an electron-opaque material frequently encircling pathogen hyphae abundantly through much of the tissues. The host roots were signalled to defend themselves through the rapid stimulation of a general cascade of non-specific defence responses. In other studies, the non-pathogenic *Fusarium* displayed the ability to colonise the outer root tissues without inducing cell damage, which is known to occur in several host–Fusarium sp. interactions (Benhamou and Theriault 1992). This property implies that at least small amounts of cell wall hydrolytic enzymes, such as pectinases and cellulases, were produced by non-pathogenic *Fusarium* to locally infringe the host cell walls, thus, facilitating spread into the root tissues. However, the regular pattern of cellulose distribution over host cell walls adjacent to invaded areas was taken as an indication that cell wall-degrading enzymes were slightly produced inside the plant (Benhamou and Garand 2001). Synthesis of extracellular lytic enzymes by non-pathogenic *Fusarium* strains has not been reported, although the possibility that these fungi may produce pectinases and cellulases as part of their enzymatic arsenal appears realistic. Pectin hydrolysis is one of the main mechanisms involved in root colonisation by non-pathogenic strains (Benhamou and Theriault 1992), evidenced by the presence of the fungus in a large number of intercellular spaces. Thus the relationship established between the host plant and non-pathogenic *F. oxysporum* appears to follow a well-defined scheme of events including proliferation along the elongating root and local penetration of the epidermis resulting in the release of pectic fragments in turn, may act as elicitors of the plant defences (Benhamou et al. 1996). Wilson (1995) called non-pathogenic strains as endophytes because they establish in cortical tissues, without causing disease symptoms. Mandeel and Baker (1991) observed that cytochemical analysis of the pattern of chitin distribution revealed a marked decrease in the amount of chitin over cell walls of non-pathogenic *Fusarium*, especially at sites where the fungus was closely appressed against the host cell wall. Collectively, these results indicate that the root cells were signalled to produce chitinases that likely accumulated extracellularly (Mandeel and Baker 1991). Understanding the molecular mechanisms regulatory circuit involved in plant gene root infection by non-pathogenic fungi represents a major challenge for future research (Benhamou and Garand 2001).

8.5 Integration of Non-pathogenic *Fusarium* with Other Methods

8.5.1 Non-pathogenic *Fusarium* and Fluorescent *Pseudomonas*

The non-pathogenic *Fusarium* has been successfully combined with *Pseudomonas* spp. to obtain effective biocontrol of plant pathogens. Couteaudier and Alabouvette (1990) suggested that the efficacy of non-pathogenic *F. oxysporum* strains in controlling *Fusarium* wilt is related to their ability to compete for carbon. Numerous studies have established that fluorescent *Pseudomonas* spp. are efficient competitors for ferric iron (Kloepper et al. 1980; Larkin and Fravel 1998). Their siderophores, called pyoverdines or pseudobactins, have a high affinity for iron (Bakker et al. 1988). Siderophore-mediated competition for iron was demonstrated or postulated to be responsible for suppression of disease development of several soil-borne pathogens, including *Fusarium* wilt, by strains of fluorescent *Pseudomonas* spp. The greater efficacy in disease suppression by the association of non-pathogenic *F. oxysporum* and fluorescent *Pseudomonas* spp. could be due to the combination of carbon and iron competition. Lemanceau et al. (1992) observed that non-pathogenic *F. oxysporum* Fo47bl0 combined with *Pseudomonas putida* WCS358 competently suppressed *Fusarium* wilt of carnations grown in soil-less culture. This suppression with this combination was considerably higher than that obtained by use of either antagonistic microorganism alone. The increased suppression obtained by Fo47bl0 combined with WCS358 only occurred when Fo47bl0 was introduced at a density high enough (at least ten times higher than that of the pathogen) to be efficient on its own. The *P. putida* isolate WCS358 had no effect on disease severity when inoculated on its own but significantly improved the control achieved with non-pathogenic *F. oxysporum* Fo47bl0. In contrast, a siderophore-negative mutant of ICS358 had no effect on disease severity even in the presence of Fo47bl0. Since the densities of both bacterial strains at the root level were similar, the difference between the wild-type yCS358 and the siderophore-negative mutant with regard to the control of *Fusarium* wilt was related to the production of pseudobactin 358. The production of pseudobactin 358 appeared to be responsible for the increased suppression by Fo47bl0 combined.

Competition for nutrients was proposed as a mechanism for suppression of *Fusarium* diseases by both non-pathogenic *F. oxysporum* and fluorescent *Pseudomonas* sp. Lemanceau et al. (1992) suggested that pseudobactin 358 production by *P. putida* WCS358 was responsible for the improved biological control of *Fusarium* wilt achieved by association of non-pathogenic *F. oxysporum* Fo47bl0 and *P. putida* WCS358 compared with the separate application of each antagonistic organism. These results suggest that pseudobactin-mediated competition for iron increases the efficacy of the antagonistic activity of non-pathogenic *F. oxysporum* Fo47bl0 against pathogenic *F. oxysporum.* Similar results were obtained with combined application of an isolate of non-pathogenic *F. oxysporum* and an isolate of fluorescent *Pseudomonas* against *Fusarium* wilt of chickpea in India (Kaur et al. 2007a, b).

8.5.2 Non-pathogenic *Fusarium* and *Serratia marcescens*

Ting et al. (2007) studied the efficacy of *F. oxysporum* isolate UPM31P1 and *Serratia marcescens* isolate UPM39B3 in suppressing *Fusarium* wilt incidence in 'Pisang Berangan'. In glasshouse trials, the treatments with UPM31 PI singly and in combination with UPM39B3 were effective in delaying the onset of *Fusarium* wilt symptoms and in reducing disease severity and incidence. Under field conditions, UPM31P1 applied singly and in combination with UPM39B3 was able to suppress wilt incidence in plants up until week 13 after planting. This amounted to a 6-week delay in the appearance of symptoms compared to untreated control plants. Treatment with UPM31P1 and UPM39B3 initially suppress the wilt incidence and encourage vegetative growth.

8.5.3 Two Non-pathogenic Strains of *Fusarium*

The biocontrol potential of two strains of non-pathogenic *F. oxysporum* strains (CS-20 and CWB318) was studied against *Fusarium* crown rot on asparagus when combined with and without NaCl. In the greenhouse and field, there were no interactions between the treatments and NaCl on root lesions, root growth or yield. The non-pathogenic *Fusarium* strain CS-20 increased root growth.

8.5.4 Non-pathogenic *Fusarium* and Fungicides

Fravel et al. (2005) assessed the compatibility of strain CS-20 with seven fungicides recommended for tomato. All the fungicides tested did not kill strain CS-20 at the concentrations tested in the in vitro experiment. Azoxystrobin (Quadris) and chlorothalonil (Bravo) were most toxic to strain CS-20 and significantly reduced growth rate and final colony size at 10 ppm a.i. or greater concentrations compared to growth on unamended medium. Thiram (thiram), mefenoxam + chlorothalonil (Ridomil Gold Bravo) significantly reduced final colony size at 30 and 50 ppm or greater, respectively. Mancozeb (Manzate) and mancozeb + copper (Mankocide) reduced final colony size only at 100 ppm, whilst mefenoxam (Ridomil Gold) and mefenoxam + copper (Ridomil Gold Copper) did not affect growth of strain CS-20.

8.6 Production, Formulation and Delivery

8.6.1 Production

Having selected an efficient strain and determined basic information on its biology and ecology, it is necessary to mass-produce and formulate it in such a way that it can be stored and applied easily. The efficacy of a biological control agent greatly depends on the quality of the inoculant, which is function of both production and formulation (Lewis 1991; Whipps 1996). *F. oxysporum* is easily grown in liquid or solid fermentation. Both processes are being used to produce the strains Fo47 and CS-20 for large-scale experiments. Because of their role in survival, chlamydospores are considered the most desirable propagules for formulation, and production methods generally focus on increasing the percentage of chlamydospores. From studies in other biocontrol systems, we know that production methods that yield the greatest number of propagules in the shortest period of time do not necessarily yield the most efficacious propagules (Hebbar et al. 1997, 1998). Thus, bioassays are necessary to verify the 'quality' of the inoculum. Similarly, viability after storage does not necessarily guarantee biocontrol efficacy, and a bioassay is needed to confirm biocontrol ability. Development of a biochemical assay obviating the need for these bioassays would be very useful.

The strain Fo47 can be produced by solid-state fermentation either in a sterilised peat or in calcinated clay enriched with an appropriate nutrient solution. In peat, no matter what the initial concentration of inoculum, strain Fo47 will reach, at the plateau, a concentration $>1 \times 10^7$ propagules $g-1$. Both the peat and the calcinated clay provide a carrier for the inoculum; there is no need for further formulation. This inoculant can be stored at 4°C or even at room temperature without losing its density or its activity for up to 18 months. Efficacy of the inoculant can be improved by choice of a specific food base that will favour the growth of the biocontrol agent after release. Steinberg et al. (1997) compared the population kinetics and the biological efficacy of several formulations of Fo47. A formulation made of microgranules enriched with a food base provided a better survival and a better biocontrol efficacy than the traditional talc formulation used in the laboratory.

Because most industrial microbial fermentation is currently done in liquid, liquid fermentation is generally preferred. One disadvantage of liquid fermentation is that unless the final formulation is liquid, it is necessary to separate the fungal biomass from the fermentation liquid before formulation. Biomass can be separated by drying, filtration or centrifugation. Hebbar et al. (1997), working with a mycoherbicidal strain of *F. oxysporum*, increased the percentage of chlamydospores in liquid fermentation by use of selected substrates, such as soy hull fibre, and by lowering the dissolved oxygen concentration. Methods developed for this mycoherbicide have been successfully used with strain CS-20. Strain Fo47 has been produced in a 400 l fermentor, in an appropriate growth medium with control of O_2 supply and pH (Durand et al. 1989). The propagules, mainly 'bud cells', are mixed with talcum, which is then dried for 48 h at 18–20°C. This inoculant can be stored at 4°C for long periods, since the inoculum concentration was only decreased by 23 and 35 % after 1 and 2 years, respectively (Alabouvette and Couteadier 1992).

8.6.2 Formulation

Commercial Formulations of Non-pathogenic Fusarium

Biofox C:

- *Target pathogen*: *F. oxysporum* and/or *F. moniliformae*
- *Crops*: Basil, carnation, cyclamen, tomato
- *Formulation*: Dust or alginate granules
- *Method of application*: Seed treatment or soil incorporation
- *Country registered*: Italy
- *Manufacturers and suppliers*: SIAPA

Fusaclean:

- *Target pathogen*: *F. oxysporum*
- *Crops*: Asparagus, basil, carnation, cyclamen, tomato, gerbera
- *Formulation*: Liquid formulation
- *Method of application*: Incorporate into potting mixture; in rows
- *Country registered*: France
- *Manufacturers and suppliers*: Natural plant protection

8.6.3 Delivery

Successful biocontrol depends on having the biocontrol agent delivered to the right place, at the right time, in the appropriate physiological state. In addition to these considerations, application must be compatible with the production system. For example, because fresh market tomatoes are a transplant crop, Larkin and Fravel (1998) applied strain CS-20 as a drench whilst the plants were in the glasshouse before transplanting to the field. To protect cyclamen from Fusarium wilt, Fo47 is introduced in the potting mixture used to grow the plant. Thus, it colonises the substrate before accidental introduction of the pathogen.

8.7 Future Research

Additional research is needed in several areas including yield studies and integration into production systems, genetics, mechanisms of action, risk assessment and genetic improvement of biocontrol agents.

The use of non-pathogenic strains of *F. oxysporum* to control Fusarium wilt has been reported for many crops including banana (Gerlach et al. 1999), basil (Fravel and Larkin 2002), carnation (Garibaldi et al. 1986), cucumber (Mandeel and Baker 1991), cyclamen (Minuto et al. 1997b), flax (Alabouvette et al. 1993), gladiolus (Magie 1980), melon (Rouxel et al. 1979), tomato (Lemanceau and Alabouvette 1991), spinach, strawberry (Tezuka and Makino 1991) and watermelon (Larkin et al. 1996). Amongst these papers, only a few presented data that were obtained under commercial production conditions.

The greatest reason for lack of adoption of biological control is the lack of consistency of biological control performance based on the application of a strain of non-pathogenic *F. oxysporum*. In addition to field studies integrating biological control into commercial production systems, a more thorough understanding of the genetics, biology and ecology of biocontrol agents and their modes of action will enable optimal exploitation of these fungi for disease control. These issues are interrelated since molecular techniques can provide insight in each of these areas. Molecular genetics offers new tools for unravelling mechanisms and understanding genetic relationships amongst *Fusaria*.

For example, an approach using transposon mutagenesis has recently been used to produce mutants of *F. oxysporum melonis* affected in their pathogenicity (Migheli et al. 1999) and mutants of Fo47 showing either an increased or a decreased biocontrol capacity (Trouvelot et al. 2002). Recently, the early physiological responses of flax cells challenged with conidia of strains of pathogenic and non-pathogenic *Fusarium oxysporum* were compared. The results observed with the pathogenic strain were typical of those described in the case of the compatible reaction, and results observed with Fo47 were similar to those observed in the case of the incompatible reaction when a pathogen is interacting with a resistant cultivar. The non-pathogenic strain elicits early defence reactions restricting its growth into the root. These preliminary results open the way to new research in the field of the plant–microbial interactions leading to the identification of the biochemical pathways triggered by the non-pathogenic strains.

It is conceivable that understanding what is triggered in the induced resistance discussed above may lead to being able to trigger this in the absence of a biocontrol agent, thus avoiding the problems inherent in production, formulation and delivery of a living agent.

More research using new tools of molecular genetics is needed to determine the genetic relationships amongst pathogenic, biocontrol and saprophytic Fusaria, as well as to elucidate the genetic determinants of pathogenicity and biological control ability. Identification of genes involved in biological control should assist in making biological control more consistent and in optimization of control; identification of genes involved in biological control may also facilitate screening for new biocontrol agents or genetic improvement of current biocontrol agents.

One issue that has received little attention is the safety of releasing non-pathogenic *F. oxysporum*. Cook (1993) list four areas of concern for the release of any biological control agents. These are (1) displacement of non-target microorganisms, (2) allergenicity to humans and other animals, (3) toxigenicity to non-target organisms and (4) pathogenicity to non-target organisms. They point out that, based on experience, any adverse effects from biological control are likely to be short term and can be eliminated by terminating use of the biocontrol agent, whilst agriculture as it is currently practised has produced significant long-term adverse affects. Gullino et al. (1995) point out the need for data on which to base informed decisions about the risk involved in releasing non-pathogenic *F. oxysporum*. They found release of non-pathogenic *F. oxysporum* to have negligible non-target effects with respect to persistence and survival, effect on indigenous microbial communities, and genetic stability and transfer. Because some *Fusaria* produce mycotoxins, it is important to establish that those used as biological control agents do not produce toxins. With *F. oxysporum* in particular, the concern arises as to whether the biocontrol agent is truly non-pathogenic, or whether it may be pathogenic on a species of plant on which it has not yet been tested. Further, given the lack of understanding about how new races of the pathogen arise, there is some concern from the public that the biocontrol agent could become pathogenic. Genetic data are needed to allay these fears. But progress in the development of non-pathogenic *Fusaria* as biological control agents needs a cooperative research effort in all these fields from basic understanding of the mechanisms to practical field studies of the environmental conditions influencing the efficacy of the control.

8.8 Conclusions

Biological control of *Fusarium* wilt diseases has become an increasingly popular disease management consideration in recent years, given its environmentally friendly nature and the discovery of novel mechanisms of plant protection associated with certain microorganisms. From a crop protection point of view, non-pathogenic strains of *F. oxysporum* represent a key component of the soil microbiota in soils suppressive to *Fusarium* wilt. Better knowledge of the mechanisms involved in the protection of plants by biocontrol agents is a requirement to the development of successful biocontrol strategies for use under commercial conditions. Also, before planning the large-scale use of non-pathogenic strains of *F. oxysporum* as biocontrol agents of *Fusarium* wilt, their behaviour and potential impact on soil ecosystems should be carefully studied as part of risk assessment.

References

Abeysinghe S (2009) Use of nonpathogenic *Fusarium oxysporum* and rhizobacteria for suppression of Fusarium root and stem rot of *Cucumis sativus* caused by *Fusarium oxysporum* f sp *radicis-cucumerianum*. Arch Phytopath Plant Prot 42:73–82

Alabouvette C (1990) Biological control of fusarium wilt pathogens in suppressive soils. In: Hornby D (ed) Biological control of soil borne plant pathogens. CAB International, Wallingford, pp 27–43

Alabouvette C, Couteadier Y (1992) Biological control of plant diseases: progress and challenges for the future. In: Tjamos EC, Papavizas GC, Cook RJ (eds) Biological control of plant diseases. Plenum Press, New York, pp 415–426

Alabouvette C, Lemanceau P, Steinberg C (1993) Recent advances in biological control of fusarium wilts. Pestic Sci 37:365–373

Alabouvette C, Schippers B, Lemanceau P, Bakker PAHM (1998) Biological control of fusarium wilts: towards the development of commercial products. In: Boland GJ, Kuykendall R (eds) Plant microbe interaction and biological control. Marcel Dekker, New York, pp 15–36

Anonymous (2010) INDEX forum 2010. http://www.indexfungorum.org/names/asp. Retrieved 22 Apr 2010

Appel DJ, Gordon TR (1994) Local and regional variation in population of *Fusarium oxysporum* from agricultural field soils. Phytopathology 84:786–791

Bakker PAHM, Weisbeek PJ, Schippers B (1988) Siderophore production by plant growth-promoting *Pseudomonas* sp. J Plant Nutr 11:925–933

Bao JR, Lazarovits G (2001) Differential colonization of tomato roots by nonpathogenic and pathogenic *Fusarium oxysporum* strains may influence Fusarium wilt control. Phytopathology 91:449–456

Benhamou N, Garand C (2001) Cytological analysis of defense-related mechanisms induced in pea root tissues in response to colonization by nonpathogenic *Fusarium oxysporum* Fo47. Phytopathology 91:730–740

Benhamou N, Theriault G (1992) Treatment with chitosan enhances resistance of tomato plants to the crown and root rot pathogen *Fusarium oxysporum* f sp *radicis-lycopersici*. J Physiol Mol Plant Pathol 41:33–52

Benhamou N, Charest PM, Jarvis WR (1989) Biology and host-parasite relationships of *Fusarium oxysporum*. In: Tjamos EC, Beckman CH (eds) Vascular wilt diseases of plants. Basic studies and control, NATO ASI series. Springer, Berlin, pp 95–105

Benhamou N, Belanger RR, Paulitz TC (1996) Induction of differential host responses by *Pseudomonas fluorescens* in Ri T-DNA-transformed pea roots after challenge with *Fusarium oxysporum* f sp *pisi* and *Pythium ultimum*. Phytopathology 86:78–114

Benhamou N, Garand C, Goulet A (2002) Ability of non-pathogenic *Fusarium oxysporum* strain Fo47 to induce resistance against *Pythium ultimum* infection in cucumber. Appl Environ Microbiol 68:4044–4060

Biles CL, Martyn RD (1989) Local and systemic resistance induced in watermelons by formae speciales of *Fusarium oxysporum*. Phytopathology 79:856–860

Burgess LW (1981) General ecology of the fusaria. In: Nelson PE, Toussoun TA, Cook RJ (eds) *Fusarium*: diseases, biology and taxonomy. Pennsylvania State University Press, Philadelphia, pp 225–235

Cook RJ (1993) Making greater use of microorganisms for biocontrol of plant pathogens. Annu Rev Phytopathol 31:53–80

Couteaudier Y, Alabouvette C (1990) Quantitative comparison of *Fusarium oxysporum* competitiveness in relation with carbon utilization. FEMS Microbiol Lett 74:261–267

Couteaudier Y, Steinberg C (1990) Biological and mathematical description of growth pattern of *Fusarium oxysporum* in sterilized soil. FEMS Microbiol Ecol 74:253–260

Cuppels DA, Walker S, Stobbs L, Burr T (2000) Development of a biological control agent for crown gall of grapevine. Report No. OREP-1999/07. Ontario Research Enhancement Program (OREP)

Doohan FM (1998) Using RT-PCR to determine mycotoxin production in *Fusarium* species. In: Proceedings of the COST Action 835: Agriculturally Important Toxigenic Fungi (AITF'98), East Mailing, Kent, UK, pp 102–102

Droby S, Cohen A, Weiss B, Horev B, Chalutz E, Katz H, Keren-Tzur M, Shachnai A (2002) Commercial testing of aspire: a yeast preparation for the biological control of postharvest decay of citrus. Biol Control 12:97–100

Duijff BJ, Pouhair D, Olivain C, Alabouvette C, Lemanceau P (1998) Implication of systemic induced resistance in the suppression of *Fusarium* wilt of tomato by *Pseudomonas fluorescens* WCS417r and by non-pathogenic *Fusarium oxysporum* Fo47. Eur J Plant Pathol 104:903–910

Duijff BJ, Recorbet G, Bakker PAHM, Loper JE, Lemanceau P (1999) Microbial antagonism at the root level is involved in the suppression of *Fusarium* wilt by the combination of non-pathogenic *Fusarium oxysporum* Fo47 and *Pseudomonas putida* WCS358. Phytopathology 89:1073–1079

Durand A, Vergoignan C, Almanza S (1989) Studies of the survival of *Fusarium oxysporum* conidia produced by submerged culture. Biotechnol Lett 11:503–508

Edel V, Steinberg C, Avelange I, Laguerre G, Alabouvette C (1995) Comparison of three molecular methods for the characterization of *Fusarium oxysporum* strains. Phytopathology 85:579–585

Edel V, Steinberg C, Gautheron N, Alabouvette C (1997) Populations of non-pathogenic *Fusarium oxysporum* associated with roots of four plants species compared to soil borne populations. Phytopathology 87:693–697

Edel V, Steinberg C, Gautheron N, Alabouvette C (2000) Ribosomal DNA-targeted oligonucleotide probe and PCR assay specific for *Fusarium oxysporum*. Mycol Res 104:518–526

Elias KS, Schneider RW, Lear MM (1991) Analysis of vegetative compatibility groups in nonpathogenic populations of *Fusarium oxysporum* isolated from symptomless tomato roots. Can J Bot 69:2089–2094

Eparvier A, Alabouvette C (1994) Use of ELISA and GUS-transformed strains to study competition between pathogenic and non-pathogenic *Fusarium oxysporum* for root colonization. Biocontrol Sci Technol 4:35–47

Fravel DR, Larkin RP (2002) Reduction of *Fusarium* wilt of hydroponically-grown basil by *Fusarium oxysporum* strain CS-20. Crop Prot 21:539–543

Fravel DR, Olivain C, Alabouvette C (2003) *Fusarium oxysporum* and its biocontrol. New Phytol 157:493–502

Fravel DR, Deahl KL, Stommel JR (2005) Compatibility of the biocontrol fungus *Fusarium oxysporum* strain CS-20 with selected fungicides. Biol Control 34:165–169

Fuch JG, Moenne LY, Defago G (1999) Ability of non-pathogenic *Fusarium oxysporum* strain Fo47 to protect tomato against fusarium wilts. Biol Control 14:105–110

Fuchs M, Jr Mcferson, Tricoli DM, Jr Mcmaster, Deng RZ, Boeshore ML, Reynolds JF, Russell PF, Quemada HD, Gonsalves D (1997) Cantaloupe line CZW-30 containing coat protein genes of cucumber mosaic virus, zucchini yellow virus, and watermelon mosaic virus-2 is resistant to these three viruses in the field. Mol Breed 3:279–290

Garibaldi A, Brunatti F, Gullino ML (1986) Suppression of Fusarium wilt of carnation by competitive non-pathogenic strains of Fusaria. Med Fac Landbouww Rijksuniv Gent 51:633–638

Garrett SD (1970) Pathogenic root infection fungi. Cambridge University Press, Cambridge/London. ISBN 0521077869

Gerlach KS, Bentley S, Moore NY, Aitken EAB, Pegg KG (1999) Investigation of nonpathogenic strains of *Fusarium oxysporum* for suppression of Fusarium wilt of banana in Australia. In: Alabouvette C (ed) Second international Fusarium workshop. PNRA-CMSE, Dijon, France, p 54

Gordon TR, Okarnoto D (1992) Variation in mitochondrial DNA among vegetatively compatible isolates of *Fusarium oxysporum*. Exp Mycol 16:245–250

Gullino ML, Migheli Q, Mezzalama M (1995) Risk analysis in the release of biological control agents: antagonistic *Fusarium oxysporum* as a case study. Plant Dis 79:1193–1201

He CY, Hsiang T, Wolyn DJ (2002) Induction of systemic disease resistance and pathogen defense responses in *Asparagus officinalis* inoculated with nonpathogenic strains of *Fusarium oxysporum*. Plant Path 51:225–230

Hebbar KP, Lumsden RD, Poch SM, Lewis JA (1997) Liquid fermentation to produce biomass of mycoherbicidal strains of *Fusarium oxysporum*. Appl Microbiol Biotechnol 48:714–719

Hebbar KP, Lumsden RD, Lewis JA, Poch SM, Bailey BA (1998) Formulation of mycoherbicidal strains of *Fusarium oxysporum*. Weed Sci 46:501–507

Hervas A, Landa B, Jimenez-Diaz DM (1997) Influence of chickpea genotype and *Bacillus* sp on protection from *Fusarium* wilt by seed treatment with nonpathogenic *Fusarium oxysporum*. Eur J Plant Pathol 103:631–642

Hervas A, Landa B, Datnoff LE, Jimenez-Diaz RM (1998) Effects of commercial and indigenous microorganisms on *Fusarium* wilt development in chickpea. Biol Control 13:166–176

Honda M, Kawakub Y (1998) Control of *Fusarium* basal rot of rakkyo (Allium chinense) by nonpathogenic *Fusarium moniliforme* and *Fusarium oxysporum*. Soil Microorgan 51:13–18

Katsube K, Alasaka Y (1997) Control of *Fusarium* wilt of spinach by transplanting seedlings preheated with non-pathogenic *Fusarium oxysporum*. Ann Phytopathol Soc Jap 63:389–394

Kaur R (2003) Characterization of selected isolates of non-pathogenic *Fusarium oxysporum,* fluorescent pseudomonads and their efficacy against chickpea wilt. PhD thesis, Punjab Agricultural University, Ludhiana, 185 pp

Kaur R, Kaur J, Singh RS, Alabouvette C (2007a) Biological control of *Fusarium oxysporum* f sp *ciceri* by non-pathogenic *Fusarium* and fluorescent *Pseudomonas*. Int J Bot 3:114–117

Kaur R, Singh RS, Alabouvette C (2007b) Antagonistic activity of selected isolates of fluorescent *Pseudomonas* against *Fusarium oxysporum* f sp *ciceri*. Asian J Plant Sci 6:446–454

Kistler HC, Alabouvette C, Baayan RP, Bentley SB, Goddingtone A (1998) Systemic numbering of vegetative compatibility groups in the plant pathogenic fungus *Fusarium oxysporum*. Phytopathology 88:30–32

Kloepper JW, Leong J, Teintze M, Schroth MN (1980) Enhanced plant growth by siderophores produced by plant growth promoting rhizobacteria. Nature 286:885–886

Kondo N, Kodama F, Ogoshi A (1997) Vegetative compatibility groups of *Fusarium oxysporum* f sp *adzukicola* and nonpathogenic *Fusarium oxysporum* on adzuki bean isolated from adzuki bean fields in Hokkaido. Ann Phytopathol Soc Jap 63:8–12

Kroon BA, Scheffer RJ, Elgersma DM (1992) Induced resistance in tomato plants against *Fusarium* wilt involved by *Fusarium oxysporum* f sp *dianthi*. Neth J Plant Pathol 97:401–408

Larkin RP, Fravel DR (1996) Ecological characteristics of biological control of *Fusarium* wilt of tomato using non-pathogenic *Fusarium* spp. Phytopathology 86:89–99

Larkin RP, Fravel DR (1998) Efficacy of various fungal and bacterial biocontrol organisms for control of Fusarium wilt of tomato. Plant Dis 82:1022–1028

Larkin RP, Fravel DR (1999a) Field efficacy of selected nonpathogenic *Fusarium* spp and other biocontrol agents for the management of *Fusarium* wilt of muskmelon. Biol Cult Tests 14:161–161

Larkin RP, Fravel DR (1999b) Mechanisms of action and dose response relationships governing biological control of *Fusarium* wilt of tomato by nonpathogenic *Fusarium* spp. Phytopathology 89:1152–1161

Larkin RP, Fravel DR (2002a) Effects of varying environmental conditions on biological control of *Fusarium* wilt of tomato by nonpathogenic *Fusarium* spp. Phytopathology 92:1160–1166

Larkin RP, Fravel DR (2002b) Reduction of *Fusarium* wilt of hydroponically-grown basil by *Fusarium oxysporum* strain CS-20. Crop Prot 21:539–543

Larkin RP, Hopkins DL, Martin FN (1993) Ecology of *Fusarium oxysporum* f sp *niveum* in soils suppressive and conducive to *Fusarium* wilt of watermelon. Phytopathology 83:1105–1116

Larkin RP, Hopkins DL, Martin FN (1996) Suppression of *Fusarium* wilt of watermelon by nonpathogenic *Fusarium oxysporum* and other microorganisms recovered from a disease-suppressive soil. Phytopathology 86:812–819

Lemanceau P, Alabouvette C (1991) Biological control of fusarium diseases by fluorescent *Pseudomonas* and non-pathogenic *Fusarium*. Crop Prot 10:279–286

Lemanceau P, Baker PAHM, De Kogel WJ, Alabouvette C, Schippers B (1992) Effect of *Pseudobactin* 358 production by *Pseudomonas putida* WCS358 on suppression of *Fusarium* wilt of carnation by nonpathogenic *Fusarium oxysporum* Fo47. Appl Environ Microbiol 58:2978–2982

Lemanceau P, Baker PAHM, De Kogel WJ, Alabouvette C, Schippers B (1993) Antagonistic effect on nonpathogenic *Fusarium oxysporum* strain Fo47 and pseudobactin 358 upon pathogenic *Fusarium oxysporum* f sp *dianthi*. Appl Environ Microbiol 59:74–82

Lewis JA (1991) Formulation and delivery systems of biocontrol agents with emphasis on fungi. In: Keister DL, Cregan PB (eds) The rhizosphere and plant growth. Kluwer, Dordrecht, pp 279–287

Louvet J, Rouxel F, Alabouvette C (1976) Recherches sur la résistance des sols aux maladies. I – Mise en évidence de la nature microbiologique de la résistance d'un sol au développement de la fusariose vasculaire du melon. Annales de Phytopathologie 8:425–436

Magie RO (1980) *Fusarium* disease of gladioli controlled by inoculation of conns with non-pathogenic *Fusaria*. Proc Fla State Hortic Soc 93:172–175

Mandeel Q, Baker R (1991) Mechanisms involved in biological control of *Fusarium* wilt of cucumber with strains of nonpathogenic *Fusarium oxysporum*. Phytopathology 81:462–469

Matta A (1989) Induced resistance to *Fusarium* wilt diseases. In: Tjamos EC, Beckman CH (eds) Vascular wilt diseases of plants- basic studies and control. Springer, Berlin, pp 175–196

Migheli Q, Lauge R, Daviere JM, Gerlinger C, Kaper F, Langin T, Daboussi MJ (1999) Transposition of the autonomous *Fot1* element in the filamentous fungus *Fusarium oxysporum*. Genetics 151:1005–1013

Minuto A, Migheli Q, Garabaldi A (1997a) Evaluation of antagonistic strains of *Fusarium* spp in the biological and integrated control of fusarium wilt of cyclamen. Crop Prot 14:221–226

Minuto A, Minuto G, Migheli Q, Mocioni M, Gullino ML (1997b) Effect of antagonistic *Fusarium* spp and of different commercial biofungicide formulations on Fusarium wilt of basil (*Ocimum basilicum* L). Crop Prot 16:765–769

Mishra PK, Fox RTV, Culham A (2003) Development of a PCR based assay for rapid and reliable identification of pathogenic *Fusaria*. FEMS Microbiol Lett 218:329–332

Nagao H, Couteaudier Y, Alabouvette C (1990) Colonization of sterilized soil and flax roots by strains of *Fusarium oxysporum* and *Fusarium solani*. Symbiosis 9:343–354

Nel B, Steinberg C, Labuschagne N, Viljoen A (2006) Isolation and characterization of nonpathogenic *Fusarium oxysporum* isolates from the rhizosphere of healthy banana plants. Plant Pathol 55:207–216

Ogawa K, Komada H (1984) Biological control of *Fusarium* wilt of sweet potato by nonpathogenic *Fusarium oxysporum*. Ann Phytopathol Soc Jap 50:1–9

Ogawa K, Watanabe K, Komada H (1996) Formulation of nonpathogenic *Fusarium oxysporum*, a biocontrol agent for commercial use. In: Proceedings of the 1st international workshop on Fusarium Biocontrol, 28–31 Oct, Bletsville, p 34

Olivain C, Alabouvette C (1997) Colonization of tomato root by a nonpathogenic strain of *Fusarium oxysporum*. New Phytol 137:481–494

Olivain C, Alabouvette C (1999) Process of tomato root colonization by a pathogenic strain of *Fusarium oxysporum* f sp *lycopersici* discussed in comparison to a nonpathogenic strain. New Phytol 141:497–510

Olivain C, Steinberg C, Alabouvette C (1995) Evidence of induced resistance in tomato inoculated by nonpathogenic strains of *Fusarium oxysporum*. In: Manka M (ed) Environmental biotic factors in integrated plant disease control. Polish Phytopathological Society, Poznan, pp 427–430

Park CS, Paultiz TC, Baker R (1988) Biological control of *Fusarium* wilt of cucumber resulting from interaction between *Pseudomonas putida* and nonpathogenic isolates of *Fusarium oxysporum*. Phytopathology 78:190–194

Patil S, Sriram S, Savitha MJ, Arulmani N (2011) Induced systemic resistance in tomato by non-pathogenic *Fusarium* species for the management of *Fusarium* wilt. Arch Phytopathol Plant Prot 44:1621–1634

Paul J, Singh RS, Kaur J, Alabouvette C (1999) Effect of inoculum density of nonpathogenic *Fusarium* in biological control of chickpea wilt caused by *Fusarium oxysporum* f sp *ciceri*. In: Proceedings of symposium on Biology control based pest management for quality crop protection in the current millennium, 18–19 July, PAU, Ludhiana, pp 97–98

Paulitz TC, Park CS, Baker R (1987) Biological control of *Fusarium* wilt of cucumber with nonpathogenic isolates of *Fusarium oxysporum*. Can J Microbiol 33:349–353

Postma J, Luttikholt AJG (1996) Colonization of carnation stems by a nonpathogenic isolate of *Fusarium oxysporum* and its effect on *Fusarium oxysporum* f sp *dianthi*. Can J Bot 74:1841–1851

Postma J, Rattink H (1992) Biological control of *Fusarium* wilt of carnation with a nonpathogenic isolate of *Fusarium oxysporum*. Can J Bot 70:1199–1205

Puhalla JE (1985) Classification of strains of *Fusarium oxysporum* on the basis of vegetative compatibility. Can J Bot 63:179–183

Recorbet G, Bestel-Corre G, Dumas-Gaudot E, Gianinazzi S, Alabouvette C (1998) Differential accumulation of β-1, 3-glucanase and chitinase isoforms in tomato roots in response to colonization by either pathogenic or non-pathogenic strains of *Fusarium oxysporum*. Microbiol Res 153:257–263

Robert DA, Kraft JM (1971) A rapid technique for studying fusarium wilt of peas. Phytopathology 61:342–343

Rouxel F, Alabouvette C, Louvet J (1979) Recherches sur la resistance des sols aux maladies. IV- Mise en evidence du role des Fusarium autochtones dans la resistance d unel a la Fusariose vasculairedu Melon. Annu Phytopathol 11:199–207

Schneider RW (1984) Effects of nonpathogenic strains of *Fusarium oxysporum* on celery root infection by *Fusarium oxysporum* f sp *apii* and a novel use of the line weaver-burk double reciprocal plot technique. Phytopathology 74:646–653

Shimizu B, Fujimori A, Miyagawa H, Ueno T, Watanabe K, Ogawa K (2000) Resistance against *Fusarium* wilt induced by non-pathogenic *Fusarium ipomoea tricolor*. J Pesticide Sci 25:365–372

Singh RS, Kaur J, Kaur R, Alabouvette C (2002a) Effect of amendment with farm yard manure on biocontrol potentiality of non-pathogenic *Fusarium* against chickpea wilt. Plant Dis Res 17:207–207

Singh RS, Kaur J, Kaur R, Alabouvette C (2002b) The role of nonpathogenic *Fusarium* in biocontrol of *Fusarium* wilts. University of Mysore, Mysore, pp 150–151

Smith SN, Snyder WC (1971) Relationship of inoculums density and soil types to severity of *Fusarium* wilt of sweet potato. Phytopathology 61:1049–1051

Sneh B (1998) Use of non-pathogenic or hypovirulent fungal strains to protect plants against closely related fungal pathogens. Biotechnol Adv 16:1–32

Steinberg C, Edel V, Alabouvette C (1997) Influence of formulation process on survival and antagonistic activity of biocontrol agents against *Fusarium* wilt. Cryptogamic Mycol 18:139–143

Tamietti G, Alabouvette C (1986) Resistance de sols aux maladies: XIII-Role des *Fusarium oxysporum* non pathogenes dans les mecanismes de resistance dun sol de noirmoutier aux fuseieses vasculaires. Agronomie 6:541–548

Tamietti G, Pramotton R (1990) La receprivite des sols aux fusarioses vasculaires: Rapports entre resistance et microllore aotuchtone avec reference particuliere aux *Fusarium* non pathogens. Agronomie 10:69–76

Tamietti G, Ferraris L, Matta A, Gentile A (1993) Physiological responses of tomato plants grown in fusarium suppressive soil. J Phytopathol 138:66–76

Tezuka N, Makino T (1991) Biological control of *Fusarium* wilt of strawberry by nonpathogenic *Fusarium oxysporum* isolated from strawberry. Annu Phytopathol Soc Jap 57:506–511

Ting ASY, Sariah M, Kadir J, Gurmit S (2007) Field evaluation of non-pathogenic *Fusarium oxysporum* isolates UPM31PL and UPM39B3 for the control of Fusarium wilt in Tisang Berangan' (MUSA, AAA). In: Proceedings of the international symposium on recent advances in banana crop protection for sustainable production and improved livelihoods, White River, South Africa, Sept 2007, ISHS Acta Horticulturae, pp 139–144

Toussoun TA (1975) *Fusarium*-suppressive soils. In: Bruehl GW (ed) Biology and control of soil-borne plant pathogens. American Phytopathological Society, St Paul, pp 145–151

Toyota K, Kitamura M, Kimura M (1995) Suppression of *Fusarium oxysporum* f sp *raphani* PEG-4 in soil following colonization by other *Fusarium* spp. Soil Biol Biochem 27:41–46

Trouvelot S, Olivain C, Recorbet G, Migheli Q, Alabouvette C (2002) Recovery of *Fusarium oxysporum* Fo47 mutants affected in their antagonistic activity after transposon mutagenesis. Phytopathology 92:936–9945

Whipps JM (1996) Developments in the biological control of soil-borne plant pathogens. Adv Bot Res 26:1–134

Wilson D (1995) Endophyte: the evolution of a term and clarification of its use and definition. Oikos 73:274–276

9 Plant Defence Activators

Abstract

Plant defence activators are of two types, namely, biological and chemical. Plants are endowed with several defence genes which are involved in synthesis of antifungal, antibacterial and antiviral compounds like, pathogenesis-related (PR) proteins, phenolics, phytoalexins, lignin, callose and terpenoids conferring resistance against plant pathogens. Most of the defence genes are sleeping genes (quiescent in healthy plants) which require specific signals to activate them. Several antagonistic organisms have been shown to provide signals, which activate the defence genes.

A number of natural and synthetic compounds induce plant defences against pathogens and herbivores and act at different points in plant defence pathways (Karban R, Baldwin IT, Induced responses to herbivory. University of Chicago Press, London, 31 pp, 1997; Gozzo F, Outlooks on pest management, pp 20–23, 2004). The non-protein amino acid DL-β-aminobutyric acid (BABA) is a potent inducer of plant resistance and is effective against a wide range of biotic and abiotic stresses. BABA is rarely found naturally in plants, but, when applied as a root drench or foliar spray, it has been shown to protect against viruses, bacteria, oomycetes, fungi, and phytopathogenic nematodes, as well as abiotic stresses such as drought and extreme temperatures (Jakab G, Cottier V, Touquin V et al., Eur J Plant Pathol 107:29–37, 2001).

Plants usually do not possess an immune system as humans and animals, but a plant definitely has its own defence system against pests and pathogens. Plant defence activators are of two types, namely, biological and chemical. During the pathogen attack, plants generally resist through their natural self-defence mechanism. But it is important to note that only a few mechanisms are triggered by biological or chemical agents without any harmful side effects. The best example one can quote in this area of research is systemic acquired resistance which is naturally activated by biotic agents (Kuk 1984). Nowadays, the induced disease resistance becomes a hot spot discussion for sustainable agriculture. In relation to biochemical resistance, it was stated that acibenzolar-S-methyl (ASM) which is a derivative of benzothiadiazole (BTH) is the first commercially available product that activates the same defence

P.P. Reddy, *Recent Advances in Crop Protection*,
DOI 10.1007/978-81-322-0723-8_9, © Springer India 2013

reaction as the biological inducer of systemic acquired resistance with the same capacity of protection and with the same change of biochemical mechanisms (Kuk 1987). Amongst different activating agents, a chemical activator plays a significant role for activating the plant defence mechanism against pests and diseases (Malamy and Klessig 1992).

9.1 Biological Plant Defence Activators

Lipopolysaccharides (LPS) are constituents of the outer membrane of many bacterial antagonists, and they trigger defence mechanisms of the host. LPS of the bacterial antagonist *Pseudomonas fluorescens* strain WCS417r act as elicitors and induce resistance against different diseases. The O-antigenic side chain of the outer membrane LPS of the strain WCS417r appears to be the main elicitor for induction of resistance to *Fusarium* wilt in radish and carnation (Leeman et al. 1995; Van Peer and Schippers 1992).

Chitin oligomers and glucans are important elicitors of host origin, and they are released by chitinases and ß-1, 3-glucanases. The antagonists *Trichoderma* strains have been shown to produce chitinases (Krishnamurthy et al. 1999), and *P. fluorescens* strains produce ß-1, 3-glucanases. Application of *P. fluorescens* Pf1, which produces ß-1, 3-glucanases, induces activity of ß-1, 3-glucanases, chitinases and phenylalanine ammonia-lyase in rice plants (Meena et al. 1999). These strains induced resistance in plants against pathogens.

Pseudomonas fluorescens strain CHA0-induced systemic resistance in tobacco against tobacco necrosis virus and a siderophore (pyoverdine) produced by this strain triggers defence mechanisms of the host (Maurhofer et al. 1994, 1998). Leeman et al. (1996) reported that the purified siderophore, pseudobactin, from *P fluorescens* strain WCS374 induced resistance to Fusarium wilt in radish.

Some enzymes produced by fungal antagonists act as elicitors. Xylanases produced by *T. viride* elicit several defence mechanisms (induce synthesis of phytoalexins and pathogenesis-related proteins, ethylene production, tissue necrosis, lipid peroxidation, electrolyte leakage and cell death) of the host (Dean and Anderson 1990).

A low molecular weight protein, oligandrin, has been obtained from culture filtrates of *Pythium oligandrum*, another antagonistic organism. This protein induces plant defence reactions that contribute to restrict stem cell invasion by *Phytophthora nicotianae* var. *parasitica* in tomato. Oligandrin is similar to elicitins. Oligandrin-treated plants show reduced disease incidence in tomato caused by *P. nicotianae* var. *parasitica*. Application of *P. oligandrum* effectively induces resistance and controls Fusarium crown and root rot in tomato caused by *Fusarium oxysporum* f. sp. *radicis-lycopersici* (Pharand et al. 2002). The induced resistance was due to the formation of physical barriers at sites attempted fungal penetration. These structures included callose-enriched wall appositions and osmophilic material (Pharand et al. 2002).

Salicylic acid has been reported as one of the important signal molecules, which act locally in intracellular signal transduction. Salicylic acid may enhance release of H_2O_2 and H_2O_2-derived active oxygen and induce activities of defence-related genes (Shirasu et al. 1997). The induced resistance may be due to production of salicylic acid by several *P. fluorescens* strains. Several fluorescent pseudomonads are known to synthesise salicylic acid. Salicylic acid biosynthetic genes expressed in *P. fluorescens* strain P3 improve the induction of systemic resistance in tobacco against tobacco necrosis virus. Production of salicylic acid has been related to efficacy of *P. aeruginosa* strain in inducing resistance in bean.

A strain of *P. fluorescens* strain WCS374 and its mutant derivatives, 374PSB and 374OA, produced high amount of salicylic acid, and all of them induced resistance and controlled the radish Fusarium wilt caused by *F. oxysporum* f. sp. *raphani*. Wild strain of *P. aeruginosa* was capable of producing salicylic acid induced systemic resistance against *Colletotrichum lindemuthianum* in bean.

Induction of H_2O_2 production due to *P. fluorescens* treatment in plants has also been reported, and H_2O_2 is known as a second messenger, which triggers synthesis of defence chemicals.

Table 9.1 Efficacy of *Pseudomonas fluorescens* Pf1 commercial formulation in inducing resistance against rice diseases

Treatment	Blight incidence (%)	Sheath blight incidence (%)
Pseudomonas fluorescens Pf1	8.0	8.6
Carbendazim	9.0	6.4
Control	50.0	16.4

Table 9.2 Efficacy of *Pseudomonas fluorescens* Pf1 commercial formulation in inducing resistance against peanut diseases

	Disease severity (0–9 scale)		
Treatment	Late leaf spot	Rust	Seed yield (kg/ha)
Seed treatment with *P. fluorescens*	4.6	4.3	1,480
Seed treat. + foliar appln of *P. fluorescens*	3.7	3.5	1,823
Carbendazim + mancozeb	3.9	3.1	1,650
Control	8.1	7.3	1,100

These biological plant activators are highly useful antagonistic organisms and have been developed as potential biocontrol agents in control of a wide range of crop diseases caused by fungi, bacteria and viruses. Several *Pseudomonas* strains isolated from plant rhizosphere (rhizobacterial strains) have been shown to induce plant resistance and control rice diseases. The biological plant activator *P. fluorescens* Pf1 strain has been isolated at Tamil Nadu Agricultural University, Coimbatore, and developed as a commercial powder formulation (Vidhyasekaran et al. 1997). It induces resistance in rice, tomato and peanut against bacterial, fungal and viral diseases. When rice seeds were treated with *P. fluorescens* Pf1 product and sown, the seedlings showed resistance to the bacterial blight pathogen, *Xanthomonas oryzae* pv. *oryzae* inoculated 30 days after sowing (Vidhyasekaran et al. 2001). When lower leaves were sprayed with the Pf1 product, uppermost leaves showed resistance to *X. oryzae* pv. *oryzae* in rice. Seed treatment of *P. fluorescens* followed by foliar spray controlled the bacterial blight, better than the standard streptocycline (Vidhyasekaran et al. 2001). *P. fluorescens* Pf1 strain induced resistance and effectively controlled rice blast (*Pyricularia oryzae*) and sheath blight (*Rhizoctonia solani*) (Table 9.1) (Vidhyasekaran and Muthamilan 1999; Vidhyasekaran et al. 1997).

Seed treatment with commercial formulation of *P. fluorescens* Pf1 product provided long-term protection (induced systemic resistance) in the fields against airborne late leaf spot (*Cercospora personatum*) and rust (*Puccinia arachidis*) diseases on peanut (Meena et al. 2000, 2002). Seed treatment followed by foliar application of *P. fluorescens* Pf1 provided most effective control of both the diseases similar to the standard carbendazim + mancozeb mixtures and increased crop yield (Table 9.2) (Meena et al. 2002).

Pseudomonas fluorescens Pf1 strain induced resistance in tomato and controlled the wilt disease caused by *Fusarium oxysporum* f. sp. *lycopersici* (Ramamoorthy et al. 2002).

Pseudomonas fluorescens Pf1 induced increased activities of several defence-related enzymes, such as peroxidase, phenyl alanine ammonia-lyase and coumarate: CoA ligase in treated rice plants tissues (Vidhyasekaran et al. 2001). It induced increased synthesis of lignin in rice (Meena et al. 1999; Vidhyasekaran et al. 2001). Several pathogenesis-related proteins such as PR-2, PR-3 and PR-5 accumulated in *P. fluorescens* Pf1-treated tomato plants (Ramamoorthy et al. 2002). Application of *P. fluorescens* Pf1 product induced accumulation of a 23-kDa PR-5 protein and a 30-kDa ß-1, 3-glucanase (PR-2 protein) in peanut. It also induced increase in activity of phenylalanine ammonia-lyase, resulting in increased phenolic content (Meena et al. 2000).

9.2 Chemical Plant Defence Activators

DL-β-aminobutyric acid (BABA)-induced resistance (BABA-IR) can provide effective protection for crop plants in many botanical families, including legumes, cereals, brassicas and Solanaceae (Jakab et al. 2001). Unlike other chemical inducers (e.g. INA and BTH), BABA does not directly activate the plant's defence arsenal and therefore does not cause direct trade-off effects on plant growth due to energetically demanding investment in defence mechanisms. Instead, BABA appears to condition the plant for a faster and stronger activation of defence responses once the induced plant is exposed to stress, a process known as 'sensitisation' or 'priming' (Conrath et al. 2002).

BABA was shown to be a unique inducer of plant defence. It is a simple non-protein amino acid which, when sprayed onto the leaf surface or drenched into the soil, induced SAR against various foliar and root pathogens. BABA provided almost complete control of late blight in tomato plants without being fungitoxic. It has instantaneous action, even when applied post-infection. This feature bears a significant advantage over BTH [benzo (1, 2, 3) thiadiazole-7-carbothioic acid S-methyl ester] which has to be applied before the appearance of the disease (Cohen 2001).

BTH is strongly effective against *Peronospora tabacina*, causative agent of blue mould, the most important worldwide distributed tobacco disease. Applied in minimal amounts (around 50 $g\,ha^{-1}$), BTH provides field protection lasting until flowering without negative influence on growth, development and yield of tobacco. BTH appears more efficient than metalaxyl, the commonly used blue mould fungicide. It ensures 90% disease reduction on the 17th day after its application versus only 46 % for metalaxyl (Tally et al. 1999). It is noteworthy that BTH is an effective inducer of resistance in tobacco not only against fungal pathogens but also against viruses and bacteria (Tally et al. 1999). BTH was also found to be effective in inducing SAR in wheat (Görlach et al. 1996), pea (Dann and Deverall 2000), potato (Bokshi et al. 2003), cotton against *Alternaria* leaf spot, bacterial blight and *Verticillium* wilt (Colson-Hanks et al. 2000), and tomato against bacterial canker (*Clavibacter michiganensis* sub. sp. *michiganensis*) (Soylu et al. 2003).

Both benzothiadiazole (BTH-Actigard 50 WG was applied at 52 g/ha 4 days prior to pathogen inoculation) and BTH in combination with the PGPR product [mixing of BioYield concentrate (product contains 4.1×10^4 endospores of two bacterial strains, *Bacillus subtilis* GB03 and *B. amyloliquefaciens* IN937A, per cm^3) evenly into the potting mix at planting (1.2 kg/m^3)] significantly reduced bacterial speck (*Pseudomonas syringae* pv. *tomato*) incidence. Synergy between BTH and the PGPR product was observed. Plants that received BTH treatment consistently exhibited less severe symptoms than untreated control and PGPR-treated plants ($P<0.05$). BTH+PGPR-treated plants were less severely infected than those treated with BTH alone. BTH (SAR-inducing compound) effectively reduced bacterial speck incidence and severity, both alone and in combination with the ISR-inducing product. Application of BTH also led to elevated activation of salicylic acid and ethylene-mediated responses, based on real-time polymerase chain reaction analysis of marker gene expression levels. In contrast, the ISR-inducing product (made up of plant growth-promoting rhizobacteria) inconsistently modified defence gene expression and did not provide disease control to the same level as did BTH (Herman et al. 2008).

Synergistic effects of BABA and BTH were successfully applied in crop protection. Moreover, synergistic interactions of BABA with fungicides were reported, namely, with metalaxyl, controlling blue mould in tobacco, and mancozeb in controlling late blight (*Phytophthora infestans*) of potato (Baider and Cohen 2003).

Application of O-acetylsalicylic acid, 2-chloronicotinic acid, 5-nitrosalicylic acid and 4-chlorosalicylic acid and salicylic acid were found to be superior to carbofuran in reducing

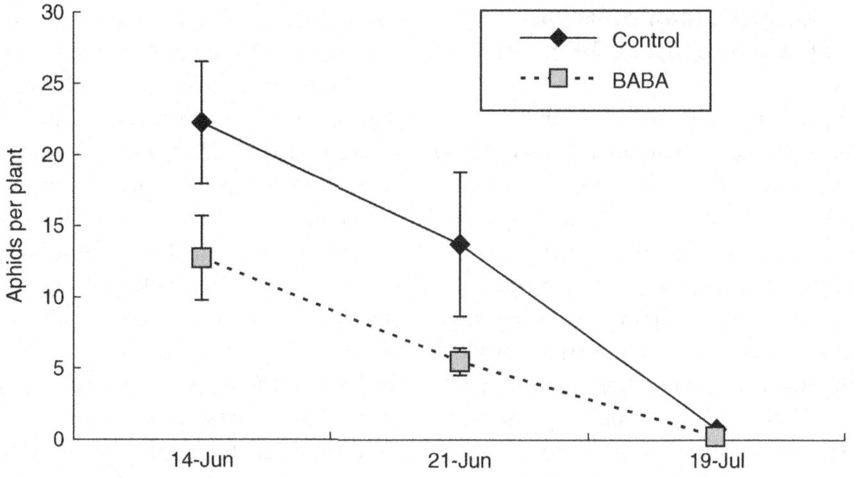

Fig. 9.1 The impact of BABA on aphid–plant interactions in the laboratory and field. Number of carrot-willow aphids (*Cavariella aegopodii*) on carrot plants (*Daucus carota*) during a preliminary 2004 field trial. Plots were either sprayed once with 25 mM BABA on 7 June or unsprayed (controls)

root-knot nematode (*Meloidogyne incognita*) population in soil and root as well as galling and enhancing plant growth and flower yield of chamomile (*Matricaria recutita*) (Pandey and Kalra 2005).

9.2.1 BABA Induces Resistance to Aphids

Despite extensive research showing that BABA enhances plant resistance to pathogens, there have been no published studies exploring effects on plant–insect interactions. Data from Imperial College, London, shows that BABA induces very effective resistance against aphids (Hodge et al. 2005). In laboratory studies, BABA application to five legume species caused significant reductions in *Aphis pisum* fitness (Hodge et al. 2005), and the performance of other pest aphid species was reduced on cereals (*Sitobion avenae*), brassicas (*Myzus persicae*) and sugar beet (*Aphis fabae* and *M. persicae*). In a preliminary field study, application of 25 mM BABA (as a foliar spray) to carrot plots significantly reduced numbers of the willow-carrot aphid (*Cavariella aegopodii*) present on plants, compared to unsprayed control plots (Fig. 9.1).

9.3 Synergistic Manipulation of Plant and Insect Defences

9.3.1 Background

A number of natural or synthetic chemicals may act as plant activators or primers, that is, chemicals that induce plant resistance by, for example, up-regulating a plant's secondary chemistry to produce more of a toxic compound. One such chemical is the amino acid BABA, which does not directly activate a plant's defence but instead allows the plant to produce a faster and stronger defence response when needed.

Metabolic enzymes are present in insects to enable them to metabolise plant xenobiotics, but selection pressure from pesticides can result in greatly enhanced activity and therefore insecticide resistance. Some synergists work by inhibiting these enzyme systems, thus allowing pesticides to work more efficiently. If time is allowed between an application of synergist and an application of pesticide, mortality of the insect can be brought about by much lower concentrations of pesticide, typically 1,000-fold less.

The mortality should be reached by utilising the plants natural xenobiotics, particularly if they have been activated or primed by chemical means.

9.3.2 Synergistic Manipulations of Plant and Insect Defences

Aphids are the most important group of insect crop pests in temperate regions (Tatchell 1989) and achieve their unique pest status by combining extremely high rates of population increase with a propensity for extracting phloem sap and transmitting plant viruses. Cyclical parthenogenesis allows aphids to combine extremely high rates of reproductive increase (asexual reproduction) with annual recombination (sexual reproduction), leading to the rapid evolution of insecticide resistance. This problem, coupled with the revocation of many insecticides, has created an urgent need for novel approaches to aphid control. One such approach involves enhancing plant resistance to aphids, whilst another acts on the resistance mechanism of the aphids itself.

This is an alternative approach for developing sustainable aphid resistance in plants; augmenting the plant's natural defence mechanisms by inhibiting the metabolic enzymes of the aphids.

9.3.3 Temporal Synergism Defeats Insect-Resistance Mechanisms

Pesticide resistance is often due to the enhancement of metabolic enzyme systems within the insect, particularly non-specific esterases and microsomal oxidases. These enzymes are present in insects to enable them to metabolise plant xenobiotics, but selection pressure from pesticides can result in greatly enhanced activity and therefore insecticide resistance. Inhibitors of these enzyme systems (synergists) can result in increased potency of insecticides. Although such regimes have been in place for many years, the synergists have only been used as mixtures. Recent work has demonstrated that for many insects, the synergist has to be applied several hours before maximum inhibition of these metabolic enzymes is realised (Moores et al. 2005; Young et al. 2005). Bespoke formulations allowing a burst release of synergist piperonyl butoxide (PBO) followed by a second burst release of insecticide several hours later have resulted in an increase of efficacy of around 1,000-fold. Thus, 1,000-times less insecticide was found to be needed to give equivalent control (Young et al. 2006). A similar regime against *Myzus persicae* resulted in 10,000-fold less pirimicarb being required to confer full control of even high resistant populations (Bingham and Moores 2008).

Since such a small concentration of insecticide is necessary to confer kill upon the insects after the lowering of their metabolic enzymes, the increase of plant xenobiotics resulting from the BABA priming will be sufficient to result in an increase of aphid mortality due to feeding on the plant. If such mortality rates are found to be insufficient for adequate control of vector and virus, then the regime will be supported by a minimum of conventional chemical control.

Insecticide-resistant *Myzus persicae* possess an enhanced non-specific esterase, E4, that can hydrolyse or/and sequester insecticides and plant xenobiotics. For optimal effects of the secondary metabolites from the crops to affect aphid mortality, this enzyme should be fully inhibited prior to exposure. This inhibition is to be obtained using the synergist piperonyl butoxide (PBO), a chemical that was originally thought to be a specific inhibitor of microsomal oxidases, but has now been shown to also inhibit esterases (Moores and Bingham 2008).

Results using BABA as a root drench on plants immediately after a PBO spray demonstrated extremely positive results. This was carried out with *B. tabaci* on cotton and tomatoes, *M. persicae* on potatoes and peppers, and *B. nigra* on black mustard (Moores and Bingham 2008).

Thus, with *Bemisia tabaci* on cotton, *B. tabaci* on tomatoes, *Myzus persicae* on potatoes, *M. persicae* on peppers and *Brevicoryne brassicae* on black mustard, results followed a similar pattern. Application of synergist (PBO 30) alone gave a significant decrease in insect pest numbers when applied to cotton and potatoes. This is perhaps indicative of the toxic alkaloids found within these plants. A root drench of BABA alone resulted in reduction of insect pest numbers with cotton, tomatoes, potatoes, peppers and black mustard. Application of PBO and BABA resulted in a significant reduction in pest numbers in all

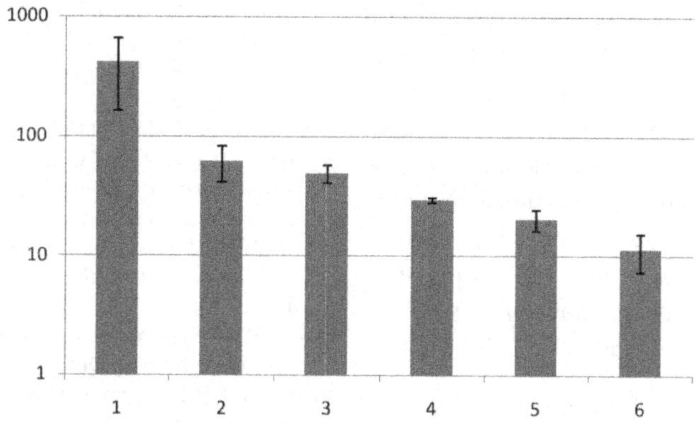

Fig. 9.2 Efficacy of plant defence activators and insecticide on whitefly control. *1* Control, *2* PBO, *3* BABA, *4* PBO+BABA, *5* imidacloprid, *6* imidacloprid+PBO. X-axis is the treatment as described above; Y-axis is the number of whitefly eggs counted after treatment as a % of the egg numbers counted before treatment

cases and with cotton and tomatoes resulted in almost complete absence. Plant health was not compromised at these doses of synergist and plant activator (Moores and Bingham 2008).

The insecticide resistance of the population is seen in the results, as field application of pirimicarb has only resulted in a reduction of less than 50% of the aphids. Even supplementing the insecticide with a synergist has left around 20% of the aphids present, although this effect would be dramatically increased if a time delay between synergist and insecticide had been applied. The effect of the insecticide has not been significantly increased with the addition of plant activator (BABA), and the increase seen with the addition of synergist to insecticide has also not been enhanced with the addition of BABA. However, on the very positive side, BABA reduced aphid numbers as efficiently as pirimicarb, and BABA+PBO worked better than pirimicarb or pirimicarb+BABA (Moores and Bingham 2008).

Thus, when dealing with an insecticide-resistant population that contains both metabolic and target-site resistance and is thus very difficult to control with conventional insecticides, the application of plant activator and synergist can be seen as an alternative control strategy that will give comparable decrease in insect pest numbers. However, there does not appear to be any synergistic interaction between plant activator and insecticide.

In protected tomatoes to *Bemisia tabaci*, there was no significant difference in the PBO alone and BABA alone, both giving significant reduction in egg numbers when compared to the control plot. There was a further significant reduction in numbers with BABA+PBO. Indeed, egg numbers were comparable with those counted using imidacloprid. The most effective treatment, unsurprisingly, was that of imidacloprid+PBO (Moores and Bingham 2008). In the control, plot numbers increased fourfold (400 %) (Fig. 9.2).

Using a plant activator such as BABA, to 'prime' a crop in conjunction with an appropriate synergist (e.g. PBO) to abrogate the metabolic defences of pest insects, can result in good control of even insecticide-resistant insect pests, without the need of an insecticide.

Other plant activators such as *cis*-jasmone and other insecticide synergists such as propyl gallate could be used in combination to give comparable results. It is expected that any crop possessing relatively toxic secondary chemistry would give similar results.

9.4 Integration of Biological and Chemical Plant Activators

The isolates of *P. fluorescens* systemically induced resistance against *Fusarium* wilt of chickpea caused by *F. oxysporum* f. sp. *ciceri*

(*Foc* Rs1) and significantly ($P = 0.05$) reduced the wilt disease by 26–50 % as compared to control. Amongst chemical inducers, salicylic acid showed the highest protection of chickpea seedlings against wilting. The reduction in disease was more pronounced when chemical inducers were applied with *P. fluorescens*. Fifty-two to sixty-four percent reduction of wilting was observed in soil treated with isolate *Pf*4–92 along with chemical inducers (Saikia et al. 2003).

References

Baider A, Cohen Y (2003) Synergistic interaction between BABA and mancozeb in controlling *Phytophthora infestans* in potato and tomato and *Pseudoperonospora cubensis* in cucumber. Phytoparasitica 31:399–409

Bingham GV, Moores GD (2008) Temporal synergism can enhance carbamate and nicitinoid insecticidal activity against resistant crop plants. Pest Manag Sci 64:81–85

Bokshi AI, Morris SC, Deverall BJ (2003) Effects of benzothiadiazole and acetylsalicylic acid on β-1, 3-glucanase activity and disease resistance in potato. Plant Pathol 52:22–27

Cohen Y (2001) The BABA story of induced resistance. Phytoparasitica 29:375–378

Colson-Hanks ES, Allen SJ, Deverall BJ (2000) Effect of 2, 6- dichloroisonicotinic acid or benzothiadiazole on Alternaria leaf spot, bacterial blight and Verticillium wilt in cotton under field conditions. Austr Plant Pathol 29:170–177

Conrath U, Pieterse CMJ, Mauch-Mani B (2002) Priming in plant-pathogen interactions. Trends Plant Sci 7:210–216

Dann EK, Deverall BJ (2000) Activation of systemic disease resistance in pea by an avirulent bacterium or a benzothiadiazole, but not by a fungal leaf spot pathogen. Plant Pathol 49:324–332

Dean JFD, Anderson JD (1990) Ethylene biosynthesis inducing xylanase II. Purification and physical characterization of the enzyme produced by Trichoderma viride. Plant Physiol 94:1849–1854

Görlach J, Volrath S, Knauf Beiter G, Hengy G, Beckhove U, Kogel KH, Oostendorp M, Staub T, Ward E, Kessman H, Ryals J (1996) Benzothiadiazole, a novel class of inducers of systemic acquired resistance, activates gene expression and disease resistance in wheat. Plant Cell 8:629–643

Gozzo F (2004) Outlooks on pest management, Feb 2004, pp 20–23

Herman MAB, Davidson JK, Smart CD (2008) Induction of plant defense gene expression by plant activators and *Pseudomonas syringae* pv *tomato* in greenhouse-grown tomatoes. Phytopathology 98:1226–1232

Hodge S, Thompson GA, Powell G (2005) Application of DL-β-aminobutyric acid (BABA) as a root drench to legumes inhibits the growth and reproduction of the pea aphid *Acyrthosiphon pisum* Harris. Bull Ent Res 95:449–455

Jakab G, Cottier V, Touquin V et al (2001) Eur J Plant Pathol 107:29–37

Karban R, Baldwin IT (1997) Induced responses to herbivory. University of Chicago Press, London, 31 pp

Krishnamurthy J, Samiyappan R, Vidyasekaran P, Nakeeran S, Rajeshwari E, Raja JAJ, Balasubramanian P (1999) Efficacy of *Trichoderma* chitinases against *Rhizoctonia solani*, the rice sheath blight pathogen. J Biosci 24:207–213

Kuk J (1984) Systemic plant immunization. Tagungsbericht Akad Landw Wiss DDR 222:189–198

Kuk J (1987) Plant immunization and its application for disease control. In: Chet I (ed) Innovative approaches to plant disease. Wiley, New York, pp 255–274

Leeman M, Den Ouden FM, Van Pelt JA, Dirkx FPM, Steijl H, Bakker PAHM, Schippers B (1996) Iron availability affects induction of systemic resistance to *Fusarium* wilt of radish *Pseudomonas fluorescens*. Phytopathology 86:149–155

Leeman M, Van Pelt JA, Den Ouden FM, Heinsbroek M, Bakker PAHM, Schippers B (1995) Induction of systemic resistance against *Fusarium* wilt of radish by lipopolysaccharides of *Pseudomonas fluorescens*. Phytopathology 85:1021–1027

Malamy J, Klessig DF (1992) Salicylic acid and plant resistance. Plant J 2:643–654

Maurhofer M, Hase C, Meuwly P, Metraux JP, Defago G (1994) Induction of systemic resistance of tobacco to tobacco necrosis virus by the root colonizing *Pseudomonas fluorescens* strain CHA0: influence of the *gag*A gene and of pyoverdine production. Phytopathology 84:678–684

Maurhofer M, Reimann C, Sacherer SP, Heeb S, Haas D, Defago G (1998) Salicylic acid biosynthetic genes expressed in *Pseudomonas fluorescens* strain P3 improve the induction of systemic resistance in tobacco against tobacco necrosis virus. Phytopathology 88:139–146

Meena B, Radhajeyalakshmi R, Marimuthu T, Vidhyasekaran P, Sabitha D, Velazhahan R (2000) Induction of pathogenesis related proteins, phenolics and phenylalanine ammonia lyase in groundnut by *Pseudomonas fluorescens*. J Plant Dis Prot 107:514–527

Meena B, Radhajeyalakshmi R, Marimuthu T, Vidhyasekaran P, Velazhahan R (2002) Biological control of groundnut late leaf spot and rust by seed and foliar applications of a powder formulation of *Pseudomonas fluorescens*. Biocont Sci Technol 12:195–204

Meena B, Radhajeyalakshmi R, Vidhyasekaran P, Velazhahan R (1999) Effect of foliar application of *Pseudomonas fluorescens* on activities of phenylalanine ammonia lyase, chitinase, and ß-1, 3-glucanase and accumulation of phenolics in rice. Acta Phytopathologie Entomologica Hungaricae 34:307–315

Moores GD, Bingham GV (2008) Compositions and methods for synergistic manipulation of plant and insect defences. International PCT Application No. PCT/GB2008/001419

Moores GD, Bingham G, Gunning RV (2005) Outlook Pest Manag 16(1):7–9

Pandey R, Kalra A (2005) Chemical activators: a novel and sustainable management approach for *Meloidogyne incognita* (Kofoid and White) Chitwood in *Chamomilla recutita* L. Arch Phytopathol Plant Prot 38:107–111

Pharand B, Carisse O, Benhamou N (2002) Cytological aspects of compost-mediated induced resistance against Fusarium crown and root rot in tomato. Phytopathology 92:424–438

Ramamoorthy V, Raguchander T, Samiyappan R (2002) Induction of defense-related proteins in tomato roots treated with *Pseudomonas fluorescens* Pf1 and *Fusarium oxysporum* f sp *lycopersici*. Plant Soil 239:55–68

Saikia R, Singh T, Kumar R, Srivastava J, Srivastava AK, Singh K, Arora DK (2003) Role of salicylic acid in systemic resistance induced by *Pseudomonas fluorescens* against *Fusarium oxysporum* f sp *ciceri* in chickpea. Microbiol Res 158:203–213

Shirasu K, Nakajima H, Rajasekhar VK, Dixon RA, Lamb C (1997) Salicylic acid potentialities an agonist-dependent gain control that amplifies pathogen signals in the activation of defense mechanisms. Plant Cell 9:261–270

Soylu S, Baysal O, Soylu EM (2003) Induction of disease resistance by the plant activator, acibenzolar-S methyl (ASM) against bacterial canker (*Clavibacter michiganensis* sub sp *michiganensis*) in tomato seedlings. Plant Sci 165:1069–1076

Tally A, Oostendorp M, Lawton K, Staub T, Bassy B (1999) Commercial development of elicitors of induced resistance to pathogens. In: Agrawal AA, Tuzun S, Bent E (eds) Inducible plant defenses against pathogens and herbivores: biochemistry, ecology, and agriculture. American Phytopathological Society Press, St Paul, pp 357–369

Tatchell GM (1989) Crop Prot 8:25–29

Van Peer R, Schippers B (1992) Lipopolysaccharides of plant growth promoting *Pseudomonas* sp strain WCS 417r induce resistance in carnation to *Fusarium* wilt. Netherlands J Plant Path 98:129–139

Vidhyasekaran P, Kamala N, Ramanathan A, Rajappan K, Paranidharan V, Velzhahan R (2001) Induction of systemic resistance by *Pseudomonas fluorescens* Pf1 against *Xanthomonas oryzae* pv *oryzae* in rice leaves. Phytoparasitica 29:155–166

Vidhyasekaran P, Muthamilan M (1999) Evaluation of *Pseudomonas fluorescens* for controlling rice sheath blight. Biocontrol Sci Technol 9:67–74

Vidhyasekaran P, Sethuraman K, Rajappan K, Vasumathi K (1997) Powder formulation of *Pseudomonas fluorescens* to control pigeonpea wilt. Biol Contr 8:166–171

Young SJ, Gunning RV, Moores GD (2005) Pest Manag Sci 61:397–401

Young SJ, Gunning RV, Moores GD (2006) Pest Manag Sci 62:114–119

Plant Growth-Promoting Rhizobacteria (PGPR) 10

Abstract

Plant growth-promoting rhizobacteria (PGPR) colonize the roots of plants following inoculation onto seed before planting and enhance plant growth and/or reduce disease, nematode or insect damage. There has been much research interest in PGPR and there is now an increasing number of PGPR being commercialized for crops. Organic growers may have been promoting these bacteria without knowing it. The addition of compost and compost teas promote existing PGPR and may introduce additional helpful bacteria to the field. The absence of pesticides and the more complex organic rotations likely promote existing populations of these beneficial bacteria. However, it is also possible to inoculate seeds with bacteria that increase the availability of nutrients, including solubilizing phosphate, potassium, oxidizing sulphur, fixing nitrogen, and chelating iron and copper.

PGPR such as *Pseudomonas* and *Bacillus* species have attracted much attention for their role in reducing plant diseases. The work to date is very promising and may offer organic growers with some of their first effective control of serious plant diseases. Some PGPR use scarce resources, and thereby prevent or limit the growth of pathogenic microorganisms. Even if nutrients are not limiting, the establishment of benign or beneficial organisms on the roots limits the chance that a pathogenic organism that arrives later will find space to become established. Numerous rhizosphere organisms are capable of producing compounds that are toxic to pathogens like HCN.

Plant growth in agricultural soils is influenced by a myriad of abiotic and biotic factors. Whilst growers routinely use physical and chemical approaches to manage the soil environment to improve crop yields, the application of microbial products for this purpose is less common. An exception to this is the use of rhizobial inoculants for legumes to ensure efficient nitrogen fixation, a practice that has been occurring in North America for over 100 years (Smith 1997). The region around the root, the rhizosphere, is relatively rich in nutrients, due to the loss of as much as 40% of plant photosynthates from the roots (Lynch and Whipps 1991). Consequently, the rhizosphere supports large and active microbial populations capable of exerting beneficial, neutral or detrimental effects on plant growth. The importance of rhizosphere microbial populations for

P.P. Reddy, *Recent Advances in Crop Protection*,
DOI 10.1007/978-81-322-0723-8_10, © Springer India 2013

maintenance of root health, nutrient uptake and tolerance of environmental stress is now recognised (Bowen and Rovira 1999; Cook 2002). These beneficial microorganisms can be a significant component of management practices to achieve the attainable yield, which has been defined as crop yield limited only by the natural physical environment of the crop and its innate genetic potential (Cook 2002).

The prospect of manipulating crop rhizosphere microbial populations by inoculation of beneficial bacteria to increase plant growth has shown considerable promise in laboratory and greenhouse studies, but responses have been variable in the field (Bowen and Rovira 1999). The potential environmental benefits of this approach, leading to a reduction in the use of agricultural chemicals and the fit with sustainable management practices, are driving this technology. Recent progress in our understanding of the biological interactions that occur in the rhizosphere and of the practical requirements for inoculant formulation and delivery should increase the technology's reliability in the field and facilitate its commercial development.

10.1 Characteristics of an Ideal PGPR

- High rhizosphere competence
- High competitive saprophytic ability
- Enhanced plant growth
- Ease for mass multiplication
- Broad spectrum of action
- Excellent and reliable control
- Safe to environment
- Compatible with other rhizobacteria
- Should tolerate desiccation, heat, oxidising agents and UV radiations (Jeyarajan and Nakkeeran 2000)

10.2 Ways that PGPR Promote Plant Growth

- Increasing nitrogen fixation in legumes
- Promoting free-living nitrogen-fixing bacteria
- Increasing supply of other nutrients, such as phosphorus, sulphur, iron and copper
- Producing plant hormones
- Enhancing other beneficial bacteria or fungi
- Controlling diseases, nematodes and insect pests

Introduction of PGPR for increasing plant growth promotion (Table 10.1) during the 1950s from the research findings opened new vistas to use PGPR as an alternate to chemical pesticides for the management of soil-borne pathogens (Kloepper 1993). Application of PGPR either as single strain or strain-mixture-based formulations checked pest and disease spread besides increasing growth and yield.

10.3 Strains of PGPR

Plant growth-promoting rhizobacteria (PGPR) were first defined by Kloepper and Schroth (1978b) to describe soil bacteria that colonise the roots of plants following inoculation onto seed and that enhance plant growth. The following are implicit in the colonisation process: ability to survive inoculation onto seed, to multiply in the spermosphere (region surrounding the seed) in response to seed exudates, to attach to the root surface and to colonise the developing root system (Kloepper 1993). The ineffectiveness of PGPR in the field has often been attributed to their inability to colonise plant roots (Benizri et al. 2001; Bloemberg and Lugtenberg 2001). A variety of bacterial traits and specific genes contribute to this process, but only a few have been identified (Benizri et al. 2001; Lugtenberg et al. 2001). These include motility, chemotaxis to seed and root exudates, production of pili or fimbriae, production of specific cell surface components, ability to use specific components of root exudates, protein secretion and quorum sensing (Lugtenberg et al. 2001). The generation of mutants altered in expression of these traits is aiding our understanding of the precise role each one plays in the colonisation process (Lugtenberg et al. 2001; Persello-Cartieaux et al. 2003). Progress in the identification of new, previously uncharacterised genes is being made using nonbiased screening strategies that rely on gene fusion technologies. These strategies employ reporter transposons (Roberts et al. 1998) and in vitro

Table 10.1 Efficacy of PGPR formulations against plant diseases and growth promotion

Formulation	Crop	Results	Reference
Talc-based *P. fluorescens*	Potato	Significant plant growth promotion	Kloepper and Scroth (1981a)
Talc-based *P. fluorescens*	Winter wheat	Significant plant growth promotion	De Freitas and Germida (1992)
Peat-based *P. fluorescens*	Cotton	Significant reduction of seedling diseases	Hagedorn et al. (1993)
Talc-based *P. fluorescens*	Chickpea	Significant increase in grain yields and controlled fusarial wilt under field conditions	Vidhyasekaran and Muthamilan (1995)
Talc-based *P. fluorescens*	Pigeonpea	Control of wilt and sign increase in grain yield	Vidhyasekaran et al. (1997)
Chitosan-based *B. pumilus*	Tomato	Induced resistance against *F. oxysporum*	Benhamou et al. (1998)
Methyl cellulose and talc-based *P. fluorescens*	Rice	Suppressed blast both in nursery and field conditions	Krishnamurthy and Gnanamanickam (1998)
B. subtilis strain LS213 (commercial product)	Watermelon, muskmelon	Increased plant growth and improved yield	Vavrina (1999)
B. subtilis formulations	Cucumber, watermelon, squash, ornamentals, vegetables, pepper, tobacco, loblolly pine, lodge pine	Significant induction of resistance against different pathogens	Reddy et al. (1999), Kenney et al. (1999), Martinez-Ochoa et al. (1999), Ryu et al. (1999), Yan et al. (1999) and Zhang et al. (1999)
Chitosan-based *B. subtilis* strain LS213 (commercial product)	Tomato, tobacco, cucumber, pepper	Reduced the incidence of bacterial spot and late blight of tomato, angular leaf spot of cucumber and blue mould of tobacco	Reddy et al. (1999)
Talc-based formulation of *P. fluorescens* (CHAO and Pf1)	Sugarcane	Increased germination of seeds, plant growth besides the suppression of damping off	Viswanthan and Samiyappan (1999)
Vermiculite-based *P. fluorescens*	Sugar beet	Significant control of damping off	Moenne-Loccoz et al. (1999)
Talc-based *P. fluorescens*	Rice	Significant reduction of sheath blight under field conditions	Vidhyasekaran and Muthamilan (1999) and Nandakumar et al. (2000)
Talc-based *P. fluorescens*	Banana	Significant reduction of panama wilt	Raguchander et al. (2000)
Vermiculite- and kaolin-based *B. subtilis*	Lettuce	Suppressed root rot caused by *Pythium aphanidermatum* and increased fresh weight	Amer and Utkhede (2000)
Vermiculite-based *P. putida*	Cucumber	Significantly reduced root rot caused by *Fusarium oxysporum* f. sp. *cucurbitacearum*	Amer and Utkhede (2000)
Talc-based *P. fluorescens* (Pf1)	Urdbean, sesame	Increased growth promotion and reduced root rot caused by *M. phaseolina*	Jayashree et al. (2000)
Talc-based rhizobacterial mixtures of fluorescent pseudomonads	Rice	Significant plant growth promotion and suppression of sheath blight	Nandakumar et al. (2001)

(continued)

Table 10.1 (continued)

Formulation	Crop	Results	Reference
Peat-based *B. subtilis* supplemented with chitin	Groundnut, pigeonpea	Significant control of groundnut root rot and pigeonpea wilt	Manjula and Podile (2001)
Chitosan-based mixed formulation of *Paenibacillus macerans* and *B. subtilis* (LS255)	Rice	Increased plant growth and yield in cultivars, IR24, IR50 and Jyothi	Vasudevan et al. (2002)
Chitin-based formulation of *B. subtilis* strain GB03 + *B. pumilus* strain INR7 (LS256) and *B. subtilis* strain GB03 + *B. subtilis* strain IN937b	Tomato, pepper	Increased yield of pepper and tomato	Burelle et al. (2002)
Talc-based *P. aeruginosa* strain 78	Mung bean	Reduced the incidence of root-knot and population density of *Meloidogyne javanica* under field conditions	Ali et al. (2002)
Talc-based fluorescent pseudomonads	Sugarcane	Significant increase in sett germination, increased cane growth and reduced red rot incidence	Viswanathan and Samiyappan (2002)
Talc-based *P. fluorescens*	Rice	Significant reduction of sheath blight, leaf folder and increased yield. Increased population of insect parasites and predators	Radja Commare et al. (2002)
Talc-based *P. fluorescens*	Groundnut	Significant reduction of leaf spot and rust	Meena et al. (2002)
Talc-based formulation of *B. subtilis* and *P. chlororaphis* (PA23)	Tomato	Increased growth promotion and significant reduction of damping off	Kavitha et al. (2003)
Chitosan-based mixed formulation of *B. subtilis* strain GB03 + *B. pumilus* strain INR7(LS256) and *B. subtilis* strain GB03 + *B. pumilus* strain T4 (LS257)	Pearl millet	Reduced downy mildew and increased plant growth promotion	Niranjan Raj et al. (2003)
Talc-based *P. fluorescens* FP7 supplemented with chitin	Mango	Significant reduction of anthracnose coupled with increase in fruit yield and quality	Vivekananthan et al. (2004)
Talc-based *B. subtilis* (BSCBE4) and *P. chlororaphis* (PA23)	Turmeric	Significant reduction of rhizome rot and yield increase of rhizomes	Nakkeeran et al. (2004)
Talc-based *B. subtilis* (BSCBE4), *P. chlororaphis* (PA23) and *P. fluorescens* (ENPF1)	*Phyllanthus amarus*	Significant reduction of stem blight caused by *Corynespora cassicola* under field conditions	Mathiyazhagan et al. (2004)
Talc-based *P. putida*	Muskmelon	Effective control of wilt caused by *Fusarium oxysporum* f. sp. *melonis*	Bora et al. (2004)

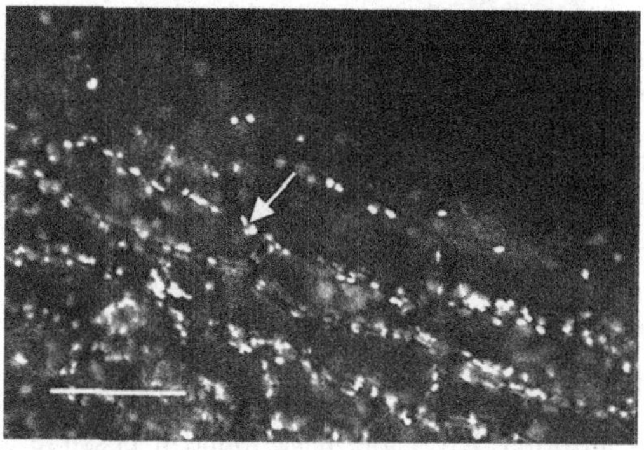

Fig. 10.1 Confocal laser scanning micrograph of a 5-day-old canola root colonised by *Pseudomonas putida* strain 6-8 labelled with green fluorescent protein (as indicated by the *arrow*). The bar is equal to 60 μm

expression technology (IVET) (Rainey 1999) to detect genes expressed during colonisation.

Using molecular markers such as green fluorescent protein or fluorescent antibodies it is possible to monitor the location of individual rhizobacteria on the root using confocal laser scanning microscopy (Bloemberg et al. 2000; Bloemberg and Lugtenberg 2001; Sorensen et al. 2001) (Fig. 10.1). This approach has also been combined with an rRNA-targeting probe to monitor the metabolic activity of a rhizobacterial strain in the rhizosphere and showed that bacteria located at the root tip were most active (Lubeck et al. 2000; Sorensen et al. 2001).

An important aspect of colonisation is the ability to compete with indigenous microorganisms already present in the soil and rhizosphere of the developing plant. Our understanding of the factors involved in these interactions has been hindered by our inability to culture and characterise diverse members of the rhizosphere community and to determine how that community varies with plant species, plant age, location on the root and soil properties. Phenotypic and genotypic approaches are now available to characterise rhizobacterial community structure. Phenotypic methods that rely on the ability to culture microorganisms include standard plating methods on selective media, community level physiological profiles (CLPP) using the BIOLOG system (Garland 1996), phospholipid fatty acid (PLFA) (Tunlid and White 1992) and fatty acid methyl ester (FAME) profiling (Germida et al. 1998). Culture-independent molecular techniques are based on direct extraction of DNA from soil and 16S-rRNA gene sequence analysis, bacterial artificial chromosome or expression cloning systems (Rondon et al. 1999). These are providing new insight into the diversity of rhizosphere microbial communities, the heterogeneity of the root environment and the importance of environmental and biological factors in determining community structure (Baudoin et al. 2002; Berg et al. 2002; Smalla et al. 2001). These approaches can also be used to determine the impact of inoculation of plant growth-promoting rhizobacteria on the rhizosphere community (Ciccillo et al. 2002; Steddom et al. 2002).

It is possible to inoculate seeds with bacteria that increase the availability of nutrients, including solubilising phosphate, potassium, oxidising sulphur, fixing nitrogen, chelating iron and copper. Phosphorus (P) frequently limits crop growth in organic production. Nitrogen-fixing bacteria are miniature of urea factories, turning N_2 gas from the atmosphere into plant available amines and ammonium via a specific and unique enzyme they possess called nitrogenase. Although there are many bacteria in the soil that 'cycle' nitrogen from organic material, it is only this small group of specialised nitrogen-fixing bacteria that can 'fix' atmospheric nitrogen in the soil. Arbuscular mycorrhizal fungi (AMF) are root symbiotic fungi improving plant stress resistance to abiotic factors such as phosphorus deficiency or dehydration.

The fourth major plant nutrient after N, P and K is sulphur (S). Although elemental sulphur, gypsum and other sulphur-bearing mined minerals are approved for organic production, the sulphur must be transformed (or oxidised) by bacteria into sulphate before it is available for plants. Special groups of microorganisms can make sulphur more available and do occur naturally in most soils. One of the most common ways that PGPR improve nutrient uptake for plants is by altering plant hormone levels. This changes root growth and shape by increasing root branching, root mass, root length and/or the amount of root hairs. This leads to greater root surface area, which in turn, helps it to absorb more nutrients.

In the last few decades, a large array of bacteria including species of *Pseudomonas, Azospirillum, Azotobacter, Klebsiella, Enterobacter, Alcaligenes, Arthrobacter, Burkholderia, Bacillus, Rhizobium* and *Serratia* have been reported to enhance plant growth.

10.3.1 *Pseudomonas fluorescens*

It belongs to PGPR, the important group of bacteria, which play a major role in the plant growth promotion, induced systemic resistance, biological control of pathogens, etc. PGPR are known to enhance plant growth promotion and reduce severity of many nematode diseases.

Most research is dedicated to improve crop yields, and plant growth with microbial inoculants has focused on symbiotic bacteria. However, free-living soil microorganisms, such as *Azospirillum* and various pseudomonas, also might be used to promote plant growth. PGPR are rhizobacteria that have the ability to promote the growth of plants following inoculation onto seeds or subterranean plant parts. PGPR mediate improved plant growth either directly by stimulation of the plant or indirectly through control of pathogens or triggering and inducting of host defence mechanism.

Pseudomonas fluorescens produce the antibiotics 2, 4-diacetylphloroglucinol (DAPG), phenazine-1-carboxylic acid (PCA), phenazine (Phe), antimicrobial metabolites such as pyoluteorin (Plt), hydrogen cyanide (HCN), siderophores pyoverdine (Pvd), salicylic acid (Sal) and pyochelin (Pch) which inhibit a broad spectrum of plant pathogenic fungi, bacteria and nematodes and control a variety of root and seedling diseases. There are many evidences substantiating the importance of DAPG in biological control. The production of DAPG is governed by *Phl* gene. The population size of DAPG producers in the rhizosphere correlated with the disease suppressiveness of the soil and in situ antibiotic production. The diverse DAPG-producing *Pseudomonas* spp. have been isolated from the rhizosphere of various crop plants, and their role in promoting plant growth and inhibiting root diseases are the subjects of ongoing investigations worldwide.

10.3.2 *Pseudomonas chitinolytica*

Pseudomonas chitinolytica, a bacterium isolated from crustacean shell amended soil was evaluated for the control of *Meloidogyne javanica* on tomatoes. Greenhouse, screen house and microplot tests all indicated the improved growth and yield and reduced nematode numbers in treated plants.

Van Gundy and associates at the University of California, Riverside, have found it possible to stimulate growth of rhizosphere bacteria by adding nutrients (e.g. peptone) through a trickle irrigation system or to distribute the rhizosphere bacteria through such a system. There is a need to search for bacteria that would be active in the rhizosphere and produce compounds lethal to nematodes, inhibitory to nematode movement or development and repellent to nematodes on roots.

10.3.3 *Bacillus subtilis*

This prevalent inhabitant of soil is widely recognised as a powerful biocontrol agent. In addition, its effect is due to its broad host range, its ability to form endospores and produce different biologically active compounds with a broad spectrum of activity.

10.3.4 *Bacillus megaterium*

Bacillus megaterium from tea rhizosphere is able to solubilise phosphate and produce IAA, siderophore and antifungal metabolites, and, thus, it helps in the plant growth and reduces disease intensity (Chakraborty et al. 2006).

10.3.5 *Bacillus thuringiensis* and *B. sphaericus*

These two species have the ability to solubilise inorganic phosphates and help in the control of the lepidopteron pests (Seshadri et al. 2007).

10.4 Disease Management Using PGPR

PGPR have attracted much attention for their role in reducing plant diseases (Table 10.2).

10.5 Nematode Management Using PGPR

10.5.1 Fruit Crops

10.5.1.1 Banana

Three *P. fluorescens* strains and the type strain *P. putida* inhibited invasion of *Radopholus similis*

Table 10.2 Management of horticultural crop diseases using PGPR

Horticultural crop	Disease/pathogen/s	Potential PGPR
Banana	Panama wilt, *Fusarium oxysporum* f. sp. *cubense*	*Pseudomonas fluorescens*, *T. viride* + *P. fluorescens*-sucker treat
Mulberry	Leaf spot, *Cercospora moricola*	*P. fluorescens*
Mango	Bacterial canker, *Xanthomonas campestris* pv. *mangiferaeindicae*	*Bacillus coagulans*
Apple	Collar rot, *Phytophthora cactorum*	*B. subtilis*-soil treatment
Pear	Fire blight, *Erwinia amylovora*	*P. fluorescens*-foliar spray
Peach	Brown rot, *Monilinia fructicola*	*B. subtilis* (B-3)-post-harvest fruit line spray at 5×10^8 cfu/g; *P. syringae*-post-harvest fruit dipping in 10^7 cfu/ml
Strawberry	Grey mould, *Botrytis cinerea*	*B. pumilus*, *P. fluorescens*
French bean	Dry root rot, *Macrophomina phaseolina*	*P. cepacia* UPR5C-seed treat
Pea	Root rot, *Aphanomyces euteiches*	*P. fluorescens* PRA25-seed treat, *P. cepacia* AMMD-seed treat
	Damping off, *Pythium ultimum*	*P. cepacia* AMMD-seed treat, *P. putida* NIR-seed treat
Tomato	Damping off, *Pythium aphanidermatum*	*P. aeruginosa* 7NSK2
	Wilt, *Fusarium oxysporum* f. sp. *lycopersici*	*P. fluorescens* strains Pf1, *P. putida*-seed treat
Potato	Wilt, *Ralstonia solanacearum*	*B. cereus*, *B. subtilis*
Brinjal	Collar rot, *Sclerotinia sclerotiorum*	*B. subtilis*-soil treat
Radish	Wilt, *F. oxysporum* f. sp. *raphani*	*P. fluorescens* strains WCS374, WCS417r – soil treat
Beet root	Damping off, *Pythium debaryanum*, *P. ultimum*	*Penicillium* spp. + *P. fluorescens*-seed treat
Cucumber	Wilt, *Fusarium oxysporum* f. sp. *cucumerinum*, *R. solani*	*P. putida* 89B-27, *Serratia marcescens*
	Cucumber mosaic virus	*P. fluorescens* strain 89B-27
Carnation	Wilt, *F. oxysporum* f. sp. *dianthi*	*P. fluorescens* strain WCS 417r-soil appln; *P. putida* WCS 358r-root dip treat; *Bacillus* sp; *S. liquefaciens*
Areca nut	Fruit rot, *Phytophthora arecae*, *Colletotrichum capsici*	*P. fluorescens*
Black pepper	Foot rot, *Phytophthora capsici*	*P. fluorescens*, *Bacillus* sp.-foliar spray
	Anthracnose, *Colletotrichum gloeosporioides*	*P. fluorescens*-foliar spray
Ginger	Rhizome rot, *Pythium aphanidermatum*, *P. myriotylum*	*P. fluorescens*-seed treat
	Bacterial wilt, *Ralstonia solanacearum*	*P. fluorescens*-soil treat
Fenugreek	Root rot, *R. solani*	*P. fluorescens*-seed treat
Vanilla	Root rot, *Phytophthora meadii*, *F. oxysporum* f. sp. *vanillae*	*P. fluorescens*-soil treat

Table 10.3 Management of plant parasitic nematodes in horticultural crops using PGPR

Antagonistic bacteria	Crop	Nematode	Reference
Pseudomonas fluorescens	Citrus (sweet orange, lemon)	*T. semipenetrans*	Shanthi et al. (1999)
	Citrus (acid lime)	*T. semipenetrans*	Parvatha Reddy et al. (2000)
	Banana	*R. similis and M. incognita*	Aalten and Gowen (1998)
	Grapevine	*M. incognita*	Shanthi et al. (1998)
	Tomato	*M. incognita*	Verma et al. (1999), Shanthi and Sivakumar (1995)
		R. reniformis	Niknam and Dhawan (2001a)
	Potato	*M. incognita*	Mani et al. (1998)
		G. rostochiensis	Mani et al. (1998)
	Okra	*M. incognita*	Sobita Devi and Dutta (2002)
	Sugar beet	*H. schachtii*	Oostendorp and Sikora (1989)
	Black pepper	*M. incognita*	Eapen et al. (1997)
P. stutzeri	Black pepper	*M. incognita*	Eapen et al. (1997)
	Turmeric	*M. incognita*	Seenivasan et al. (2001)
	Tomato	*M. incognita*	Khan and Tarannum (1999)
Bacillus subtilis	Tomato	*M. incognita*	Khan and Tarannum (1999)
		R. reniformis	Niknam and Dhawan (2001b)
	Pigeonpea	*H. cajani*	Siddiqui and Mahmood (1995)
B. thuringiensis	Tomato	*M. incognita*	Khan and Tarannum (1999)
	Pea	*M. incognita*	Chahal and Chahal (1991, 2003)

in banana roots. *P. fluorescens* strain kept *R. similis* numbers significantly lower in banana roots after the initial invasion stage. All strains also showed an in vitro repellent effect towards the *R. similis*. *P. fluorescens* significantly reduced *R. similis* invasion by 50%.

10.5.1.2 Grapevine

Application of *P. fluorescens* at 4 g/plant significantly reduced root galling (40%) and egg mass production (60%) and increased root colonisation by *P. fluorescens* and fruit yield in grapevine (166% over control) (Shanthi et al. 1998).

10.5.2 Vegetable Crops

10.5.2.1 Potato

DAPG-producing *P. fluorescens* F113 is proposed as a potential biocontrol inoculant for the protection of potato crop against the cyst nematode, *Globodera rostochiensis*.

10.5.2.2 Tomato

Shanthi and Sivakumar (1995) reported the increased plant growth of tomato with *P. fluorescens* strain Pf-1. Bacterial treatment also reduced the level of infestation by the root-knot nematode, which was concentration dependent. The suppressing ability of the bacterial strains appeared to be related to their root colonising ability.

Three *P. fluorescens* strains and the type strain *P. putida* inhibited invasion of *Meloidogyne* spp. in tomato roots. All strains also showed an in vitro repellent effect towards *Meloidogyne* spp.

10.5.2.3 Sugar Beet

P. fluorescens had induced systemic resistance and inhibited early root penetration of *Heterodera schachtii,* the cyst nematode in sugar beet (Oostendorp and Sikora 1989).

10.5.3 Spice Crops

10.5.3.1 Turmeric

Turmeric rhizome treatment with *P. fluorescens* at 10 g/kg increased plant growth characters (58 and 31% increase in pseudostem height and number of tillers over control) and decreased root-knot nematode population (40%) and gall indices (25%) (Seenivasan et al. 2001).

PGPR effective for the management of plant parasitic nematodes have been presented in Table 10.3.

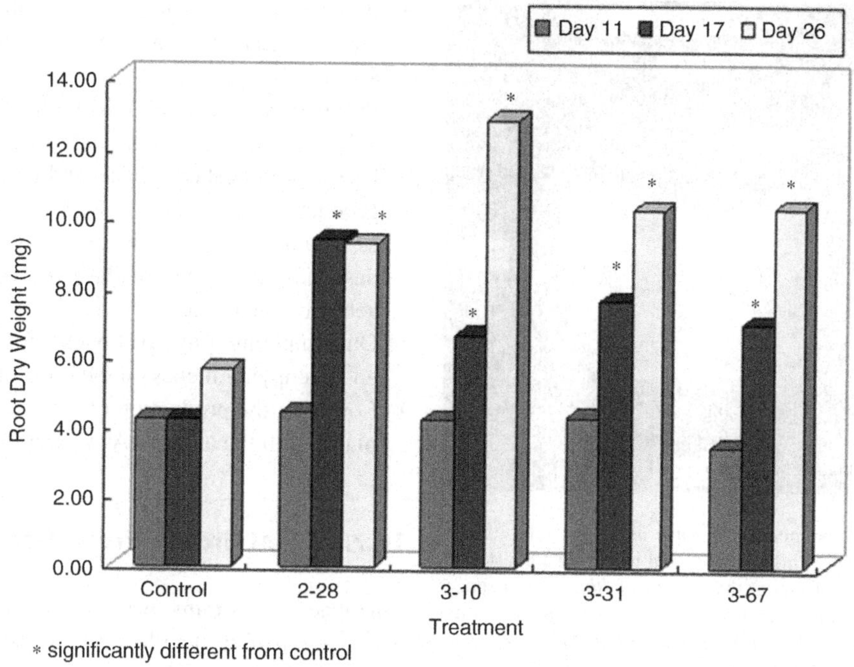

Fig. 10.2 Example of growth promotion of lentil following inoculation with PGPR isolates, 2–28, 3–10, 3–31 and 3–67. Plants were grown in containers in a growth chamber and sampled 11, 17 and 26 days following inoculation

10.6 Mode of Action

PGPR enhance plant growth by direct and indirect means, but the specific mechanisms involved have not all been well characterised (Glick 1995; Kloepper 1993). Direct mechanisms of plant growth promotion by PGPR can be demonstrated in the absence of plant pathogens (Fig. 10.2) or other rhizosphere microorganisms, whilst indirect mechanisms involve the ability of PGPR to reduce the deleterious effects of plant pathogens on crop yield. PGPR have been reported to directly enhance plant growth by a variety of mechanisms: fixation of atmospheric nitrogen that is transferred to the plant, production of siderophores that chelate iron and make it available to the plant root, solubilisation of minerals such as phosphorus and synthesis of phytohormones (Glick 1995). Direct enhancement of mineral uptake due to increases in specific ion fluxes at the root surface in the presence of PGPR has also been reported (Bashan and Levanony 1991; Bertrand et al. 2000). PGPR strains may use one or more of these mechanisms in the rhizosphere. Molecular approaches using microbial and plant mutants altered in their ability to synthesise or respond to specific phytohormones have increased our understanding of the role of phytohormone synthesis as a direct mechanism of plant growth enhancement by PGPR (Glick 1995; Persello-Cartieaux et al. 2003). PGPR that synthesise auxins and cytokinins or that interfere with plant ethylene synthesis have been identified (Garcia de Salamone et al. 2001; Glick 1995; Persello-Cartieaux et al. 2003).

PGPR that indirectly enhance plant growth via suppression of phytopathogens do so by a variety of mechanisms. These include the ability to produce siderophores that chelate iron, making it unavailable to pathogens; the ability to synthesise antifungal metabolites such as antibiotics (Fig. 10.3), fungal cell wall-lysing enzymes, or hydrogen cyanide, which suppress the growth of fungal pathogens; the ability to successfully compete with pathogens for nutrients or specific niches on the root; and the ability to induce systemic resistance (Bloemberg and Lugtenberg 2001; Glick 1995; Persello-Cartieaux et al. 2003). Biochemical and molecular approaches are providing new insight into the genetic basis of these traits, the biosynthetic pathways involved,

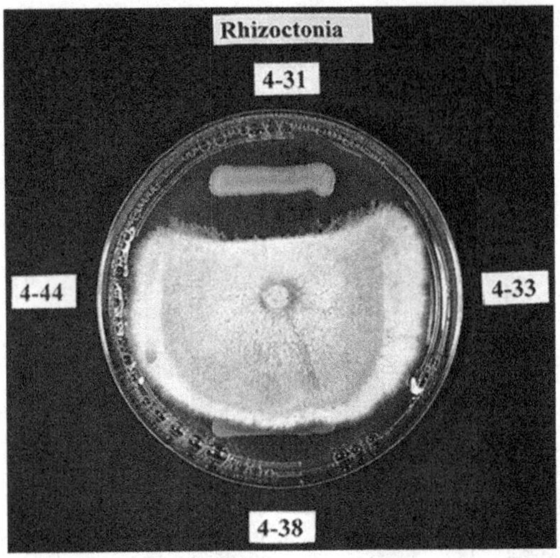

Fig. 10.3 Example of in vitro assay for inhibition of fungal growth. Different bacterial isolates were tested for their ability to inhibit the growth of *Rhizoctonia* spp., a soil-borne plant pathogen of legumes. A zone of inhibition can be observed around isolate 4–31 in the *upper* quadrant of the plate

their regulation and importance for biological control in laboratory and field studies (Bloemberg and Lugtenberg 2001; Bowen and Rovira 1999; Glick 1995; Persello-Cartieaux et al. 2003).

10.7 Biocontrol Mechanisms of PGPR

PGPR exhibit multiple numbers of mechanisms to promote plant growth and to serve as potential biocontrol agents. Generally, PGPR traits associated with the biocontrol of plant pathogens include:

- Atmospheric nitrogen fixation and its supply to plants
- Synthesising various phytohormones including auxins and cytokinins
- Providing mechanisms for the solubilisation of minerals such as phosphorus
- Antibiotic synthesis
- Secretion of iron binding siderophores to obtain soluble iron from the soil and provide it to a plant, thereby depriving fungal pathogens in the vicinity, of soluble iron
- Production of low molecular weight metabolites such as hydrogen cyanide with antifungal activity
- Production of enzymes including chitinase, β-1, 3-glucanase, protease and lipase which can lyse some fungal cells (Chet and Inbar 1994)
- Production of oxidative stress enzymes such as catalases, superoxide dismutases, peroxidase and polyphenol oxidases for scavenging active oxygen species
- Outcompeting phytopathogens for nutrients and occupying niches on the root surface
- Lowering the production of stress ethylene in plants with the enzyme ACC deaminase

10.7.1 Plant Growth Promotion

Rhizobacterial strains were found to increase plant growth after inoculation in seeds and therefore called 'plant growth-promoting rhizobacteria' (Kloepper et al. 1980). The mechanisms of growth promotion by these PGPR are complex and appear to comprise both changes in the microbial balance in the rhizosphere and alterations in host plant physiology. PGPR, including fluorescent pseudomonads, are capable of surviving and colonising the rhizosphere of all field crops. They promote plant growth by secreting auxins, gibberellins and cytokinins. PGPR has a significant impact on plant growth and development in either indirect or direct ways. Indirect promotion of plant growth occurs when bacteria prevent some of the deleterious effects of a phytopathogenic organism by one or more mechanisms. On the other hand, the direct promotion of plant growth by PGPR generally entails providing the plant with compound that is synthesised by the bacterium or facilitating the uptake of nutrients from the environment (Glick 1995). Plant growth benefits due to the addition of PGPR include increase in germination rates; root growth; yield including grain, leaf area, chlorophyll content, magnesium, nitrogen and protein content; hydraulic activity; tolerance to drought and salt stress; shoot and root weights; and delayed leaf senescence. Seed treatment with PGPR resulted in increased yield and growth in potato under field conditions (Kloepper et al. 1980). Van Peer and Schippers (1989) documented the increased root and shoot fresh weight of tomato, cucumber, lettuce and potato as a result of bacterisation with Pseudomonas strains.

Amongst PGPR isolates tested against some seed-borne fungal diseases, *B. pumilus* (SE-34), *B. pasteurii* (T4) and *B. subtilis* (IN 937-6 and GB-03) strains stood first in the improvement of crop, both in greenhouse and field conditions. Potential strains increased the biomass of plants, total number of leaves, fruits, length, girth and biomass of the fruit. The colonisation of these bacterial strains reduced the incidence of seed mycoflora which indirectly enhanced the per cent seed germination and vigour index of seedlings. Seed treatment with four isolates of rhizobacteria (namely, RS29, RS39, RS41 and RP3) isolated from different rhizotic zones of pigeonpea resulted in 90% seed germination in contrast with 50% obtained in untreated control after 72 h of incubation, and the isolates RS34, ER17, RP7 and RS41 increased shoot height and shoot dry biomass as compared to uninoculated control, whereas the isolates RS45, RS36, RS37, ER23 and RP24 influenced root dry biomass significantly.

10.7.2 Siderophore Production

Siderophores are low molecular weight, extracellular compounds with a high affinity for ferric iron that are secreted by microorganisms to take up iron from the environment, and their mode of action in suppression of disease was thought to be solely based on competition for iron with the pathogen. Fluorescent pseudomonas are characterised by the production of yellow-green pigments termed pyoverdines which fluoresce under UV light and function as siderophores. The siderophores produced by fluorescent pseudomonads promoted plant growth. The siderophores of fluorescent pseudomonads were later reported to be implicated in the suppression of plant pathogen's competition for iron between pathogens, and siderophores of fluorescent pseudomonads have been implicated in the biocontrol of wilt diseases caused by *Fusarium*, damping off of cotton caused by *Pythium ultimum* and *Pythium* root rot of wheat. Pyoverdines chelate iron in the rhizosphere and deprive pathogens of iron which is required for their growth and pathogenesis. Rhizobacteria produce various types of siderophores (pseudobactin and ferrooxamine B) that chelate the scarcely available iron and thereby prevent pathogens from acquiring iron.

The involvement of siderophore production in disease suppression by Pseudomonas strain WCS 358 was studied on carnation, radish and flax using *F. oxysporum* f. sp. *dianthi*, *F. oxysporum* f. sp. *raphani* and *F. oxysporum* f. sp. *lini*, respectively. Siderophore production by a plant growth-promoting fluorescent *Pseudomonas* sp. RBT 13 was found effective against several fungal and bacterial pathogens. Five strains of fluorescent pseudomonads exhibited growth promotion of lentil and biocontrol of wilt caused by *F. oxysporum* f. sp. *lini* with siderophore production as the mechanism. Several strains of siderophore producing *P. fluorescens* have been shown to inhibit *F. oxysporum* f. sp. *cubense*, *F. oxysporum* f. sp. *vasinfectum*, *Rhizoctonia solani* and *Acrocylindrium oryzae*.

Pseudomonas culture and purified siderophores showed good antifungal activity against the plant deleterious fungi, namely, *Aspergillus niger*, *A. flavus*, *A. oryzae*, *F. oxysporum* and *Sclerotium rolfsii*. Though siderophores are part of primary metabolism (iron is an essential element), on occasions, they also behave as antibiotics which are commonly considered to be secondary metabolites. The siderophores exerted maximum impact on *F. oxysporum* than on *Alternaria* sp. and *Colletotrichum capsici*.

The fluorescent pseudomonads had the property to form ferric siderophores complex which prevent the availability of iron to the microorganisms. Ultimately, this led to iron starvation and prevented the survival of the microorganisms including nematodes. *Pseudomonas aeruginosa* strain IE-6 and its streptomycin resistant strain IE-6+ markedly suppressed nematode population densities in root and subsequent root-knot development. The iron concentration in soil was lowered by the addition of an iron chelator.

10.7.3 Hydrogen Cyanide (HCN) Production

The cyanide ion is exhaled as HCN and metabolised to a lesser degree into other compounds. HCN first inhibits the electron transport, and the energy supply to the cell is disrupted, leading to the death of the organisms. It inhibits proper functioning of enzymes and natural receptor's reversible mechanism of inhibition, and it is also known to inhibit the action of cytochrome oxidase. HCN is produced by many rhizobacteria and is postulated to play a role in biological control of pathogens. Production of HCN by certain strains of fluorescent pseudomonads has been involved in the suppression of soil-borne pathogens. Suppression of black root rot of tobacco and take all of wheat by *P. fluorescens* strain CHAO was attributed to the production of HCN. *P. fluorescens* HCN inhibited the mycelial growth of *Pythium* in vitro. The cyanide producing strain CHAO stimulated root hair formation, indicating that the strain induced and altered plant physiological activities. Four of the six PGPR strains that induced systemic resistance in cucumber against *Colletotrichum orbiculare* produced HCN. Fluorescent Pseudomonas strain RRS1 isolated from Rajanigandha (tuberose) produced HCN, and the strain improved seed germination and root length. Pessi and Haas (2000) reported that low oxygen concentrations are a prerequisite for the activity of the transcription factor *ANR* which positively regulates HCN biosynthesis.

HCN from *P. fluorescens* strain CHAO not repressed by fusaric acid played a significant role in disease suppression of *F. oxysporum* f. sp. *radicis-lycopersici* in tomato. HCN is a broad-spectrum antimicrobial compound involved in biological control of root disease by many plant-associated fluorescent pseudomonads. Further, the enzyme HCN synthase is encoded by three biosynthetic genes (*henA, henB* and *henC*).

10.7.4 Indole-3-Acetic Acid (IAA) Production

IAA is a phytohormone which is known to be involved in root initiation, cell division and cell enlargement. This hormone is very commonly produced by PGPR and implicated in the growth promotion by PGPR. However, the effect of IAA on plants depends on the plant sensitivity to IAA and the amount of IAA produced from plant-associated bacteria and induction of other phytohormones. The bacterial IAA from *P. putida* played a major role in the development of host plant root system. Similarly, IAA production in *P. fluorescens* HP 72 correlated with suppressing of creeping bent grass brown patch.

10.7.5 Antibiosis

Antibiotics are generally considered to be organic compounds of low molecular weight produced by microbes. Antibiosis plays an active role in the biocontrol of plant disease, and it often acts in concert with competition and parasitism. Antibiosis has been postulated to play an important role in disease suppression by rhizobacteria. A variety of secondary metabolites listed in below are produced by fluorescent pseudomonads that exhibit antimicrobial properties (Table 10.4).

Fluorescent pseudomonad strains are known to reduce fungal growth in vitro by the production of one or more antifungal antibiotics that may also have activity in vivo. Several strains of *Pseudomonas* spp. have been shown to produce wide array of antibiotics which include 2, 4-diacetyl phloroglucinol, hydrogen cyanide, kanosamine, phenazine-1-carboxylic acid, pyoluteorin, oomycin A, pyrrolnitrin, pyocyanin and viscosinamide as well as several other uncharacterised moieties. Phloroglucinols, phenazines, pyoluteorin, pyrrolnitrin and cyclic lipopeptides all of which are diffusible, and hydrogen cyanide is volatile in nature.

Root-associated fluorescent pseudomonads produce and excrete secondary metabolites which are inhibitory to plant pathogenic organisms including fungi, bacteria and nematodes. Amongst these metabolites, the polyketide compound, DAPG, has received particular attention because of its broad-spectrum antifungal, antibacterial and antihelminthic activity. Phenazines (PHZ) are N-containing heterocyclic pigments synthesised by species of *Pseudomonas, Streptomyces, Burkholderia* and *Brevibacterium*. Pyrrolnitrin

Table 10.4 Secondary metabolites produced by fluorescent pseudomonads effective in biocontrol of pathogens

Metabolite	*Pseudomonas* sp.	Pathogen controlled
Pyoluteorin	*P. fluorescens* Pf-5	*Pythium*
Pyrrolnitrin	*P. fluorescens* Pf-5	*Pythium*
	P. cepacia	*Rhizoctonia solani*
	P. fluorescens PEM-2	*R. solani*
	P. fluorescens BL915	*R. solani*
Phenazine-1- carboxylic acid (PCA)	*P. fluorescens* 2-79	*Gaeumannomyces*
	P. aeruginosa	*R. solani*
2,4 Diacetylphloroglucinol (DAPG)	*P. aureofasciens*	*Gaeumannomyces*
	P. cepacia	*R. solani*
	P. fluorescens Q2-87	*Gaeumannomyces*
Oomycin A	*P. fluorescens*	*Pythium ultimum*

(PRN) is a broad-spectrum antifungal metabolite produced by many fluorescent and non-fluorescent strains of the genus *Pseudomonas*. A phenylpyrrol derivative of PRN has been developed as an agricultural fungicide. Pyrrolnitrin persists actively in the soil for at least 30 days. Pyoluteorin (PLT) is an aromatic polyketide antibiotic consisting of a resorcinol ring derived through polyketide biosynthesis. PLT is produced by several *Pseudomonas* spp. including strains that suppress plant diseases caused by phytopathogenic fungi (Maurhofer et al. 1994). PLT mainly inhibits the oomycetous fungi including *Pythium ultimum* against which it is strongly active when applied to seeds. PLT-producing pseudomonads decrease the severity of *Pythium* damping off. *P. fluorescens* strain CHAO, and its antibiotic overproducing derivative CHAO/PME 3424 repeatedly reduced *M. incognita* galling in tomato, brinjal, mung and soybean in early growth stage. A strong negative correlation existed between rhizobacteria colonisation and nematode invasion.

10.7.6 Induced Systemic Resistance (ISR)

Induced resistance is defined as an enhancement of the plant's defensive capacity against a broad spectrum of pathogens and pests that is acquired after appropriate stimulation. The resulting elevated resistance due to an inducing agent upon infection by pathogen is called induced systemic resistance (ISR) or systemic acquired resistance (SAR). The induction of systemic resistance by rhizobacteria is referred to as ISR, whereas that by other agencies is called SAR. Once resistance is induced, it will afford non-specific protection against pathogenic fungi, bacteria, nematodes and viruses, as well as against insect pests.

A large number of defence enzymes that have been associated with ISR include phenylalanine ammonia lyase (PAL), chitinase, β-1, 3-glucanase, peroxidase (PO), polyphenol oxidase (PPO), superoxide dismutase (SOD), catalase (CAT), lipoxygenase (LOX), ascorbate peroxidase (APX) and proteinase inhibitors. These enzymes also bring about liberation of molecules that elicit the initial steps in induction of resistance, phytoalexins and phenolic compounds.

Induced systemic resistance by PGPR has been achieved in large number of crops including *Arabidopsis*, cucumber, tomato, potato, radish, carnation, sugarcane, chilli, brinjal, rice and mango against broad spectrum of pathogens including fungi, bacteria, nematodes and viruses. Seed treatment and seedling root dipping induced early and enhanced levels of PO in rice plants. Two peroxidase isoforms were induced in the PGPR-treated rice plants inoculated with the sheath blight pathogen, *R. solani* (Nandakumar et al. 2001). High level expression of PO was reported in *P. fluorescens* Pf1 treated chilli plants challenged with *C. capsici*.

Similarly, increased activity of PPO was observed in PGPR-treated tomato plants challenged with *F. oxysporum* f. sp. *lycopersici* (Ramamoorthy et al. 2002). Plants treated with Pseudomonas strains initially showed higher level of PAL compared to control. Seedling dip with talc-based formulation of *P. fluorescens* induced the activity of PAL in finger millet leaves against blast disease. The inoculation of PGPR strains *P. putida* 89B-27 and *Serratia marcescens* 90-166 and the pathogen, *F. oxysporum* f. sp. *cucumerinum*, on separate halves of roots of cucumber seedlings exhibited that both PGPR strains induced systemic resistance against the *Fusarium* wilt as expressed by delayed disease symptom development and reduced number of dead plants. The same PGPR strains also induced systemic resistance in cucumber against bacterial angular leaf spot caused by *Pseudomonas syringae* pv. *lachrymans*.

Maize plants raised from *P. fluorescens*-treated seeds showed higher activity of peroxidase, polyphenol oxidase and PAL, when leaf sheaths were inoculated with the pathogen, *R. solani*. The bacterised seeds with *P. fluorescens* lead to accumulation of higher phenolic compounds and higher activity of PO, PPO and PAL that may play a role in defence mechanism in plants against pathogen. The control of nematode diseases in tomato and bell pepper was achieved by treatments with PGPR strains through induction of systemic resistance. Siddiqui and Shaukat (2002) observed that the application of PGPR strains to one half of the split root system of tomato caused a significant reduction (42%) in nematode penetration in the other half of the split root system, and this was attributed to ISR activity of the strain.

Amongst the PGPR strains, *B. subtilis* strain GB3 was the most effective in providing significant suppression of bacterial spot and was well correlated with increased activity of defence-related enzymes, namely, peroxidase and PAL. PGPR that were effective in greenhouse were also able to induce resistance in tomato against bacterial spot under field conditions.

10.8 Challenges in PGPR Research

10.8.1 Selection and Characterisation of PGPR

One of the challenges in developing PGPR for commercial application is ensuring that an effective selection and screening procedure is in place, so that the most promising organisms are identified and brought forward. In the agricultural chemical industry, thousands of prospective compounds are screened annually in efficient high-throughput assays to select the best one or two compounds for further development. Similar approaches are not yet in place for PGPR. Effective strategies for initial selection and screening of rhizobacterial isolates are required. It may be important to consider host plant specificity or adaptation to a particular soil, climatic conditions or pathogen in selecting the isolation conditions, and screening assays (Bowen and Rovira 1999; Chanway et al. 1989). The spermosphere model, an enrichment technique that relies on seed exudates as the nutrient source, has been used for selection and isolation of promising N_2-fixing rhizosphere bacteria from rice (Thomas-Bauzon et al. 1982). One approach for selection of organisms with the potential to control soil-borne phytopathogens is to isolate from soils that are suppressive to that pathogen (Weller et al. 2002). Other approaches involve selection based on traits known to be associated with PGPR such as root colonisation (Silva et al. 2003), 1-aminocyclopropane-1-carboxylate (ACC) deaminase activity (Cattelan et al. 1999; Glick 1995) and antibiotic (Giacomodonato et al. 2001) and siderophore production (Cattelan et al. 1999). The development of high-throughput assay systems and effective bioassays will facilitate selection of superior strains (Mathre et al. 1999; McSpadden Gardener and Fravel 2002).

10.8.2 Field Application of PGPR

One of the challenges of using PGPR is natural variation. It is difficult to predict how an organism may respond when placed in the field (compared to the controlled environment of a laboratory). Another challenge is that PGPR are living organisms. They must be able to be propagated artificially and produced in a manner to optimise their viability and biological activity until field application. Like rhizobia, PGPR bacteria will not live forever in a soil, and over time growers will need to re-inoculate seeds to bring back populations.

The application of PGPR for control of fungal pathogens in greenhouse systems shows considerable promise (Paulitz and Belanger 2001), due in part to the consistent environmental conditions and high incidence of fungal disease in greenhouses. Achieving consistent performance in the field where there is heterogeneity of abiotic and biotic factors and competition with indigenous organisms is more difficult. Knowledge of these factors can aid in determination of optimal concentration, of timing and placement of inoculant and of soil and crop management strategies to enhance survival and proliferation of the inoculant (Bowen and Rovira 1999; McSpadden Gardener and Fravel 2002). The concept of engineering or managing the rhizosphere to enhance PGPR function by manipulation of the host plant, substrates for PGPR, or through agronomic practices, is gaining increasing attention (Bowen and Rovira 1999; Mansouri et al. 2002). Development of better formulations to ensure survival and activity in the field and compatibility with chemical and biological seed treatments is another area of focus; approaches include optimisation of growth conditions prior to formulation and development of improved carriers and application technology (Bashan 1998; Bowen and Rovira 1999; Date 2001; Mathre et al. 1999; Yardin et al. 2000).

10.8.3 Commercialisation of PGPR

Prior to registration and commercialisation of PGPR products, a number of hurdles must be overcome (Fravel et al. 1999; Mathre et al. 1999; McSpadden Gardener and Fravel 2002). These include scale-up and production of the organism under commercial fermentation conditions whilst maintaining quality, stability and efficacy of the product. Formulation development must consider factors such as shelf life, compatibility with current application practices, cost and ease of application. Health and safety testing may be required to address such issues as non-target effects on other organisms including toxigenicity, allergenicity and pathogenicity, persistence in the environment and potential for horizontal gene transfer. The product claim, whether as a fertiliser supplement or for biological control, will determine to which federal agency applications for registration should be addressed. Capitalisation costs and potential markets must be considered in the decision to commercialise. McSpadden Gardener and Fravel (2002) estimated that a minimum capitalisation of US $1 million is required to register a biopesticide product in North America.

PGPR with wide scope for commercialisation includes *Pseudomonas fluorescens, P. putida, P. aeruginosa, Bacillus subtilis* and other *Bacillus* spp. The potential PGPR isolates are formulated using different organic and inorganic carriers either through solid or liquid fermentation technologies. They are delivered either through seed treatment, bio-priming, seedling dip, soil application, foliar spray, fruit spray, hive insert, sucker treatment and sett treatment. Application of PGPR formulations with strain mixtures perform better than individual strains for the management of pest and diseases of crop plants, in addition to plant growth promotion. Supplementation of chitin in the formulation increases the efficacy of antagonists. More than 33 products of PGPR have been registered for commercial use in greenhouse and field in North America. Though PGPR has a potential scope in commercialisation, the threat of certain PGPR (*P. aeruginosa*, *P. cepacia* and *B. cereus*) to infect human beings as opportunistic pathogens has to be clarified before large scale acceptance, registration and adoption of PGPR for pest and disease management.

10.8.4 Problems Encountered in Commercialisation

The challenges encountered in PGPR development and utilisation includes:

- Lack of suitable screening protocol for the selection of promising candidate of PGPR
- Lack of sufficient knowledge on the microbial ecology of PGPR strains and plant pathogens
- Optimisation of fermentation technology and mass production of PGPR strains
- Inconsistent performance and poor shelf life
- Lack of patent protection
- Prohibitive registration cost (Schisler and Slininger 1997; Fravel et al. 1998, 1999)
- Awareness, training and education shortfalls
- Lack of multidisciplinary approach

- Technology constraints (Sabitha et al. 2001)

Amongst several PGPR strains, *Bacillus*-based products gain momentum for commercialisation. Because, *Bacillus* spp. produce endospores tolerant to extremes of abiotic environments such as temperature, pH, pesticides and fertilisers (Backman et al. 1997). Owing to the potentiality of *Bacillus* spp., 18 different commercial products of *Bacillus* origin are sold in China to mitigate soil-borne diseases (Backman et al. 1997). The registered commercial products of PGPR are listed in Table 10.5.

10.9 Development of Formulations

PGPR formulation development is essential for easy application, storage, commercialisation and field use.

10.9.1 Features of an Ideal Formulation

- Increased shelf life.
- Not phytotoxic to crop plants.
- Dissolve well in water and should release bacteria.
- Tolerate adverse environmental conditions.
- Cost-effective and should give reliable control of plant diseases.
- Compatible with other agrochemicals.
- Carriers must be cheap and readily available for formulation development (Jeyarajan and Nakkeeran 2000).

10.9.2 Carriers

Carriers should support the survival of bacteria for a considerable length of time. Carriers may be either organic or non-organic. They should be economical and easily available.

10.9.2.1 Organic/Non-organic Carriers

The organic carriers used for formulation development include peat, turf, talc, lignite, kaolinite, pyrophyllite, zeolite, montmorillonite, alginate, press mud, sawdust and vermiculite. Carriers increase the survival rate of bacteria by protecting it from desiccation and death of cells (Heijnen et al. 1993). The shelf life of bacteria varies, depending upon bacterial genera, carriers and their particle size. Survival of *P. fluorescens* (2-79RN10, W4F393) in montmorillonite, zeolite and vermiculite with smaller particle size increased the survival rate than in kaolinite, pyrophyllite and talc with bigger particle size. The carriers with smaller particle size have increased surface area, which increases resistance to desiccation of bacteria by the increased coverage of bacterial cells (Dandurand et al. 1994).

10.9.3 Formulations

Formulations of fluorescent pseudomonads were developed through liquid fermentation technology. The fermentor biomass was mixed with different carrier materials (talc/peat/kaolinite/lignite/vermiculite) and stickers (Vidhyasekaran and Muthamilan 1995). Krishnamurthy and Gnanamanickam (1998) developed talc-based formulation of *P. fluorescens* for the management of rice blast caused by *Pyricularia grisea*, in which methyl cellulose and talc were mixed at 1:4 ratio and blended with equal volume of bacterial suspension at a concentration of 10^{10} cfu/ml. Nandakumar et al. (2001) developed talc-based strain mixture formulation of fluorescent pseudomonads. It was prepared by mixing equal volume of individual strains and blended with talc as per Vidhyasekaran and Muthamilan (1995). Talc-based strain mixtures were effective against rice sheath blight and increased plant yield under field conditions than the application of individual strains. Talc- and peat-based formulations of *P. chlororaphis* and *B. subtilis* were prepared and used for the management of turmeric rhizome rot (Nakkeeran et al. 2004).

One school of thought explains that CMC is added as a sticker at 1:4 ratio to talc. Though it is effective in disease management, it would lead to the increase in the production cost, which would prevent the growers to adopt the technology. Moreover, another school of thought explain that CMC and talc should be used at 1:100 ratios.

Table 10.5 Commercial products of PGPR in plant disease management

Product	Target pathogens/diseases	Crops recommended	Manufacturer
Biosave 10, 11, 100, 110, 1000 – *P. syringae* ESC-100	*Botrytis cinerea, Penicillium* spp., *Mucor pyroformis, Geotrichum candidum*	Pome fruit (Biosave 100), citrus (Biosave 1000)	Eco Science Corp, Orlando, FL, USA
Blight Ban A506 – *P. fluorescens* A 506	*Erwinia amylovora and* russet-inducing bacteria	Almond, apple, apricot, blueberry, cherry, peach, pear, potato, strawberry, tomato	Plant Health Technologies, USA
Cedomon – *P. chlororaphis*	Leaf stripe, net blotch, *Fusarium* sp., spot blotch, leaf spot and others	Barley and oats, potential for wheat, other cereals	Bio Agri AB, Sweden
Campanion – *B. subtilis* GB03	*Rhizoctonia, Pythium, Fusarium, Phytophthora*	Horticultural crops, turf	Growth Products, USA
Conquer – *P. fluorescens*	*P. tolassii*	Mushrooms	Mauri Foods, Australia
Victus – *P. fluorescens*	*P. tolassii*	Mushrooms	Mauri Foods, Australia
BioJect Spotless – *P. aureofaciens*	Dollar spot, anthracnose and *P. aphanidermatum*	Turf, other crops	Eco Soil Systems, San Diego, CA, USA
BioJet – *Pseudomonas* sp. + *Azospirillum*	Brown blotch, dollar spot disease	Turf, other crops	Eco Soil Systems, San Diego, CA, USA
Deny – *Burkholderia Cepacia (Pseudomonas cepacia)*	*Rhizoctonia, Pythium, Fusarium*, diseases caused by lesion, spiral, lance and sting nematodes	Alfalfa, barley, beans, clover, cotton, peas, sorghum, vegetable crops, wheat	Stine Microbial Products, Shawnee, KS, USA
Intercept – *P. cepacia*	*Rhizoctonia solani, Fusarium* sp., *Pythium* sp.	Maize, vegetables, cotton	Soil Technologies Corp, USA
Kodiak, Kodiak HB, Epic, Concentrate, Quantum 4000 and System 3 – *B. subtilis* GB03	*Rhizoctonia solani, Fusarium* spp., *Alternaria* spp., *Aspergillus* spp.	Cotton, legumes	Gustafson Inc, Dallas, USA
EcoGuard – *Bacillus licheniformis* SB3086	Dollar spot	Turf	Novozymes A/S, Denmark
Green Releaf – *B. licheniformis* SB3086	Leaf spot, blight	Ornamental turf, lawns, golf courses, ornamental plants, conifers and tree seedlings in outdoor, greenhouse and nursery sites	Novozymes A/S, Denmark
Bio Yield – *B. subtilis* + *B. amyloliquefaciens*	Broad-spectrum action against greenhouse pathogens	Tomato, cucumber, Pepper, tobacco	Gustafson Inc, Dallas, USA

(continued)

Table 10.5 (continued)

Product	Target pathogens/diseases	Crops recommended	Manufacturer
Taegro, Tae-Technical – *B. subtilis* var *amyloliquefaciens* FZB24	*Rhizoctonia*, *Fusarium*	Only in greenhouses and other indoor sites on shade and forest tree seedlings, ornamentals, shrubs	Earth Biosciences Inc, USA
Rhizo-Plus – *B. subtilis* strain FZB24	*R. solani*, *Fusarium* spp., *Sclerotinia*, *Verticillium*	Greenhouse grown crops, forest tree seedlings, ornamentals, shrubs	KFZB Biotechnik GMBH, Berlin, Germany
Serenade – *B. subtilis* strain QWT713. Available as wettable powder	Powdery mildew, downy mildew, Cercospora leaf spot, early blight, late blight, brown rot, fire blight	Cucurbits, grapes, hops, vegetables, peanuts, pome fruits, stone fruits	AgraQuest Inc, Davis, USA
Rhapsody – *B. subtilis* strain QST713. Aqueous suspension formulation	Powdery mildew, sour rot, downy mildew, early leaf spot, early blight, late blight, bacterial spot and walnut blight	Cherries, cucurbits, grapes, leafy vegetables, peppers, potatoes, tomatoes, walnuts	AgraQuest Inc, Davis, USA
Subtilex – *B. subtilis* MB1600	*Fusarium* spp., *Rhizoctonia* spp. and *Pythium* spp.	Ornamentals, vegetable crops	Becker Underwood, Ames, IA
HiStick N/T – *B. subtilis* MBI600	*Fusarium* spp., *Rhizoctonia* spp., *Aspergillus* spp.	Soybean, alfalfa, dry/snap beans, peanuts	Becker Underwood, Ames, IA, USA
GB 34 Concentrate – *B. pumilus*	*Rhizoctonia*, *Fusarium* which attack developing soybean roots	Soybean	Gustafson TX, USA
Sonata ASO – *B. pumilus* QST 2808	Fungal pests such as moulds, mildews, blights, rusts, oak death syndrome	Used in nurseries, landscapes, oak trees, greenhouse crops	AgraQuest Inc, Davis, USA
Ballad – *B. pumilus* QST2808	Rust	Soybean	AgraQuest Inc, Davis, CA, USA
YieldShield – *B. pumilus* GB34	Soil-borne fungal pathogens causing root diseases	Soybean	Gustafson Inc, TX, USA
System 3 – *B. subtilis* GB03 and chemical pesticides	Seedling pathogens	Barley, beans, cotton, peanut, pea, rice, soybean	Helena Chemical Co, Memphis, USA
AtEze – *P. chlororaphis* 63-28	*Pythium* spp., *Rhizoctonia solani*, *Fusarium oxysporum*	Ornamentals, vegetables	EcoSoil Systems Inc, San Diego, CA
Pix plus plant regulator, *B. cereus* BPO1 technical – *B. cereus* UW85	Used as growth regulator	Cotton	Micro Flo Company, Lakeland, FL, USA
Biosave 10LP, 110 – *P. syringae*	*Botrytis cinerea*, *Penicillium* spp., *Geotrichum candidum*	Pome fruit, citrus, cherries, potatoes	Eco Science Corp, FL, USA

Hence, feasibility of the technique and shelf life of the product has to be evaluated to make the technology as a viable component in disease management so as to promote organic farming.

10.9.3.1 Talc Formulation

Talc is a natural mineral referred as steatite or soapstone composed of various minerals in combination with chloride and carbonate. Chemically, it is referred as magnesium silicate [$Mg_3Si_4O_3(OH)_2$] and available as powder form from industries suited for wide range of applications. It has very low moisture equilibrium, relative hydrophobicity, chemical inertness and reduced moisture absorption and prevents the formation of hydrate bridges that enable longer storage periods. Owing to the inert nature of talc and easy availability as raw material from soapstone industries, it is used as a carrier for formulation development. Kloepper and Schroth (1981a) demonstrated the potentiality of talc to be used as a carrier for formulating rhizobacteria. The fluorescent pseudomonads did not decline in talc mixture with 20% xanthum gum after storage for two months at 4 °C. *P. fluorescens* isolate Pf1 survived up to 240 days in storage. The initial population of Pf1 in talc-based formulation was 37.5×10^7 cfu/g and declined to 1.3×10^7 cfu/g after 8 months of storage (Vidhyasekaran and Muthamilan 1995). Amendment of sucrose (0.72 M) in King's B medium increased population and shelf life of *P. fluorescens* (P7NF, TL3) in talc-based formulation up to 12 months (Caesar and Burr 1991). *P. putida* strain 30 and 180 survived up to 6 months in talc-based formulations. The population load at the end of sixth month was 10^8 cfu/g of the product (Bora et al. 2004).

10.9.3.2 Peat Formulation

Peat (turf) is a carbonised vegetable tissue formed in wet conditions by decomposition of various plants and mosses. It is formed by the slow decay of successive layers of aquatic and semi-aquatic plants, for example, sedges, reeds, rushes and mosses. Peat soils are used as carrier materials to formulate PGPR. Though peat carriers are cheap to use, it harbours lot of contaminants. The quality of peat is variable and not readily available worldwide. Sterilisation of peat through heat releases toxic substances to the bacteria and thereby reduce bacterial viability (Bashan 1998). Peat-based formulation of *Azospirillum brasilense* had a shelf life up to 4 months. The population load after 4 months of storage was 10^7 cfu/g of the product (Bashan 1998). This population was sufficient for successful plant inoculation. Vidhyasekaran and Muthamilan (1995) reported that the shelf life of *P. fluorescens* in peat-based formulation was maintained up to 8 months (2.8×10^6 cfu/g). Shelf life of *P. chlororaphis* (PA23) and *B. subtilis* (CBE4) in peat carriers was retained for more than six months (Kavitha et al. 2003; Nakkeeran et al. 2004).

10.9.3.3 Press Mud Formulation

Press mud is a by-product of sugar industries. It was composted using vermin-composting technique and later used as a carrier for *Azospirillum* spp. This carrier maximises the survival of *Azospirillum* spp. than lignite, which is predominantly used as a carrier material in India (Muthukumarasamy et al. 1999).

10.9.3.4 Vermiculite Formulation

Vermiculite is a light mica-like mineral used to improve aeration and moisture retention. It is widely used as potting mixture and used as a carrier for the development of formulations for harbouring microbial agents. Vermiculite-based formulation of *P. fluorescens* (Pf1) retained shelf life for a period of 8 months. The viable load of bacteria in the formulation was 1×10^6 cfu/g (Vidhyasekaran and Muthamilan 1995). Shelf life of *Azospirillum* in vermiculite-based formulation was retained up to 10 months. The viable cells after 44 weeks of storage were 1.3×10^7 cfu/g (Saleh et al. 2001).

10.9.3.5 Microencapsulation

Microcapsules of rhizobacteria consist of a cross-linked polymer deposited around a liquid phase, where bacteria are dispersed. Microparticles are characterised based on the distribution of particle size, morphology and bacterial load. The process of microencapsulation involves mixing of gelatin polyphosphate polymer pair (81:19 w/w) at acidic pH with rhizobacteria suspended in oil (Charpentier et al. 1999). Though rhizobacteria has been formulated through microencapsulation method, its shelf life declines at a faster rate since polymers serve as a barrier for oxygen. This was later improved by developing microcapsules by spray drying. The release of *P. fluorescens–P. putida* from the microencapsulated pellets occurred after 15 min immersion in aqueous buffer. It showed that water served as triggering material for the bacterial release (Charpentier et al. 1999).

Though microencapsulation aids in formulating bacteria, still the technology has to be well refined for early release of bacterial cells and for the establishment in the infection court to counter attack the establishment of pathogens. Most of the experiments on microencapsulation have been restricted only to lab. The technology should be standardised for the industrial application so that the technical feasibility could be assessed to popularise the same for field use.

10.10 Modes of Delivery

It is delivered through seed, soil, foliage, rhizomes and setts or through combination of several methods of delivery.

10.10.1 Seed Treatment

Seed treatment with cell suspensions of PGPR was effective against several diseases. Delivering of *Serratia marcescens* strain 90-166 as seed dip before planting and soil application of 100 ml of the same at the rate of 10^8 cfu/ml to the sterilised soil-less planting mix after seeding reduced bacterial wilt of cucumber and controlled cucumber beetles besides increasing the fruit weight (Zhender et al. 2001). Transfer of technology for commercial use could be possible if PGPR strains are made available as a product. After realisation of the same, several carriers were used for formulation development. Talc-based formulation of *P. fluorescens* Pf1 was coated onto seeds at the rate of 4 g/kg (10^7 cfu/g) of chickpea seeds (cv Shoba) for the management of chickpea wilt. Sowing of treated chickpea seeds resulted in establishment of rhizobacteria on chickpea rhizosphere (Vidhyasekaran and Muthamilan 1995). Treatment of cucumber seeds with strain mixtures comprising of *Bacillus pumilus* (INR7), *B. subtilis* (GB03) and *Curtobacterium flaccumfaciens* (ME1) with a mean bacterial density of 5×10^9 cfu/seed reduced intensity of angular leaf spot and anthracnose equivalent to the synthetic elicitor Actigard and better than seed treatment with individual strains (Raupach and Kloepper 1998). Treatment of pigeonpea seeds with talc-based formulation of *P. fluorescens* (Pf1) effectively controlled fusarial wilt of pigeonpea under greenhouse and field conditions (Vidhyasekaran et al. 1997). Soaking of rice seeds in water containing 10 g of talc-based formulation of *P. fluorescens* consisting mixture of Pf1 and Pf2 (10^8 cfu/g) for 24 h controlled rice sheath blight under field conditions (Nandakumar et al. 2001). Seed treatment of lettuce with either vermiculite- or kaolin-based carrier of *B. subtilis* (BACT-0) significantly reduced root rot caused by *P. aphanidermatum*, and it also increased the fresh weight of lettuce under greenhouse conditions. Seed treatment with vermiculite-based *P. putida* reduced Fusarium root rot of cucumber and increased the yield and growth of cucumber (Amer and Utkhede 2000). Treatment of tomato seeds with powder formulation of PGPR (*B. subtilis, B. pumilus*) reduced symptom severity of TMV and increased the fruit yield (Murphy et al. 2000).

10.10.2 Bio-priming

A successful antagonist should colonise rhizosphere during seed germination (Weller 1983). Priming with PGPR increases germination and improves seedling establishment. It initiates the physiological process of germination but prevents the emergence of plumule and radical. Initiation of physiological process helps in the establishment and proliferation of PGPR on the spermosphere (Taylor and Harman 1990). Bio-priming of seeds with bacterial antagonists increase the population load of antagonist to a tune of tenfold on the seeds, thus protecting rhizosphere from the ingress of plant pathogens (Callan et al. 1990). Chickpea seeds treated with talc-based formulation of Pf1 was primed by incubating the treated seeds for 20 h at 25°C over sterile vermiculite moistened with sterile water. Population of Pf1 increased up to 100% in the rhizosphere, indicating that it provides a congenial microclimate for proliferation and establishment of bacterial antagonist (Vidhyasekaran and Muthamilan 1995). Drum priming is a commercial seed treatment method followed to treat seeds with pesticides. Drum priming of carrot and parsnip seeds with *P. fluorescens* Pf CHAO proliferated well on the seeds and could be explored for realistic scale up of PGPR (Wright et al. 2003).

10.10.3 Seedling Dip

PGPR is delivered through various means for the management of crop diseases based on the survival nature of the pathogen. In several crops, pathogens gain entry into plants either through seed, root or foliage. In rice, sheath blight incited by *Rhizoctonia solani* is a major obstacle in rice production. As the pathogen is soil borne, it establishes host–parasite relationships by entering through root. Hence, protection of rhizosphere region by prior colonisation with PGPR will prevent the establishment of host–parasite relationship. Delivering of *P. fluorescens* strain mixtures by dipping the rice seedlings in bundles in water containing talc-based formulation of strain mixtures (20 g/l) for 2 h and later transplanting it to the main field suppressed sheath blight incidence (Nandakumar et al. 2001). Similarly, dipping of rice seedlings in talc-based formulation of *P. fluorescens* (PfALR1) prior to transplanting reduced sheath blight severity and increased yield in Tamil Nadu, India (Rabindran and Vidhyasekaran 1996). Dipping of strawberry roots for 15 min in bacterial suspension of *P. putida* (2×10^9 cfu/ml) isolated from strawberry rhizosphere reduced Verticillium wilt of strawberry by 11% compared to untreated control (Berg et al. 2001). Dipping of *Phyllanthus amarus* seedlings in talc-based formulation of *B. subtilis* (BSCBE4) or *P. chlororaphis* (PA23) for 30 min prior to transplanting reduced stem blight of *P. amarus* (Mathiyazhagan et al. 2004).

10.10.4 Soil Application

Soil being as the repertoire of both beneficial and pathogenic microbes, delivering of PGPR strains to soil will increase the population dynamics of augmented bacterial antagonists and thereby would suppress the establishment of pathogenic microbes onto the infection court. Vidhyasekaran and Muthamilan (1995) stated that soil application of peat-based formulation of *P. fluorescens* (Pf1) at the rate of 2.5 kg of formulation mixed with 25 kg of well-decomposed farm yard manure, in combination with seed treatment, increased rhizosphere colonisation of Pf1 and suppressed chickpea wilt caused by *F. oxysporum* f. sp. *ciceris*. Broadcasting of talc-based formulation of strain mixtures (Pf1 and FP7) by blending 2.5 kg of formulation with 50 kg of sand after 30 days of transplanting paddy seedlings to main field significantly reduced sheath blight and increased yield under field conditions (Nandakumar et al. 2001). Incorporation of commercial chitosan-based formulations LS254 (comprising of *Paenibacillus macerans* + *B. pumilus*) and LS255 (comprising of *P. macerans* + *B. subtilis*) into soil at the ratio of 1:40 (formulation: soil) increased bio-matter production by increasing both root and shoot length and yield (Vasudevan et al. 2002). Soil application of the strain mixture formulations LS256 and LS257 comprising of two different *Bacillus* spp. was better than seed treatment and suppressed downy mildew under greenhouse and field conditions (Niranjan Raj et al. 2003).

10.10.5 Foliar Spray

The efficacy of biocontrol agents for foliar diseases is greatly affected by fluctuation of microclimate. Phyllosphere is subjected to diurnal and nocturnal and cyclic and non-cyclic variation in temperature, relative humidity, dew, rain, wind and radiation. Hence, water potential of phylloplane microbes will be varying constantly. It will also vary between leaves or the periphery of the canopy and on sheltered leaves. Higher relative humidity could be observed in the shaded, dense region of the plant than that of peripheral leaves.

The dew formation is greater in centre and periphery. The concentration of nutrients like amino acids, organic acids and sugars exuded through stomata, lenticels, hydathodes and wounds varies highly. It affects the efficacy and survival of antagonist in phylloplane (Andrews 1992).

Delivering of Pseudomonas to beet leaves actively compete for amino acids on the leaf surface and inhibited spore germination of *Botrytis cinerea*, *Cladosporium herbarum* and *Phoma betae* (Blakeman and Brodie 1977). Application of *B. subtilis* to bean leaves decreased incidence of bean rust (*Uromyces phaseoli*) by 75% equivalent to weekly treatments with the fungicide mancozeb (Baker et al. 1983). Application of *P. fluorescens* on to foliage (1 kg of talc-based formulation/ha) on 30, 45, 60, 75 and 90 days after sowing reduced leaf spot and rust of groundnut under field conditions (Meena et al. 2002). Preharvest foliar application of talc-based fluorescent pseudomonads strain FP7 supplemented with chitin at fortnightly intervals (5 g/l; spray volume 20 l/tree) onto mango trees from preflowering to fruit maturity stage induced flowering to the maximum, reduced the latent infection by *C. gloeosporioides* beside increasing the fruit yield and quality (Vivekananthan et al. 2004). Though seed treatment and foliar application of *P. fluorescens* reduce the severity of rust and leaf spot under field conditions, it is not technically feasible due to increased dosage and economy realised from the crop. Hence, dosage and frequency of application has to be standardised based on the crop value, which could be as a reliable and practical approach.

10.10.6 Fruit Spray

Pseudomonas syringae (10% wettable powder) in the modified packing line was sprayed at the rate of 10 g/l over apple fruit to control blue and grey moulds of apple. The population of antagonist increased in the wounds more than tenfold during 3 months in storage (Janisiewicz and Jeffers 1997). Research on the exploration of PGPR have to go a long way to explore its usage to manage post-harvest diseases.

10.10.7 Hive Insert

Honey bees and bumble bees serve as vectors for the dispersal of biocontrol agents for the management of diseases of flowering and fruit crops (Sandhu and Waraich 1985; Kevan et al. 2003). An innovative method of application of biocontrol agent right in the infection court at the exact time of susceptibility was developed. A dispenser is attached to the hive and loaded with powder formulation of the PGPR or with other desired biocontrol agent. When the foragers exit the hive, the antagonist get dusted on to bee and delivered to the desired crop whilst attempting for sucking the nectar. *Erwinia amylovora* causing fire blight of apple infects through flower and develops extensively on stigma. Colonisation by antagonist at the critical juncture is necessary to prevent flower infection. Since flowers do not open simultaneously, the biocontrol agent *P. fluorescens* has to be applied to flowers repeatedly to protect the stigma. Nectar seeking insects like *Apis mellifera* can be used to deliver *P. fluorescens* to stigma. Bees deposit the bacteria on the flowers soon after opening due to their foraging habits. Honey bees have also been used for the management of grey mould of strawberry and raspberry (Peng et al. 1992; Sutton 1995; Kovach et al. 2000).

10.10.8 Sucker Treatment

Plant growth-promoting rhizobacteria also play a vital role in the management of soil-borne diseases of vegetatively propagated crops. The delivery of PGPR varies depending upon the crop. In banana, rhizobacteria are delivered through sucker treatment or rhizome treatment. Banana suckers were dipped in talc-based *P. fluorescens* suspension (500 g of the product in 50 l of water) for 10 min after paring and pralinage. Subsequently, it was followed by capsule application (50 mg of *P. fluorescens* per capsule) on the third and fifth month after planting. It resulted in 80.6% reduction in panama wilt of banana compared to control (Raguchander et al. 2000).

10.10.9 Sett Treatment

Red rot of sugarcane is a major production constraint in sugarcane cultivation. Usage of chemical fungicides for the management of red rot was less effective to protect the crop. Since PGPR act as a predominant prokaryote in the rhizosphere, fluorescent pseudomonads were explored for the management of red rot under field conditions. Viswanathan and Samiyappan (2002) delivered fluorescent pseudomonads through sett treatment. Two budded sugarcane setts were soaked in talc formulation of *P. fluorescens* (20 g/l) for one hour and incubated for 18 h prior to planting. Planting of treated setts increased cane growth and sugar recovery and reduced red rot incidence under field conditions.

10.10.10 Multiple Delivery Systems

Plant pathogens establish host–parasite relationships by entering through infection court such as spermosphere, rhizosphere and phyllosphere. Hence, protection of sites vulnerable for the entry and infection of pathogens would offer a better means for disease management. Seed treatment of pigeonpea with talc-based formulation of fluorescent pseudomonads at the rate of 4 g/kg of seed followed by soil application at the rate of 2.5 kg/ha at 0, 30 and 60 days after sowing controlled pigeonpea wilt incidence under field conditions. The additional soil application of talc-based formulation improved disease control and increased yield compared to seed treatment alone (Vidhyasekaran et al. 1997). Delivering of *P. fluorescens* as seed treatment followed by three foliar applications suppressed rice blast under field conditions (Krishnamurthy and Gnanamanickam 1998). Combined application of talc-based formulation of fluorescent pseudomonads comprising of Pf1 and FP7 through seed treatment, seedling dip, soil application and foliar spray suppressed rice sheath blight and increased plant growth better than application of the same strain mixture either through seed, seedling dip or soil (Nandakumar et al. 2001). Application of strain-mixture-based formulation of Pf1 and FP7 with or without chitin through seed, seedling dip and foliar spray suppressed leaf folder damage and sheath blight in rice under field conditions (Radja Commare et al. 2002). Seed and foliar application of talc-based fluorescent pseudomonads reduced leaf spot and rust of groundnut under field conditions (Meena et al. 2002). The increased efficacy of strain mixtures through combined application might be due to increase in the population of fluorescent pseudomonads in both rhizosphere and phyllosphere (Viswanthan and Samiyappan 1999). Delivering of rhizobacteria through combined application of different delivery systems will increase the population load of rhizobacteria and thereby might suppress the pathogenic propagules.

10.11 Future Prospects

Over the years, the PGPR have gained worldwide importance and acceptance for agricultural benefits. These microorganisms are the potential tools for sustainable agriculture and the trend for the future. Scientific researches involve multidisciplinary approaches to understand adaptation of PGPR to the rhizosphere, mechanisms of root colonisation, effects on plant physiology and growth, biofertilisation, induced systemic resistance, biocontrol of plant pathogens, production of determinants, etc. Biodiversity of PGPR and mechanisms of action for the different groups (diazotrophs, bacilli, pseudomonads, Trichoderma, AMF, rhizobia, phosphate-solubilising bacteria and fungi, lignin degrading, chitin degrading, cellulose-degrading bacteria and fungi) are demonstrated. Effects of physical, chemical and biological factors on root colonisation and the proteomics perspective on biocontrol and plant defence have also shown positive results. Visualisation of interactions of pathogens and biocontrol agents on plant roots using autofluorescent protein markers has provided more understanding of biocontrol processes with overall positive consequences.

As our understanding of the complex environment of the rhizosphere, of the mechanisms of action of PGPR and of the practical aspects of inoculant formulation and delivery increases, we can expect to see new PGPR products becoming available. The success of these products will depend on our ability to manage the rhizosphere to enhance survival and competitiveness of these beneficial microorganisms (Bowen and Rovira 1999). Rhizosphere management will require consideration of soil and crop cultural practices as well as inoculant formulation and delivery (Bowen and Rovira 1999; McSpadden Gardener and Fravel 2002). Genetic enhancement of PGPR strains to enhance colonisation and effectiveness may involve addition of one or more traits associated with plant growth promotion (Bloemberg and Lugtenberg 2001; Glick 1995; Lubeck et al. 2000). Genetic manipulation of host crops for root-associated traits to enhance establishment and proliferation of beneficial microorganisms (Mansouri et al. 2002; Smith and Goodman 1999) is being pursued. However, regulatory issues and public acceptance of genetically engineered organisms may delay their commercialisation. The use of multi-strain inocula of PGPR with known functions is of interest as these formulations may increase consistency in the field (Jetiyanon and Kloepper 2002; Siddiqui and Shaukat 2002). They offer the potential to address multiple modes of action, multiple pathogens and temporal or spatial variability.

PGPR offer an environmentally sustainable approach to increase crop production and health. The application of molecular tools is enhancing our ability to understand and manage the rhizosphere and will lead to new products with improved effectiveness.

References

Aalten PM, Gowen SR (1998) Entomopathogenic nematodes and fluorescent *Pseudomonas* rhizosphere bacteria inhibiting *Radopholus similis* invasion in banana roots. Brighton Crop Prot Conf Pests Disease 2:675–680

Ali NI, Siddiqui IA, Shahid J, Shaukat S, Zaki MJ (2002) Nematicidal activity of some strains of *Pseudomonas* spp. Soil Biol Biochem 34:1051–1058

Amer GA, Utkhede RS (2000) Development of formulations of biological agents for management of root rot of lettuce and cucumber. Can J Microbiol 46:809–816

Andrews JH (1992) Biological control in the phyllosphere. Annu Rev Phytopathol 30:603–635

Backman PA, Wilson M, Murphy JF (1997) Bacteria for biological control of plant diseases. In: Rechcigl NA, Rechecigl JE (eds) Environmentally safe approaches to crop disease control. Lewis Publishers, Boca Raton, pp 95–109

Baker SC, Stavely JR, Thomas CA, Sasser M, Mac Fall SJ (1983) Inhibitory effect of *Bacillus subtilis* on *Uromyces phaseoli* and on development of rust pustules on bean leaves. Phytopathology 73:1148–1152

Bashan Y (1998) Inoculants of plant growth-promoting bacteria for use in agriculture. Biotechnol Adv 16:729–770

Bashan Y, Levanony H (1991) Alterations in membrane potential and in proton efflux in plant roots induced by *Azosprillum brasilense*. Plant Soil 137:99–103

Baudoin E, Benizri E, Guckert A (2002) Impact of growth stage on the bacterial community structure along maize roots, as determined by metabolic and genetic finger printing. Appl Soil Ecol 19:135–145

Benhamou N, Kloepper JW, Tuzun S (1998) Induction of resistance against Fusarium wilt of tomato by combination of chitosan with an endophytic bacterial strain: Ultrastructural and cytochemistry of the host response. Planta 204:153–168

Benizri E, Baudoin E, Guckert A (2001) Root colonization by inoculated plant growth promoting rhizobacteria. Biocontrol Sci Technol 11:557–574

Berg G, Fritze A, Roskot N, Smalla K (2001) Evaluation of potential biocontrol rhizobacteria from different host plants of *Verticillium dahliae*. Kleb J Appl Microbiol 91:963–971

Berg G, Roskot N, Steidle A, Eberl L, Zock A, Smalla K (2002) Plant-dependent genotypic and phenotypic diversity of antagonistic rhizobacteria isolated from different *Verticillium* host plants. Appl Environ Microbiol 68:3328–3338

Bertrand H, Plassard C, Pinochet X, Toraine B, Normand P, Cleyet-Marel JC (2000) Stimulation of the ionic transport system in *Brassica napus* by a plant growth-promoting rhizobacterium (*Achromobacter* sp). Can J Microbiol 46:229–236

Blakeman JP, Brodie IDS (1977) Competition for nutrients between epiphytic microorganisms and germination of spores of plant pathogens on beet root leaves. Physiol Plant Pathol 10:29–42

Bloemberg GV, Lugtenberg BJJ (2001) Molecular basis of plant growth promotion and biocontrol by rhizobacteria. Curr Opin Plant Biol 4:343–350

Bloemberg GV, Wijfjes AHM, Lamers GEM, Stuurman N, Lugtenberg BJJ (2000) Simultaneous imaging of *Pseudomonas fluorescens* WCS365 populations expressing three different autofluorescent proteins in the rhizosphere: New perspectives for studying microbial communities. Mol Plant Microbe Interact 13:1170–1176

Bora T, Ozaktan H, Gore E, Aslan E (2004) Biological control of *Fusarium oxysporum* f sp *melonis* by wettable powder formulations of the two strains of *Pseudomonas putida*. J Phytopathol 152:471–475

Bowen GD, Rovira AD (1999) The rhizosphere and its management to improve plant growth. Adv Agron 66:1–102

Burelle K, Vavrina CS, Rosskopf EN, Shelby RA (2002) Field evaluation of plant growth promoting rhizobacteria amended transplant mixes and soil solarization for tomato and pepper production in Florida. Plant Soil 238:257–266

Caesar AJ, Burr TJ (1991) Effect of conditioning, betaine, and sucrose on survival of rhizobacteria in powder formulations. Appl Environ Microbiol 57:168–172

Callan NW, Mathre DE, Miller JB (1990) Bio-priming seed treatment for biological control of *Pythium ultimum* pre-emergence damping-off in *sh2* sweet corn. Plant Dis 74:368–372

Cattelan AJ, Hartel PG, Fuhrmann JJ (1999) Screening for plant growth-promoting rhizobacteria to promote early soybean growth. Soil Sci Soc Am J 63:1670–1680

Chahal VPS, Chahal PPK (1991) Control of *Meloidogyne incognita* with *Bacillus thuringiensis*. In: Wright RJ et al (eds) Plant and soil interaction, pp 677–680

Chahal VPS, Chahal PPK (2003) *Bacillus thuringiensis* for the control of *Meloidogyne incognita*. In: Trivedi PC (ed) Nematode management in plants. Scientific Publishers, Jodhpur, pp 251–257

Chakraborty U, Chakraborty B, Basnet M (2006) Plant growth promotion and induction of resistance in *Camellia sinensis* by *Bacillus megaterium*. J Basic Microbiol 46(Suppl 3):186–195

Chanway CP, Nelson LM, Holl FB (1989) Cultivar-specific growth promotion of spring wheat (*Triticum aestivum* L) by co-existent *Bacillus* species. Can J Microbiol 34:925–929

Charpentier CA, Gadille P, Benoit JP (1999) Rhizobacteria microencapsulation: properties of microparticles obtained by spray drying. J Microencap 16:215–229

Chet I, Inbar J (1994) Biological control of fungal pathogens. Appl Biochem Biotechnol 48:37–43

Ciccillo F, Fiore A, Bevivino A, Dalmastri C, Tabacchioni S, Chiarini L (2002) Effects of two different application methods of *Burkholderia ambifaria* MCI 7 on plant growth and rhizospheric bacterial diversity. Environ Microbiol 4:238–245

Cook RJ (2002) Advances in plant health management in the twentieth century. Annu Rev Phytopathol 38:95–116

Dandurand LM, Morra MJ, Chaverra MH, Orser CS (1994) Survival of *Pseudomonas* spp in air dried mineral powders. Soil Biol Biochem 26:1423–1430

Date RA (2001) Advances in inoculant technology: a brief review. Austr J Exp Agric 41:321–325

de Freitas JR, Germida JJ (1992) Growth promotion of winter wheat by fluorescent pseudomonads under growth chamber conditions. Soil Biol Biochem 24:1127–1135

Eapen SJ, Ramana KV, Sarma YR (1997) Evaluation of *Pseudomonas fluorescens* isolates for control of *Meloidogyne incognita* in black pepper (*Piper nigrum* L). In: Edison S, Ramana KV, Sasikumar B, Babu KN, Eapen SJ (eds) Biotechnology of spices, medicinal & aromatic plants. Indian Institute of Spices Research, Calicut, pp 129–133

Fravel DR, Connick WJ Jr, Lewis JA (1998) Formulation of microorganisms to control plant diseases. In: Burges HD (ed) Formulation of microbial biopesticides. Kluwer Academic, Boston, pp 187–202

Fravel DR, Rhodes DJ, Larkin RP (1999) Production and commercialization of biocontrol products. In: Albajes R, Lodovica Gullino M, Van Lenteren JC, Elad Y (eds) Integrated pest and disease management in greenhouse crops. Kluwer Academic, Boston, pp 365–376

Garcia de Salamone IE, Hynes RK, Nelson LM (2001) Cytokinin production by plant growth promoting rhizobacteria and selected mutants. Can J Microbiol 47:404–411

Garland JL (1996) Patterns of potential C source utilization by rhizosphere communities. Soil Biol Biochem 28:223–230

Germida JJ, Siciliano SD, de Freitas JR, Seib AM (1998) Diversity of root-associated bacteria associated with field-grown canola (*Brassica napus* L.) and wheat (*Triticum aestivum*). FEMS Microbiol Ecol 26:43–50

Giacomodonato MN, Pettinari MJ, Souto GI, Mendez BS, Lopez NI (2001) A PCR-based method for the screening of bacterial strains with antifungal activity in suppressive soybean rhizosphere. World J Microbiol Biotechnol 17:51–55

Glick BR (1995) The enhancement of plant growth by free-living bacteria. Can J Microbiol 41:109–117

Hagedorn C, Gould WD, Bardinelli TR (1993) Field evaluation of bacterial inoculants to control seedling disease pathogens in cotton. Plant Dis 77:278–282

Heijnen CE, Burgers SLGE, van Veer JA (1993) Metabolic activity and population dynamics of rhizobia introduced into unamended and betonite amended loamy sand. Appl Environ Microbiol 59:743–747

Janisiewicz WJ, Jeffers SN (1997) Efficacy of commercial formulation of two biofungicides for control of blue mold and gray mold of apple in cold storage. Crop Prot 16:629–633

Jayashree K, Shanmugam V, Raguchander T, Ramanathan A, Samiyappan R (2000) Evaluation of *Pseudomonas fluorescens* (Pf-1) against blackgram and sesame root-rot disease. J Biol Control 14:55–61

Jetiyanon J, Kloepper JW (2002) Mixtures of plant growth-promoting rhizobacteria for induction of systemic resistance against multiple plant diseases. Biol Control 24:285–291

Jeyarajan R, Nakkeeran S (2000) Exploitation of microorganisms and viruses as biocontrol agents for crop disease management. In: Upadhyay RK et al (eds) Biocontrol potential and their exploitation in sustainable agriculture. Kluwer Academic/Plenum, New York, pp 95–116

Kavitha K, Nakkeeran S, Chandrasekar G, Fernando WGD, Mathiyazhagan S, Renukadevi P, Krishnamoorthy AS (2003) Role of antifungal antibiotics, siderophores and IAA production in biocontrol of *Pythium aphanidermatum* inciting damping off in tomato by *Pseudomonas chlororaphis* and *Bacillus subtilis*. In: Proceedings of the 6th international workshop on PGPR. Indian Institute of Spice Research, Calicut, pp 493–497

Kenney DS, Reddy MS, Kloepper JW (1999) Commercial potential of biological preparations for vegetable transplants. Phytopathology 89:S39

Kevan PG, Al-Mazrawi MS, Sutton JC, Tam L, Boland G, Broadbent B, Thompson SV, Brewer GJ (2003) Using pollinators to deliver biological control agents against crop pests. In: Downer RA, Mueninghoff JC, Volgas GC (eds) Pesticide formulations and delivery systems: meeting the PGPR formulations challenges of the current crop protection. American Society for Testing and Materials International, West Conshohocken

Khan MR, Tarannum Z (1999) Effects of field application of various micro-organisms on *Meloidogyne incognita* on tomato. Nematol Medit 27:233–238

Kloepper JW (1993) Plant growth promoting rhizobacteria as biological control agents. In: Metting FB Jr (ed) Soil microbial ecology- applications in agricultural and environmental management. Marcel Dekker, New York, pp 255–274

Kloepper JW, Schroth MN (1978) Plant growth-promoting rhizobacteria on radishes. In: Proceedings of the 4th international conference on plant pathogenic bacteria, vol 2. Station de Pathologie Vegetale et Phytobacteriologie, INRA, Angers, France, pp 879–882

Kloepper JW, Schroth MN (1981a) Development of powder formulation of rhizobacteria for inoculation of potato seed pieces. Phytopathology 71:590–592

Kloepper JW, Schroth MN (1981b) Plant growth promoting rhizobacteria and plant growth under gnotobiotic conditions. Phytopathology 71:642–644

Kloepper JW, Leong J, Teintze M, Schroth MN (1980) Enhanced plant growth by siderophores produced by plant growth promoting rhizobacteria. Nature 286: 885–886

Kovach J, Petzoldt R, Harman GE (2000) Use of honey bees and bumble bees to disseminate *Trichoderma harzianum* 1295-22 to strawberries for *Botrytis* control. Biol Control 18:235–242

Krishnamurthy K, Gnanamanickam SS (1998) Biological control of rice blast by *Pseudomonas fluorescens* strain Pf7-14: evaluation of a marker gene and formulations. Biol Control 13:158–165

Lubeck PS, Hansen M, Sorensen J (2000) Simultaneous detection of the establishment of seed-inoculated *Pseudomonas fluorescens* strain DR54 and native soil bacteria on sugar beet root surfaces using fluorescence antibody and *in situ* hybridization techniques. FEMS Microbiol Ecol 33:11–19

Lugtenberg BJJ, Dekkers L, Bloemberg GV (2001) Molecular determinants of rhizosphere colonization by pseudomonas. Annu Rev Phytopathol 38:461–490

Lynch JM, Whipps JM (1991) Substrate flow in the rhizosphere. In: Keister DL, Cregan B (eds) The rhizosphere and plant growth, vol 14, Beltsville Symposium in Agricultural Research. Kluwer, Dordrecht, pp 15–24

Mani MP, Rajeswari S, Sivakumar CV (1998) Management of the potato cyst nematodes, *Globodera* spp through plant rhizosphere bacterium *Pseudomonas fluorescens* Migula. J Biol Control 12:131–134

Manjula K, Podile AR (2001) Chitin supplemented formulations improve biocontrol and plant growth promoting efficiency of *Bacillus subtilis* AF1. Can J Microbiol 47:618–625

Mansouri H, Petit A, Oger P, Dessaux Y (2002) Engineered rhizosphere: the trophic bias generated by opine-producing plants is independent of the opine type, the soil origin, and the plant species. Appl Environ Microbiol 68:2562–2566

Martinez-Ochoa N, Kokalis-Burelle N, Rodriguez-Kabana R, Kloepper JW (1999) Use of organic amendments, botanical aromatics, and rhizobacteria to induce suppressiveness of tomato to the root-knot nematode. Meloidogyne incognita. Phytopathology 89:S49

Mathiyazhagan S, Kavitha K, Nakkeeran S, Chandrasekar G, Manian K, Renukadevi P, Krishnamoorthy AS, Fernando WGD (2004) PGPR mediated management of stem blight of *Phyllanthus amarus* (Schum and Thonn) caused by *Corynespora cassiicola* (Berk and Curt) Wei. Arch Phytopathol Plant Protect 33:183–199

Mathre DE, Cook RJ, Callan NW (1999) From discovery to use: traversing the world of commercializing biocontrol agents for plant disease control. Plant Dis 83:972–983

Maurhofer M, Hase C, Meuwly P, Metraux JP, Defago G (1994) Induction of systemic resistance of tobacco to tobacco necrosis virus by the root colonizing *Pseudomonas fluorescens* strain CHA0: influence of the *gag*A gene and of pyoverdine production. Phytopathology 84:678–684

McSpadden Gardener BB, Fravel DR (2002) Biological control of plant pathogens: research, commercialization, and application in the USA. Plant Health Progress. doi:10.1094/PHP-2002-0510-01-RV

Meena B, Radhajeyalakshmi R, Marimuthu T, Vidhyasekaran P, Velazhahan R (2002) Biological control of groundnut late leaf spot and rust by seed and foliar applications of a powder formulation of *Pseudomonas fluorescens*. Biocontrol Sci Technol 12:195–204

Moenne-Loccoz Y, Naughton M, Higgins P, Powell J, O'Connor B, O'Gara F (1999) Effect of inoculum preparation and formulation on survival and biocontrol efficacy of *Pseudomonas fluorescens* F113. J Appl Micrbiol 86:108–116

Murphy JF, Zhender GW, Schuster DJ, Sikora EJ, Polston JE, Kloepper JW (2000) Plant growth promoting rhizobacterial mediated protection in tomato against tomato mottle virus. Plant Dis 84:779–784

Muthukumarasamy R, Revathi G, Lakshminarasimhan C (1999) Diazotrophic associations in sugarcane cultivation in South India. Trop Agric 76:171–178

Nakkeeran S, Kavitha K, Mathiyazhagan S, Fernando WGD, Chandrasekar G, Renukadevi P (2004) Induced systemic resistance and plant growth promotion by *Pseudomonas chlororaphis* strain PA-23 and *Bacillus subtilis* strain CBE4 against rhizome rot of turmeric (*Curcuma longa* L). Can J Plant Pathol 26:417–418

Nandakumar R, Babu S, Viswanathan R, Raguchander T, Samiyappan R (2000) Induction of systemic resistance in rice against sheath blight disease by *Pseudomonas fluorescens*. Soil Biol Biochem 33:603–612

Nandakumar R, Babu S, Viswanathan R, Sheela J, Raguchander T, Samiyappan R (2001) A new bio-formulation containing plant growth promoting rhizobacterial mixture for the management of sheath blight and enhanced grain yield in rice. Biocontrol 46:493–510

Niknam GR, Dhawan SC (2001a) Induction of systemic resistance by *Bacillus subtilis* isolate Bs1 against *Rotylenchulus reniformis* in tomato. National Congress on Centenary of Nematology in India – Appraisal & Future Plans, Indian Agriculture Research Institute, New Delhi, pp 143–144

Niknam GR, Dhawan SC (2001b) Effect of seed bacterization, soil drench and bare root-dip application methods of *Pseudomonas fluorescens* isolate Pf1 on the suppression of *Rotylenchulus reniformis* infecting tomato. National Congress on Centenary of Nematology in India – Appraisal & Future Plans, Indian Agricultural Research Institute, New Delhi, p 144

Niranjan Raj S, Deepak SA, Basavaraju P, Shetty HS, Reddy MS, Kloepper JW (2003) Comparative performance of formulations of plant growth promoting rhizobacteria in growth promotion and suppression of downy mildew in pearl millet. Crop Prot 22:579–588

Oostendorp M, Sikora RA (1989) Seed treatment with antagonistic rhizobacteria for the suppression of *Heterodera schachtii* early root infection of sugar beet. Rev Nematol 12:77–83

Parvatha Reddy P, Nagesh M, Rao MS, Rama N (2000) Management of *Tylenchulus semipenetrans* by integration of *Pseudomonas fluorescens* with oil cakes. In: Proceedings of international symposium on citriculture, Nagpur, India, pp 830–833

Paulitz TC, Belanger RB (2001) Biological control in greenhouse systems. Ann Rev Phytopathol 39:103–133

Peng G, Sutton JC, Kevan PG (1992) Effectiveness of honey bees for applying the biocontrol agent *Gliocladium roseum* to strawberry flowers to suppress *Botrytis cinerea*. Can J Plant Pathol 14:117–188

Persello-Cartieaux F, Nussaume L, Robaglia C (2003) Tales from the underground: molecular plant-rhizobacteria interactions. Plant Cell Environ 26:189–199

Pessi G, Haas D (2000) Transcriptional control of the hydrogen cyanide biosynthetic genes *hcn*ABC by the anaerobic regulator ANR and the quorum-sensing regulators LasR and RhlR in *Pseudomonas aeruginosa*. J Bacteriol 182:6940–6949

Rabindran R, Vidhyasekaran P (1996) Development of formulation of *Pseudomonas fluorescens* PfALR2 for management of rice sheath blight. Crop Prot 15:715–721

Radja Commare R, Nandakumar R, Kandan A, Suresh S, Bharathi M, Raguchander T, Samiyappan R (2002) *Pseudomonas fluorescens* based bioformulation for the management of sheath blight disease and leaf folder insect in rice. Crop Prot 21:671–677

Raguchander T, Shanmugam V, Samiyappan R (2000) Biological control of panama wilt disease of banana. Madras Agric J 87:320–321

Rainey PB (1999) Adaptation of *Pseudomonas fluorescens* to the plant rhizosphere. Environ Microbiol 1:243–257

Raupach GS, Kloepper JW (1998) Mixtures of plant growth promoting rhizobacteria enhance biological control of multiple cucumber pathogens. Phytopathology 88: 1158–1164

Reddy MS, Rodriguez-Kabana R, Kenney DS, Ryu CM, Zhang S, Yan Z, Martinez-Ochoa N, Kloepper JW (1999) Growth promotion and induced systemic resistance (ISR) mediated by a biological preparation. Phytopathology 89:S65

Roberts DP, Yucel I, Larkin RP (1998) Genetic approaches for analysis and manipulation of rhizosphere colonization by bacterial biocontrol agents. In: Boland GJ, Kuykendall LD (eds) Plant-microbe interactions and biological control, vol 63, Books in soils, plants, and the environment. Marcel Dekker Inc., New York, pp 415–431

Rondon MR, Goodman RM, Handelsman J (1999) The earth's bounty: assessing and accessing soil microbial diversity. Trends Biotechnol 17:403–409

Ryu CM, Reddy MS, Zhang S, Murphy JF, Kloepper JW (1999) Plant growth promotion of tomato by a biological preparation (LS 213) and evaluation for protection against cucumber mosaic virus. Phytopathology 89:S87

Sabitha D, Nakkeeran S, Chandrasekar G (2001) *Trichoderma* – bioarsenal in plant disease management and its scope for commercialization. In: Proceedings of Indian Phytopathological Society, southern zone meeting, 10–12 Dec 2001, Indian Institute of Spice Research, Calicut, Kerala, pp 43–55

Saleh SA, Mekhemar GAA, Abo El-Soud AA, Ragab AA, Mikhaeel FT (2001) Survival of *Azorhizobium* and *Azospirillum* in different carrier materials: inoculation of wheat and *Sesbania rostrata*. Bull Fac Agric Univ Cairo 52:319–338

Sandhu DK, Waraich MK (1985) Yeasts associated with pollinating bees and flowers nectar. Microbial Ecol 11:51–58

Schisler DA, Slininger PJ (1997) Microbial selection strategies that enhance the likelihood of developing commercial biological control products. J Ind Microbiol Biotechnol 19:172–179

Seenivasan N, Parameswaran S, Sridar P, Gopalakrishnan C, Gnanamurthy P (2001) Application of bioagents and neem cake as soil application for the management of root-knot nematode in turmeric. National Congress on Centenary of Nematology in India – Appraisal & Future Plans, Indian Agricultural Research Institute, New Delhi, p 164

Seshadri S, Ignacimuthu S, Vadivelu M, Lakshminarasimhan C (2007) Inorganic phosphate solubilization by two insect pathogenic *Bacillus* sp. Dev Plant Soil Sci 102:351–355

Shanthi A, Sivakumar CV (1995) Biocontrol potential of *Pseudomonas fluorescens* (Migula) against root-knot nematode, *Meloidogyne incognita* (Kofoid and White, 1919) Chitwood, 1949 on tomato. J Biol Control 9:113–115

Shanthi A, Rajeswari S, Sivakumar CV (1998) Soil application of *Pseudomonas fluorescens* (Migula) for the control of root-knot nematode (*Meloidogyne incognita*) on grapevine (*Vitis vinefera* Linn). In: Usha Mehta K (ed) Nematology – challenges and opportunities in 21st century. Sugarcane Breeding Institute, Coimbatore, pp 203–206

Shanthi A, Sundarababu R, Sivakumar CV (1999) Field evaluation of rhizobacterium, *Pseudomonas fluorescens* for the management of the citrus nematode, *Tylenchulus semipenetrans*. In: Proceedings of the national symposium on rational approaches in nematode management for sustainable agriculture. Indian Agriculture Research Institute, New Delhi, pp 38–42

Siddiqui ZA, Mahmood I (1995) Some observations on the management of the wilt disease complex of pigeonpea by treatment with a vesicular arbuscular fungus and biocontrol agents for nematodes. Biosource Technol 54:227–230

Siddiqui IA, Shaukat SS (2002) Resistance against damping-off fungus *Rhizoctonia solani* systematically induced by the plant growth-promoting rhizobacteria *Pseudomonas aeruginosa* (1E-6S(+)) and *P fluorescens* (CHAO). J Phytopathol 150:500–506

Silva HSA, Romeiro RS, Mounteer A (2003) Development of a root colonization bioassay for rapid screening of rhizobacteria for potential biocontrol agents. J Phytopathol 151:42–46

Smalla K, Wieland G, Buchner A, Zock A, Parzy J, Kaiser S, Roskot N, Heuer H, Berg G (2001) Bulk and rhizosphere soil bacterial communities studied by denaturing gradient gel electrophoresis: plant-dependent enrichment and seasonal shifts revealed. Appl Environ Microbiol 67:4742–4751

Smith RS (1997) New inoculant technology to meet changing legume management. In: Elmerich C, Kondorosi A, Newton WE (eds) Biological nitrogen fixation for the 21st century. Kluwer, Dordrecht, pp 621–622

Smith KP, Goodman RM (1999) Host variation for interactions with beneficial plant-associated microbes. Annu Rev Phytopathol 37:473–491

Sobita Devi L, Dutta U (2002) Effect of *Pseudomonas fluorescens* on root-knot *(Meloidogyne incognita)* on okra plant. Indian J Nematol 32:215

Sorensen J, Jensen LE, Nybroe O (2001) Soil and rhizosphere as habitats for *Pseudomonas* inoculants: new knowledge on distribution, activity and physiological state derived from micro-scale and single-cell studies. Plant Soil 232:97–108

Steddom K, Menge JA, Crowley D, Borneman J (2002) Effect of repetitive applications of the biocontrol bacterium *Pseudomonas putida* 06909-rif/nal on citrus soil microbial communities. Phytopathology 92: 857–862

Sutton JC (1995) Evaluating of micro-organisms for biocontrol: *Botrytis cinerea* and strawberry, a case study. Adv Plant Pathol 11:73–190

Taylor AG, Harman GE (1990) Concept and technologies of selected seed treatments. Annu Rev Phytopathol 28:321–339

Thomas-Bauzon D, Weinhard P, Villecourt P, Balandreau J (1982) The spermosphere model I. Its use in growing, counting and isolating N2-fixing bacteria from the rhizosphere of rice. Can J Microbiol 28:922–928

Tunlid A, White D (1992) Biochemical analysis of biomass, community structure, nutritional status, and metabolic activity of microbial communities in soil. In: Stotzky G, Bollag JM (eds) Soil biochemistry, vol 7. Marcel Dekker, New York, pp 229–262

Van Peer R, Schippers B (1989) Plant growth responses to bacterization and rhizosphere microbial development in hydroponic culture. Can J Microbiol 35:456–463

Vasudevan P, Reddy MS, Kavitha S, Velusamy P, Avid Paul Raj RS, Purushothaman SM, Brindha Priyadarsini V, Bharathkumar S, Kloepper JW, Gnanamanickam SS (2002) Role of biological preparations in enhancement of rice seedling growth and grain yield. Curr Sci 83:1140–1144

Vavrina CS (1999) The effect of LS213 (*Bacillus pumilus*) on plant growth promotion and systemic acquired resistance in muskmelon and watermelon transplants and subsequent field performance. Proc Int Symp Stand Estab 107:111

Verma KK, Gupta DC, Paruthi IJ (1999) Preliminary trial on the efficacy of *Pseudomonas fluorescens* as seed treatment against *Meloidogyne incognita* in tomato. In: Proceedings of the national symposium on rational approaches in nematode management for sustainable agriculture, Indian Agriculture Research Institute, New Delhi, pp 79–81

Vidhyasekaran P, Muthamilan M (1995) Development of formulations of *Pseudomonas fluorescens* for control of chickpea wilt. Plant Dis 79:782–786

Vidhyasekaran P, Muthamilan M (1999) Evaluation of *Pseudomonas fluorescens* for controlling rice sheath blight. Biocontrol Sci Technol 9:67–74

Vidhyasekaran P, Sethuraman K, Rajappan K, Vasumathi K (1997) Powder formulation of *Pseudomonas fluorescens* to control pigeonpea wilt. Biol Control 8:166–171

Viswanathan R, Samiyappan R (2002) Induced systemic resistance by fluorescent pseudomonads against red rot disease of sugarcane caused by *Colletotrichum falcatum*. Crop Prot 21:1–10

Viswanthan R, Samiyappan R (1999) Management of damping off disease in sugarcane using plant growth promoting rhizobacteria. Madras Agric J 86:643–645

Vivekananthan R, Ravi M, Ramanathan A, Samiyappan R (2004) Lytic enzymes induced by *Pseudomonas fluorescens* and other biocontrol organisms mediate defence against the anthracnose pathogen in mango. World J Microbiol Biotechnol 20:235–244

Weller DM (1983) Colonization of wheat roots by a fluorescent pseudomonad suppressive to take all. Phytopathology 73:1548–1553

Wright B, Rowse HR, Whipps JM (2003) Application of beneficial microorganisms to seeds during drum priming. Biocont Sci Technol 13:519–614

Yan Z, Reddy MS, Wang Q, Mei R, Kloepper JW (1999) Role of rhizobacteria in tomato early blight control. Phytopathology 89:S87

Yardin MR, Kennedy IR, Thies JE (2000) Development of high quality carrier materials for field delivery of key microorganisms used as bio-fertilisers and bio-pesticides. Radiation Phys Chem 57:565–568

Zhang S, Reddy MS, Ryu CM, Kloepper JW (1999) Relationship between *in vitro* and *in vivo* testing of PGPR for induced systemic resistance against tobacco blue mold. Phytopathology 89:S89

Zhender GW, Murphy JF, Sikora JE, Kloepper JW (2001) Application of rhizobacteria for induced resistance. Eur J Plant Pathol 107:39–50

Soil Solarisation

11

Abstract

Solarisation is a technique that uses clear polyethylene film to cover moistened soil and trap lethal amounts of heat from solar radiation to reduce soil-borne pests. The capacity of soil solarisation to suppress propagule numbers of soil-borne pathogens relies on many factors. The temperatures obtained in the moistened soil covered by the transparent sheeting and the exposure time of the organisms to these elevated temperatures are both important characteristics of this pre-plant soil treatment. Solarisation has been effective in disease control in many geographical locations around the world. It is most successful in regions with the appropriate meteorological parameters such as high air temperatures and extended periods of high radiation.

Researchers found that solarisation could be a useful soil disinfestation method, especially in areas with hot and arid conditions during the summer months. In certain cases, the treatment has also been effective, primarily for weed management, in cooler coastal areas (Elmore CL, Stapleton JJ, Bell CE, DeVay JE, Soil solarization: a nonpesticidal method for controlling diseases, nematodes and weeds. UC DANR Pub 21377, Oakland, CA, 14pp, 1997). The pesticidal activity of solarisation was found to stem from a combination of physical, chemical and biological effects.

Soil-borne pests and pathogens, including weed propagules, nematodes, insects, fungi, bacteria and certain other agents, can be limiting factors in the production of crop plants. One of the principal strategies used by the growers of high-value horticultural crops to combat these organisms is pre-plant soil disinfestation, using chemical or physical methods. Soil fumigants are the most effective soil disinfestation chemicals, and methyl bromide (MB) is the most important soil fumigant chemical used by growers around the world. It is a broad-spectrum pesticide with excellent activity against most potential soil pests.

Scientists have been continuously working to develop usable alternatives for soil disinfestation. Apart from synthetic chemical alternatives, numerous non-chemical strategies have also been researched, field validated and in some cases implemented commercially. Amongst the potential alternative control method being touted to

P.P. Reddy, *Recent Advances in Crop Protection*,
DOI 10.1007/978-81-322-0723-8_11, © Springer India 2013

replace methyl bromide is soil solarisation that is amongst the most useful of the non-chemical disinfestation methods.

The use of clear polyethylene film to cover moistened soil and trap lethal amounts of heat from solar radiation was first reported by Katan and colleagues in Israel in the mid-1970s (Katan et al. 1976). DeVay and associates at the University of California, Davis, began an intensive research programme on the promising technique shortly thereafter, and the term 'soil solarisation' was soon coined to describe the process by cooperators in the San Joaquin Valley. Researchers found that solarisation could be a useful soil disinfestation method, especially in areas with hot and arid conditions during the summer months, such as the Central Valley and southern deserts. In certain cases, the treatment has also been effective, primarily for weed management, in cooler coastal areas (Elmore et al. 1997). The pesticidal activity of solarisation was found to stem from a combination of physical, chemical and biological effects, as described in several comprehensive reviews (Katan 1987; Chen et al. 1991; DeVay et al. 1991; Stapleton 1998, 2000).

In the United States, successful studies have been conducted in several states. A study in California showed that solarisation alone increased strawberry yield by 12% over non-treated plots (Hartz et al. 1993). In northern Florida, research showed that soil solarisation decreased densities of *Phytophthora nicotianae* and *Ralstonia solanacearum* to depths of 25 and 15 cm, respectively (Chellemi et al. 1994). In an Alabama study, temperatures attained via solar heating were 48°C at the soil surface and 34°C at 20 cm below the surface (Himelrick and Dozier 1991). In studies previously conducted at The University of Tennessee, temperatures of 49.5°C at 5 cm depth and 41.5°C at 25 cm depth have been recorded under solarisation treatments.

Although solarisation can provide excellent soil disinfestation under suitable conditions, it has significant limitations and should not be considered a cure-all or universal replacement for MB. For example, solarisation is most effective close to the surface of the soil under climatic and weather conditions of high air temperature and long days for soil heating. It will not control all pest organisms, may require that land be taken out of production for 3-to-6-week treatments during the summer months and requires disposal of used plastic film. Therefore, its practical value to the user must be assessed by several factors, including extent and predictability of pesticidal efficacy, effect on crop growth and yield, economic cost/benefit and personal pest management philosophy (Stapleton 1997).

11.1 Advantages and Disadvantages of Soil Solarisation

11.1.1 Advantages

- Non-pesticidal and simple.
- No health or safety problems associated with use.
- No registration is required.
- Crops produced are pesticide-free and may command a higher market price.
- Controls multiple soil-borne diseases and pests.
- Selects for beneficial microorganisms.
- Tends to increase soil fertility.
- Increases soluble NO_3, NH_4, Ca, Mg, K and soluble organic matter.
- May improve soil tilth.
- Can speed up in-field composting of green manure.

11.1.2 Disadvantages

- Is restricted to areas with warm to hot summers.
- May be less effective in cooler coastal areas.
- Land must be taken out of production for 4–6 weeks during the summer.
- May not fit in with some cropping cycles.
- May be difficult for those using a small amount of land intensively.
- Limited number of retail outlets for UV-inhibiting plastics.

- Disposal of plastic may be a problem.
- Large amounts of plastic cannot be currently recycled.
- Some pests are not controlled or are difficult to control.
- No pest control in the furrows between strips (if applied in strip coverage).
- High winds and animals may tear the plastic.

11.2 Method of Soil Solarisation

11.2.1 Soil Preparation

Solarisation is most effective when the plastic sheeting (tarp) is laid as close as possible to a smooth soil surface. Preparation of the soil begins by disking, rototilling or turning the soil by hand to break up clods and then smoothing the soil surface. Large rocks, weeds or any other objects or debris that will raise or puncture the plastic should be removed.

11.2.2 Plastic Mulching

11.2.2.1 Clear Versus Coloured Plastic

Transparent or clear plastic is most effective for solarisation. Black plastic, often used for mulching, does not heat the soil as well as clear plastic. It can be used for solarisation, but its main effect is reducing weed growth. In areas where solarisation is ineffective because of low solar radiation or a heavy infestation of weeds, black plastic may combine some solarisation benefit with residual weed control. It can also be used for solarising existing crops, for example, by disinfesting soil whilst establishing permanent tree or vine crops.

Since soil temperatures are lower with black plastic, the treatment time must be lengthened for best results. Other colours of plastic, such as green or brown, which allow some heating of the soil but not to the degree of clear plastic, require longer treatment times. These other colours of plastic give so much less effective solarisation that they should probably only be used as mulch.

11.2.2.2 Types of Plastic

The thinner the plastic, the greater the heating will be. Polyethylene (PE) plastic 1 mile (0.025 mm) thick is efficient and economical but not very resistant to tearing by wind or puncture by animals. Users in windy areas should consider plastic sheets that are 1.5–2 miles (0.038–0.050 mm) thick. If holes or tears occur in the plastic, they should be patched with clear patching tape. Users are encouraged to select plastic sheeting containing ITV inhibiting additives that prevent sheets from becoming brittle and difficult to remove from the field and extend the life of the plastic. Plastic sheets laid by hand can often be used more than once for solarisation, although if the plastic is dirty or dusty, reuse is less effective.

Polyethylene sheets may be modified by an additive that enables them to absorb infrared (IR) radiation and improve their capacity to retain heat. Although these are available, they have not proven to be very effective. Coloured plastic films are available that absorb light in the photosynthetic range to inhibit growth of weeds and at the same time heat the soil. These can be used for solarisation but generally do not heat soil as well as transparent films. There has been considerable interest in the development of high-density or 'impermeable' plastic sheeting to better contain fumigant chemicals in soil. These plastics may also improve the effects of solarisation by sealing in more heat and volatile compounds. Experimental work has also been done using a sprayable polymer as a replacement for plastic sheeting. Such a material would be easy to apply and less expensive to use, but to date suitable chemicals have not been found. The use of a double layer of plastic with air space between the layers mimics the greenhouse effect and raises soil temperatures from 2 to 10 °F higher than that obtained with a single layer. Using a double layer requires additional preparation time and expense, but it may make soil solarisation more feasible in areas with cooler climates.

11.2.2.3 Availability

For small applications in gardens, UV-inhibiting plastic that is 1.5–4 miles (0.038–0.100 mm) thick can be purchased from nursery, hardware or lumber establishments. These are sometimes

Fig. 11.1 Plastic sheets laid by machines in strips for soil solarisation

called 'drop cloths' and are used to catch paint drippings. For agricultural plantings, plastic can usually be purchased in rolls from 1.8 to 3.6 m wide and approximately 1,200 m long. Size will vary by source (wide range of colours and sizes including solarisation and IRT films).

11.2.2.4 Laying the Plastic

Plastic sheets may be laid by hand or machine. The open edges of the plastic sheeting should be anchored to the soil by burying the edges in a shallow trench around the treated area. Plastic is laid either in complete coverage, where the entire field or area to be planted is treated, or strip coverage, where only beds or selected portions of the field are treated.

11.2.2.4.1 Complete Coverage

In complete coverage, plastic sheeting is laid down to form a continuous surface over the entire field or area to be planted. The edges of the sheets may be joined with an ultraviolet (UV)-resistant glue or anchored by laying adjacent strips of plastic and burying both edges in soil. Anchoring the edges in the soil may be more cost-effective initially than gluing the edges together but may also result in untreated soil being close to subsequently planted crops. The ends of the sheets should be held in place by burying them in the soil. If beds are formed after complete coverage, care must be taken to avoid deep tillage that could bring untreated soil to the surface. Complete coverage is recommended if the soil is heavily infested with pathogens, nematodes or perennial weeds, since there is less chance of reinfestation by soil being moved to the plants through cultivation or furrow-applied irrigation water.

11.2.2.4.2 Strip Coverage

In strip coverage, plastic is applied in strips over preformed beds (Fig. 11.1). Strips should be a minimum of 75 cm wide; beds up to 1.5 m wide are preferred because several crop rows can be planted per bed. In some cases, strip coverage may be more practical and economical than complete coverage because less plastic is needed, and it is not necessary to join the edges of the plastic sheets together. Strip coverage effectively kills most pests and eliminates the need for deep cultivation after solarisation. It is especially effective against weeds, since the furrows are cultivated. With strip coverage, however, long-term control of soil pathogens and nematodes may be lost because pests in the untreated soil in the rows between the strips can contaminate and reinfest treated areas.

11.2.3 Irrigation

Wet soil conducts heat better than dry soil and makes soil organisms more vulnerable to heat. The soil under the plastic sheets must be saturated to at least 70% of field capacity in the upper layers and moist to depths of 60 cm for soil solarisation to be effective.

Soil may be irrigated either before or after the plastic sheets are laid. If the soil is irrigated beforehand, the plastic must be applied as soon as possible to avoid water loss; if heavy machinery is used to lay the plastic, however, the soil must be dry enough to avoid compaction. If the soil is to be irrigated after the plastic is laid, one or more hose or pipe outlets may be installed under one end of the plastic; drip lines may be installed before the plastic is laid, or irrigation water may be run underneath the plastic in furrows or in the tracks made by tractor wheels if the plastic sheets were applied by machine. Fields treated by strip coverage can be irrigated by drip lines on or in the bed.

The soil does not usually need to be irrigated again during solarisation, although if the soil is very light and sandy, or if the soil moisture is less than 50% of field capacity, it may be necessary to irrigate a second time. This will cool the soil, but because of the increased moisture, the final temperatures will be greater.

11.2.4 Duration of Treatment

The plastic sheets should be left in place for 4–6 weeks to allow the soil to heat to the greatest depth possible. To control the most resistant species, leave the plastic in place for 6 weeks. Experience has shown that there is little or no need to take the temperature of the soil. The greatest concern is to solarise the soil during a period of high solar radiation with little wind or cloud cover. Soil in the Central Valley can be solarised for 4 weeks any time from late May to September. In coastal areas, the best time may be August to September or May to June, transitional periods when fog or wind may be at a minimum.

11.2.5 Removal of the Plastic and Planting

After solarisation is complete, the plastic may be removed before planting. Or the plastic may be left on the soil as a mulch for the following crop by transplanting plants through the plastic. Clear plastic may be painted white or silver to cool the soil and repel flying insect pests in the following crop. A disadvantage of leaving the plastic on the soil is that it may degrade and be difficult to clean up in the spring.

Treated soil can be planted immediately to a fall or winter crop or left fallow without the plastic until the next growing season. If the soil must be cultivated for planting, the cultivation must be shallow (<5 cm) to avoid moving viable weed seed to the surface.

11.2.6 Disposal of Plastic Film

The disposal of plastic film after solarisation presents an additional expense and involves consideration of environmental pollution. At present, there is no programme for recycling plastic used in soil solarisation, primarily due to the relatively low amount of plastic used. A few programmes are operating in other states where a more constant supply of used agricultural plastic is available. In addition, soil adhering to used plastic makes recycling more costly. UV-treated plastics that are thicker than 4 miles (0.1 mm) may be usable for more than one season if handled carefully. Although most plastics have been put into landfills after use in solarisation, some farmers store plastic at their own sites until recycling programmes can be started.

Efforts have been made to develop a plastic film for solarisation that would degrade completely after use in a suitable or predictable amount of time. These 'biodegradable' or 'photodegradable' plastics are not currently recommended for solarisation. Photodegradable plastics degrade with exposure to UV light. Although they may be effective for solarisation, the timed degradation (6–12 weeks) has not been uniformly effective. Also, the buried part of the plastic remains in the soil until it is brought to the surface with cultivation, leaving a source of pollution in the field.

11.3 Effects of Solarisation

11.3.1 Increased Soil Temperature

The heating effect of soil solarisation is greatest at the surface of the soil and decreases with depth. The maximum temperature of soil solarised in the field is usually from 42 to 55°C at a depth of 5 cm and from 32 to 37°C at 45 cm. Control of soil pests is usually best in the upper 10–30 cm. Higher soil temperatures and deeper soil heating may be achieved inside greenhouses or by using a double layer of plastic sheeting. Soil solarised in greenhouses may reach 60°C at a depth of 10 cm and 53°C at 20 cm. Soil solarised in black plastic nursery sleeves under a single or double layer of clear plastic can exceed 70°C.

11.3.2 Improved Soil Physical and Chemical Features

Solarisation initiates changes in the physical and chemical features of soil that improve the growth and development of plants. It speeds up the breakdown of organic material in the soil, resulting in the release of soluble nutrients such as nitrogen (NO_3, NH_4^+), calcium (Ca^{++}), magnesium (Mg^{++}), potassium (K^+) and fulvic acid, making them more available to plants. Improvements in soil tilth through soil aggregation are also observed.

11.3.3 Control of Soil-Borne Pathogens

Repeated daily heating during solarisation kills many plant pathogens, nematodes and weed seed and seedlings. The heat also weakens many organisms that can withstand solarisation, making them more vulnerable to heat-resistant fungi and bacteria that act as natural enemies. Changes in the soil chemistry during solarisation may also kill or weaken some soil organisms.

Although many soil pests are killed at temperatures above 30 to 33°C, plant pathogens, weeds and other soil-borne organisms differ in their sensitivity to soil heating. Some pests that are difficult to control with soil fumigants are easily controlled by soil solarisation. Other pests are also affected but cannot be consistently controlled by solarisation. These may require additional control measures.

11.3.3.1 Fungi and Bacteria

Solarisation controls populations of many important soil-borne fungal and bacterial plant pathogens, including *Verticillium dahliae*, which causes *Verticillium* wilt in many crops; certain *Fusarium* spp. that cause *Fusarium* wilt in some crops; *Phytophthora cinnamomi*, which causes *Phytophthora* root rot; *Agrobacterium tumefaciens*, which causes crown gall disease; *Clavibacter michiganensis*, which causes tomato canker; and *Streptomyces scabies*, which causes potato scab. Other fungi and bacteria are more difficult to control with solarisation, such as certain high temperature fungi in the genera *Macrophomina*, *Fusarium* and *Pythium* and the soil-borne bacterium *Ralstonia solanacearum*.

11.3.4 Encouragement of Beneficial Soil Organisms

Fortunately, although many soil pests are killed by soil solarisation, many beneficial soil organisms are able to either survive solarisation or recolonise the soil very quickly afterwards. Important amongst these beneficials are the mycorrhizal fungi and fungi and bacteria that parasitise plant pathogens and aid plant growth. The shift in the population in favour of these beneficials can make solarised soils more resistant to pathogens than non-solarised or fumigated soil. Solarisation may also result in increased relative populations of beneficial organisms (i.e. rhizosphere bacteria such as *Bacillus* spp. and *Pseudomonas* spp., actinomycetes, *Trichoderma* spp. and mycorrhizal fungi).

11.3.4.1 Earthworms

The effect of soil solarisation on earth worms has not received much attention, but it is thought that they retreat to lower depths and escape the effects of soil heating.

11.3.4.2 Fungi

Beneficial fungi, especially *Trichoderma*, *Talaromyces* and *Aspergillus* spp., survive or even increase in solarised soil. Mycorrhizal fungi are more resistant to heat than most plant pathogenic fungi. Their populations may be decreased in the upper soil profile, but studies have shown that this is not enough to reduce their colonisation of host roots in solarised soil.

11.3.4.3 Bacteria

Populations of the beneficial bacteria *Bacillus* and *Pseudomonas* spp. are reduced during solarisation but recolonise the soil rapidly afterwards. Populations of *Rhizobium* spp., which fix nitrogen in root nodules of legumes, may be greatly reduced by solarisation and should be reintroduced by inoculation of leguminous seed. Soil-borne populations of other nitrifying bacteria are also reduced during solarisation. Population levels of actinomycetes are not greatly affected by soil solarisation. Many members of this group are known to be antagonistic to plant pathogenic fungi.

During solarisation of soil, populations of oxidise negative fluorescent pseudomonads and Gram-positive bacteria, including *Bacillus* species, may be reduced by 78–86% compared with non-solarised soil (Stapleton and DeVay 1984), whereas populations of actinomycetes may be reduced from 45 to 58% in solarised soil (Stapleton and DeVay 1984). Surprisingly, after solarisation, *Pseudomonas* species quickly recolonise the soil and their populations reach high levels. Of great significance is the change in populations of *Bacillus* species during solarisation; the percentage of colonies in solarised soil which exhibited antibiosis to *Geotrichum candidum* increased nearly 20-fold when compared with non-solarised soil (Stapleton and DeVay 1984). These bacteria are amongst those which are rhizosphere competent and are believed to contribute to the increased growth response of plants grown in solarised soil (Katan 1987). Although initial populations of *Bacillus* species are greatly reduced, they are spore formers and are a major component of the soil microflora.

In contrast to the studies in California and Israel, studies in western Australia showed that solarisation increased the total numbers of bacteria and actinomycetes in soil. However, as in the California study (Stapleton and DeVay 1984) where there was an increase in the proportion of antagonistic Gram-positive bacteria in solarised soil, the western Australia study showed that the proportion of bacteria (actinomycetes) antagonistic to *Fusarium oxysporum*, *F. solani* and *Rhizoctonia solani* was increased compared with non-solarised soil. In other studies, *Actinomyces scabies* was controlled by soil solarisation.

11.3.5 Increased Plant Growth and Yield

Plants often grow faster and produce both higher and better quality yields when grown in solarised soil. This can be attributed, in part, to improved disease and weed control, but increases in plant growth are still seen when soil apparently free of pests is solarised. A number of factors may be involved. First, minor or unknown pests may also be controlled. Second, the increase in soluble nutrients improves plant growth. Third, relatively greater populations of helpful soil microorganisms have been documented following solarisation, and some of these, such as certain fluorescent pseudomonads and *Bacillus* bacteria, are known to be biological control agents.

Solarisation increased okra biomass; the longer the duration of solarisation, the greater the increase in okra biomass (based on comparison amongst 2-, 4- and 6-week solarisation periods). The positive yield response indicated that solarisation did not impair organic matter decomposition and subsequent release of plant nutrients. Both of these studies (Ozores-Hampton et al. 2004; Seman-Varner et al. 2008) suggest that solarisation does not interfere much with beneficial soil organisms that decompose organic matter.

Plants often grow faster and produce both higher and better quality yields when grown in solarised soil. This can be attributed to the increase in soluble nutrients and relatively greater proportions of helpful soil microorganisms. Solarisation may also result in an increased growth response (as evidenced by increased trunk diameters) and yield in orchard trees, by increasing the availability of plant nutrients and changes the soil microflora to favour biological pest control. Improved growth of rice and wheat plants was observed following solarisation. Soil inorganic nitrogen levels were consistently higher as was extractable manganese.

11.3.6 Increased Availability of Nutrients

In general, availability of some nutrients can be expected to increase with solarisation, because the heat generated under clear plastic will encourage accelerated decomposition of organic matter. In a field study in southwest Florida, soil nutrients were not affected by solarisation compared to conventional treatment with methyl bromide (Ozores-Hampton et al. 2004). Solarisation coupled with compost increased soil nutrient levels more than treatments with methyl bromide or solarisation alone, both of which were combined with inorganic fertiliser. Seman-Varner et al. (2008) measured nutrient concentration in the soil and plant tissue of an okra crop following different durations of solarisation. Whilst soil potassium (K) and manganese (Mn) were higher following solarisation, Copper (Cu) and zinc (Zn) were lower. In addition, soil pH was slightly decreased by solarisation. Soil phosphorus (P), magnesium (Mg), calcium (Ca) and iron (Fe) were not affected by solarisation. Nutrients supplied to the crop were exclusively provided by chopped cowpea hay. Okra tissue concentrations of K, N, Mg and Mn were higher when grown on solarised plots. In contrast, concentrations of P and Zn were lowered by solarisation.

The increased availability of mineral nutrients following soil solarisation are particularly those tied up in organic fraction, such as NH_4-N, NO_3-N, P, Ca and Mg, as a result of the death of the microbiota. Extractable P, K, Ca and Mg sometimes have been found in greater amounts after soil solarisation. The liberation of N compounds (vapour and liquid) is a component of the mode of action; increased concentration of reduced N would then nitrify after termination of soil solarisation to provide NO_3 for increased crop growth.

Soils covered with plastic film mulches usually retain a higher level of soluble minerals. Constant moisture content, higher temperature and better aeration of the soil all tend to favour higher microbial populations in the soil, thus ensuring more complete nitrification.

Plastic mulch prevents leaching of nutrients, particularly nitrogen. The dominant advantage of using polyethylene is that it aids in the retention of nutrients within the root zone, thereby permitting more efficient nutrient utilisation by the crop.

11.3.7 Decomposition of Organic Matter

Solarisation increases the levels of available mineral nutrients in soils by breaking down soluble organic matter and increasing bioavailability. Soil solarisation also speeds up the breakdown of organic material in the soil, often resulting in the added benefit of release of soluble nutrients such as nitrogen (NO_3^-, NH_4^+), calcium (Ca^{++}), magnesium (Mg^{++}), potassium (K^+) and fulvic acid, making them more available to plants.

11.4 Solarisation Under Different Situations

Solarisation has considerable versatility, being adaptable to various agricultural production applications.

11.4.1 Protected Cultivation

Worldwide, probably the major commercial use of solarisation is in conjunction with greenhouse/glasshouse/plastic-house culture, especially in regions where the protected crops are grown only in the winter. The empty structures can be closed in the heat of the summer and plastic film laid on the soil for solarisation. This method of double insulation provides a still-air chamber above the solarised soil for added heating and greater efficacy. Solarisation in greenhouses produces significantly higher soil temperatures than solarisation in fields or gardens and can therefore be more effective in cooler weather. Greenhouse solarisation is extensively used primarily in Japan, the Near East, southern Europe and other Mediterranean countries to control diseases of strawberries, tomatoes, eggplants, cucumbers and other intensively managed crops (Stapleton 1997).

The soil surface inside the greenhouse should be levelled and irrigated before being covered with plastic sheeting. A time of year with maximum solar radiation should be chosen. To maximise the transmission of light, it may be advisable to wash the roof of the greenhouse before treatment. Once plastic is applied, the greenhouse should be tightly closed for 4 or more weeks to contain the heat.

11.4.2 Containerised Planting Media and Seed Beds

Soil solarisation has been shown to be effective for disinfesting containerised soil and soil in cold frames. Soil temperatures should be monitored closely in this planting media to assure that temperatures are high enough to control pests. Materials can be solarised either in bags or flats covered with transparent plastic or in layers 7.5–22.5 cm wide sandwiched between two sheets of plastic. In warmer areas of California, soil inside black plastic sleeves can reach 70°C during solarisation, equivalent to target temperatures for soil disinfestation by aerated steam (Stapleton et al. 1999). At these temperatures, soil is effectively solarised within a week. A double layer of plastic can increase soil temperatures by up to 50 °F.

Soil temperatures can be monitored using simple soil thermometers inserted 10–15 cm into the soil mix or by using thermocouples and a digital reading logger. Temperatures can be monitored at different locations, but the duration should be lengthened to raise the temperature at the coolest location to the desired level.

Solarisation without combination with a chemical pesticide is not suitable for open field nursery production, due to zero tolerance for surviving pathogen propagules deeper in the soil.

11.4.3 Open Field Production (Annual Crops)

Open field cultivation of row crops is the setting in which solarisation was first discovered and tested. Soil to be treated should be thoroughly moistened either by pre-irrigation or by drip irrigation beneath the clear plastic film. Soil can be solarised in fields by either complete coverage or by covering only the planting beds. This technique works best with shallow-rooted, late-season crops in warmer locations so that the solarisation can be done between crops in the summer (Elmore et al. 1997).

11.4.4 Orchards and Vineyards

Solarisation can be modified for use in managing certain weed, disease and nematode pests before or during establishment or replanting of orchard and vineyard crops, especially in warm areas such as the Central Valley. The method is most useful for managing shallowly distributed pests. It would not be expected to effectively control nematodes, fungi and other pests deep in the soil. When using solarisation in conjunction with growing plants, treatment with black, rather than clear, plastic film may be preferable. For example, early studies indicated that solarisation with clear plastic film beginning shortly after planting killed almond and apricot trees due to excessive heat. However, similar treatment with black film controlled the disease *Verticillium* wilt without damaging the trees. Other benefits included conservation of soil moisture and reducing humidity in the crop canopy (Elmore et al. 1997). This type of solarisation is mainly used on a commercial basis in the Central Valley for *Prunus* spp. and citrus.

In the orchard or vineyard, clear plastic is either laid by hand around the bases of individual trees or vines and connected to strips laid between the rows or laid in anchored strips and glued along the tree rows. For best results, solarisation should begin as soon as trees are planted. Partial shading by young trees does not prevent soil heating nor does soil solarisation appear to bother most young trees during treatment. However, solarising certain species of trees, such as herbaceous perennials, avocado and young *Prunus* trees, with clear plastic may result in plant damage, especially when trees are young. The *Prunus* trees were killed by clear plastic but not by black film.

It has been successfully used on a large scale to reduce Verticillium wilt symptoms in young pistachio orchards in California and has also been successfully used in vineyards and in avocado, stone fruit, citrus and olive orchards in the state.

In addition to killing soil-borne pests, solarisation of orchards and vineyards can greatly reduce the amount of water needed for irrigation and increase the growth, flowering and/or fruit set of the trees. In large commercial orchards, the cost of post-plant solarisation should be compared to the benefits before making a treatment decision. Experience has shown that pests that are not eradicated by solarisation may recolonise roots and soil, and pathogens and nematodes may survive in roots remaining in the soil. Periodic retreatment may be necessary.

11.4.5 Nonconventional Users

Solarisation is presently used on a relatively small scale in conventional agriculture, but its use will probably increase as methyl bromide becomes unavailable. On the other hand, solarisation has become a widespread practice for organic growers, home gardeners and other users who cannot or will not use chemical soil disinfestants.

Although solarisation can be an effective soil disinfestant, the stand-alone process, which largely depends on passive solar heating, has inherent limitations. Fortunately, it is compatible with other physical, chemical and biological methods of soil disinfestation to provide more efficacious and/or predictable treatment through integration. As MB is phased out, many current users will turn to other pesticides for soil disinfestation. Combining these pesticides (perhaps at lower dosages) with solarisation (perhaps for a shorter treatment period) may be effective. Using solarisation in conjunction proves to be the most popular option for users in warm climatic areas who want to continue using chemical soil disinfestants.

11.5 Factors Limiting Effectiveness of Solarisation

11.5.1 Location

Soil solarisation is most effective in warm, sunny locations. It also has been used successfully, but less predictably, in the cooler coastal areas and in many cooler parts during periods of highest air temperatures and clear skies. Greenhouse, nursery and seedbed (containerised) media solarisation are more effective in cooler climates than field solarisation.

11.5.2 Weather

Highest soil temperatures occur when days are long, air temperatures are high, skies are clear and there is no wind. The soil heating effect may be limited on cloudy days. Wind or air movement across the plastic will rapidly dissipate the trapped heat. Also, strong winds may lift or tear sheets.

11.5.3 Timing

The best time for solarisation of soil is from June to August, although good results may be obtained in May and September, depending on weather and location. The heat peak in many areas is around May 15. To maximise production, soil solarisation should be done during a period in crop rotations when fields are idle. Soil can be solarised during summer (when summer temperatures are too hot for many crops) and planted during fall or winter.

11.5.4 Duration of Treatment

The longer the soil is heated, the better the control of pests will be. However, heating the soil longer than required for effective control (6–8 weeks) may be deleterious to the soil. Although some pest organisms are killed within 14 days, 4–6 weeks of treatment in full sun during the summer is recommended for field application. Solarisation of containerised growth media and greenhouses may be done in a few days during the heat of summer. Some relatively heat-resistant organisms may require longer (up to 8 weeks) solarisation for control. The combination of pesticides, fertilisers and certain organic amendments with solarisation may reduce the needed treatment time.

11.5.5 Soil Preparation

A smooth seedbed is ideal for solarisation. Air pockets between the plastic and the soil greatly reduce soil heating. Solarisation will be ineffective if the seedbed is not smooth and the plastic does not rest directly on the soil.

11.5.6 Soil Moisture Content

If the soil is too dry (less than 70% of field capacity), weed seed and pathogens may not imbibe enough water to make them vulnerable to the increased heat.

11.5.7 Soil Colour

Dark soils absorb more solar radiation than lighter coloured soils and reach higher temperatures during solarisation. However, adding dark material, such as charcoal, to a light loam soil has only raised maximum temperatures 1–2°F. Organic material such as manure may give the same limited effect.

11.5.8 Orientation of Beds

The heating of soil in raised beds will be most uniform if the beds are oriented north to south rather than from east to west. More uniform heating gives better control of pests. Solarisation is most effective when there is no slope or when the slope has a south or southwest exposure. Lower temperatures and poor control of pests will occur on north-facing slopes. Cultivation after solarisation and cultivation deeper than 7.5 cm after soil solarisation should be avoided because it may bring weed seed and pathogens to the upper soil layer, causing severe weed and disease problems.

Integrity of plastic sheeting holes or tears in the plastic will adversely affect solarisation. Animals and people should be prevented or discouraged from walking on or otherwise disturbing the plastic.

11.6 Disease Management

Repeated daily heating during solarisation kills many plant pathogens, nematodes and weed seed and seedlings. The heat also weakens many organisms that can withstand solarisation, making them more vulnerable to heat-resistant fungi and bacteria that act as natural enemies. Changes in the soil chemistry during solarisation may also kill or weaken some soil organisms.

Soil solarisation can decrease the incidence and effects of disease-causing organisms that originate in the soil. Solarisation for 47–48 days in September–October in west-central Florida either reduced disease frequency or slowed disease progression of crown rot and blight caused by *Rhizoctonia* spp. and root rot caused by *Pythium* spp. when a double layer of low polyethylene mulch was used. In another case, solarisation with a single layer of plastic mulch was more effective than methyl bromide in limiting an epidemic from *Pythium* spp. in a pepper field. In contrast, there was no difference between white plastic mulch and clear plastic mulch in the suppression of *Pythium* spp. Incidence of *Phytophthora* blight (*Phytophthora capsici*) was reduced by solarising for 6–8 weeks. Furthermore, densities of *Phytophthora nicotianae* and *Ralstonia solanacearum* were reduced by solarisation in soil depths down to 25 cm (Chellemi et al. 1994). Southern blight (*Sclerotium rolfsii*) and Fusarium wilt (*Fusarium oxysporum* f. sp. *lycopersici*) can also be substantially reduced with solarisation and results are comparable to a methyl bromide control (Chellemi et al. 1997), but effects may be limited to the upper soil layer (Chellemi et al. 1994). Solarisation as one of the methods to manage Verticillium wilt (*Verticillium*

Table 11.1 Fungal pathogens controlled by soil solarisation

Scientific name	Disease caused (crop)
Didymella lycopersici	Didymella stem rot (tomato)
Fusarium oxysporum f. sp. *conglutinans*	Fusarium wilt (cucumber)
F. oxysporum f. sp. *fragariae*	Fusarium wilt (strawberry)
F. oxysporum f. sp. *lycopersici*	Fusarium wilt (tomato)
F. oxysporum f. sp. *vasinfectum*	Fusarium wilt (cotton)
Plasmodiophora brassicae	Club root (cruciferae)
Phoma terrestris	Pink root (onion)
Phytophthora cinnamomi	Phytophthora root rot (many crops)
Pyrenochaeta lycopersici	Corky root (tomato)
Pythium ultimum, *Pythium* spp.	Seed rot or seedling disease (many crops)
P. myrothecium	Pod rot (peanut)
Rhizoctonia solani	Seed rot or seedling disease (many crops)
Sclerotinia minor	Lettuce drop
Sclerotium cepivorum	White rot (garlic and onions)
S. rolfsii	Southern bight (many crops)
Thielaviopsis basicola	Black root rot (many crops)
Verticillium dahliae	Verticillium wilt (many crops)

albo-atrum) has been recommended. Several studies have concluded that bacterial wilt caused by *R. solanacearum* is not controlled by solarisation (Chellemi et al. 1994, 1997).

Solarisation controls populations of many important soil-borne fungal and bacterial plant pathogens, including *Verticillium dahliae*, which causes *Verticillium* wilt in many crops; certain *Fusarium* spp. that cause *Fusarium* wilt in some crops; *Phytophthora cinnamomi*, which causes *Phytophthora* root rot; *Agrobacterium tumefaciens*, which causes crown gall disease; *Clavibacter michiganensis*, which causes tomato canker; and *Streptomyces scabies*, which causes potato scab. Other fungi and bacteria are more difficult to control with solarisation, such as certain high temperature fungi in the genera *Macrophomina*, *Fusarium*, *Pythium* and the soil-borne bacterium *Ralstonia solanacearum* (Tables 11.1, 11.2, 11.3, 11.4, and 11.5).

Table 11.2 Bacterial pathogens controlled by soil solarisation

Scientific name	Disease caused (crop)
Agrobacterium tumefaciens	Crown gall (many crops)
Clavibacter michiganensis	Canker (tomato)
Streptomyces scabies	Scab (potato)

Table 11.3 Fungal pathogens unpredictably controlled by soil solarisation

Scientific name	Disease caused (crop)
Fusarium oxysporum f. sp. *pini*	Fusarium wilt (pines)
Macrophomina phaseolina	Charcoal rot (many crops)

Table 11.4 Bacterial pathogens unpredictably controlled by soil solarisation

Scientific name	Disease caused (crop)
Ralstonia solanacearum	Bacterial wilt (several crops)

Table 11.5 Some diseases controlled by soil solarisation

Disease	Crop
Verticillium wilt	Tomato, potato, eggplant, cotton, strawberry
Fusarium wilt	Tomato, melon, onion, cotton
Pink root rot	Onion
Southern stem rot (white mould)	Peanut
Rhizoctonia seedling disease (sore shin or damping off)	Potato, onion, bean
Crown gall	Walnut
Phytophthora root rot	Ornamentals

11.7 Nematode Management

Plant-parasitic nematodes are killed by high temperatures. Early research on *Meloidogyne javanica* (root-knot nematode) showed that movement of juveniles stopped immediately after being exposed to 50°C and there was no recovery even after returning the temperature to 25°C. Lowering the temperature resulted in longer periods required to kill juveniles; at 42°C it required 3 h. In Tanzania, egg masses of *M. javanica* were buried in soil (15 cm depth) and exposed to solarisation. Within 2–3 weeks, all eggs were dead. Soil temperatures under

Table 11.6 Hours needed to kill 100% of *Meloidogyne incognita* eggs and juveniles

	Hours to kill 100 %	
Temperature (°C)	Eggs	Juveniles
38	389.8	–
39	164.5	47.9
40	32.9	46.2
41	19.7	17.5
42	13.1	13.8

solarisation reached an average of 43°C, with maximum of 45°C.

Solarisation can be effective in Florida but may be impaired by overcast skies and rainfall during the warmest summer months. In order to counteract these effects, the solarisation time could be prolonged. Increasing the length of solarisation will cause mortality based on the accumulation of sublethal but detrimental temperatures. Solarisation of soil for 41 days mainly during October in southwest Florida, resulted in the reduction of awl (*Dolichodorus heterocephalus*) and stubby-root (*Paratrichodorus minor*) nematodes. Nematodes were exposed to average maximum temperatures of 38.4, 33.6 and 29.8°C at depths of 5, 15 and 23 cm. A recent laboratory study confirms that nematodes can be affected by an accumulation of sublethal temperatures. Eggs and juveniles of *M. incognita* were exposed to a series of temperatures from 38 to 45°C. At 44 and 45°C, juveniles were killed within 1 h. For lower sublethal temperatures, the hours required for suppression decreased with increasing temperatures (Table 11.6). Success of solarisation depends on maximum temperatures reached in the field. The higher the temperature, the shorter the duration of solarisation required to kill nematodes. Temperatures over 40°C should be the goal in order to shorten the solarisation period. In a 6-week solarisation period (Florida: July to August), temperatures high enough to kill nematodes could be accumulated.

Several nematode species were negatively affected by solarisation. It was effective in reducing populations of *M. incognita*, *D. heterocephalus*, *P. minor*, *Belonolaimus longicaudatus* (sting), *Criconemella* spp. (ring) and *Rotylenchulus reniformis* (reniform) (Chellemi et al. 1997; McSorley and Parrado 1986; Ozores-Hampton et al. 2004). However, Chellemi et al. (1997) reported that they were unable to reduce *Meloidogyne* spp. and *R. reniformis*. In some cases, solarisation was able to suppress populations of *M. incognita* or *R. reniformis* initially, but numbers recovered at the end of the cropping season (McSorley and Parrado 1986). In addition, when using strip solarisation, it is possible that fast recolonisation of plant-parasitic nematodes may occur, because row middles (occupying up to 50% of the field) will remain untreated, although this late-season recovery may not limit overall yield (Chellemi et al. 1997).

Resurgence of certain nematode species may occur to higher levels than before solarisation. The stubby-root nematode *Paratrichodorus minor* increased in numbers following 3 weeks of solarisation. A possible explanation for resurgence could be that some nematode species have population reservoirs in deeper soil layers that are larger than those found in the upper soil layers. This unusual vertical distribution in soil often occurs with stubby-root nematodes. Although solarisation reduces or eliminates these nematodes in the upper layers, recolonisation can occur quickly by drawing upon a population pool from deeper soil layers. Similar resurgence of stubby-root nematodes has also been observed when these nematodes are managed with soil fumigation (Weingartner et al. 1975).

Soil solarisation can be used to control many species of nematodes (Table 11.7). However, soil solarisation is not always as effective in controlling nematodes as it is in controlling fungal disease and weeds because nematodes are relatively mobile and can recolonise soil rapidly. Nematode management may therefore require yearly treatment. Control by solarisation is greatest in the upper 30 cm of the soil. Nematodes deeper in the soil profile may survive solarisation (Table 11.8) and damage plants with deep root systems.

Nematode control by solarisation is usually adequate to improve the growth of shallow-rooted, short-season plants. It is particularly useful for organic gardeners and home gardeners. Solarisation may also be a beneficial addition to

Table 11.7 Nematodes controlled by soil solarisation

Scientific name	Common name
Criconemella xenoplax	Ring nematode
Ditylenchus dipsaci	Stem and bulb nematode
Globodera rostochiensis	Potato cyst nematode
Helicotylenchus digonicus	Spiral nematode
Heterodera schachtii	Sugar beet cyst nematode
Meliodogyne hapla	Northern root-knot nematode
M. javanica	Javanese root-knot nematode
Paratylenchus hamatus	Pin nematode
Pratylenchus penetrans	Lesion nematode
P. thornei	Lesion nematode
P. vulnus	Lesion nematode
Tylenchulus semipenetrans	Citrus nematode
Xiphinema spp.	Dagger nematode

Table 11.8 Nematodes unpredictably controlled by soil solarisation

Scientific name	Common name
Meloidogyne incognita	Southern root-knot nematode

an integrated nematode control system. For example, excellent control of root-knot nematode (*Meloidogyne incognita*) was obtained in the San Joaquin Valley by combining solarisation with the application of composted chicken manure (Gamliel and Stapleton 1993b).

11.8 Weed Management

The use of soil solarisation for weed control is in the proven but experimental stage. We know how to make it work in the field. It has been used in the field, but there are still obstacles to its use.

Polyethylene sheeting for weed control has been widely used in agricultural and urban areas. Black polyethylene has been the dominant material where it is used as a barrier mulch. This mulch excludes light, thus prohibits photosynthesis and kills young annual weeds or even many perennial weeds if the mulch remains for a long enough period. In the early 1970s, research in Japan (Inada 1973) demonstrated that a green polyethylene mulch that would absorb light at 450-nm wavelength, reduced weed growth, though not as effectively as black polyethylene mulch. Some of these coloured mulches are still used in greenhouses and in the field, not for soil solarisation, but as a barrier for weed control.

Katan et al. (1976) described the use of clear polyethylene for soil solarisation (soil pasteurisation) to control soil pathogens and weeds. Subsequent studies for weed control have utilised this same application technology. Various synthetic mulches have been evaluated; however, polyethylene seems to have characteristics of transmittance and durability, for greatest effectiveness.

Soil solarisation for weed control is both visually dramatic and highly effective when properly conducted. Though restricted to a time interval in summer during periods of high radiation, it can effectively control species that would be a problem in the subsequent fall or spring crops. Horowitz et al. (1983) found in Israel that a 2–4-week solarisation period effectively controlled annual weeds that were appreciable after 1 year. Rubin and Benjamin (1984) reported similar results from solarisation for 4–5 weeks. In the field, solarisation has been best adapted for the control of weeds in cool seasons and for fall-seeded crops such as onions, garlic, carrots, broccoli or other Brassica crops and lettuce. In greenhouses or in the more temperate to tropical regions, solarisation can be used before planting of the spring-planted crops such as tomatoes, peppers, squash and cucurbits. Also in the field, it can be used to pre-plant strawberries. Other crops that have been evaluated include broad beans, potatoes, orchard trees and vineyards. In some studies, there has been no need for additional weed control measures such as hand weeding or cultivation in the annual crop following solarisation. Weed control has not been effective into the second or third years following treatment, though weeds are reduced. If beds are not disturbed following treatment, weed control can be effective for two crops (Bell and Elmore 1983).

The greatest difficulty in consistently achieving good weed control has been when the operator (farmer) cultivates after removing the mulch, thus bringing untreated soil (and seed) to the surface. Thus, much of the mulching for weed control has been on preformed beds rather than flat soil and then bedding following treatment.

Table 11.9 Susceptibility of winter annual weeds to soil solarisation

Weed species	Reported	Location
Anagallis coerulea	Sensitive	Israel
Avena fatua	Moderately sensitive	Israel, California (USA)
Centaurea iberica	Sensitive	Israel
Emex spinosa	Sensitive	Israel
Capsella bursa-pastoris	Sensitive	Israel, California (USA)
Lactuca scariola	Sensitive	Israel
Mercurialis annua	Sensitive	Israel
Lamium amplexicaule	Sensitive	Israel, California (USA)
Poa annua	Sensitive	California (USA), Louisiana (USA)
Raphanus raphanistrum	Sensitive	California (USA), Louisiana (USA)
Sonchus oleraceae	Sensitive	Israel, California (USA)
Senecio vulgaris	Sensitive	California (USA)
Montia perfoliata	Sensitive	California (USA)
Urtica urens	Sensitive	Israel
Erodium sp.	Sensitive	Australia

Standifer (1984) used full-tarped beds and achieved control of annual sedges and many other annual species. Rubin and Benjamin (1984) used flat solarised soil, but did not form beds, and achieved control. Solarisation can be followed by cultivating to 3 or 6 cm compared to an uncultivated area. This test was conducted from June to July or from August to September. No difference was found in germination of weeds at the 0- and 3-cm cultivations; however, a greater number of weeds germinated after 6-cm cultivation than at the uncultivated site. The difference was greater between cultivation depths in the August to September treatment than the more optimum June solarisation period. There were no differences in control of annual grass species by cultivation at 0, 5 and 20 cm following solarisation. A numerical, but not a statistical, increase in germination of *Portulaca oleracea* and *Malva parviflora* occurred at 20-cm cultivation compared to the 5 cm depth. Under 1–2-week solari sation experiments, Egley (1983) found increased weed emergence following 5-cm cultivation in solarised plots. Thus, it would seem that shallow cultivation would not be detrimental if the solarisation was conducted under optimum conditions.

In bed solarisation culture, it has been noted that more uniform control is achieved when the beds are formed in a north/south direction as compared to an east/west direction. There is always a cool north side on the east/west bed. Bed-top width solarisation of 80 cm or more was required to get a uniform temperature and control in the centre of the beds where crops were to be planted.

11.8.1 Weed Species Controlled

Winter annual weeds that germinate during short days and cool temperatures have been effectively controlled (Table 11.9). These species seem to be very temperature sensitive and require small increments of temperature increases to achieve control. Weeds that germinate predominantly in the longer day length, warmer period of the year (summer annuals) are not as susceptible (Table 11.10). Most species, however, are controlled. In some areas of marginal solar radiation, windy or cloudy conditions, etc., some of these summer annuals may escape. Predominant species that escape include *Melilotus* sp., *Medicago* sp. and to some lesser extent *Portulaca* sp. (Porter and Merriman 1983). Results from the literature may not give the true picture of control of some of these summer annual weeds. If solarisation conditions were marginal, species would be listed as 'not controlled'. Results on the perennial species *Cynodon dactylon* (Bermuda grass) and *Sorghum halepense* (Johnson grass) have varied

Table 11.10 Susceptibility of summer annual weeds to soil solarisation

Weed species	Reported	Location
Conyza canadensis	Moderately resistant	Israel
Echinochloa crusgalli	Sensitive	Louisiana (USA), California (USA), Australia
Ipomoea lacunosa	Sensitive	Mississippi (USA)
Malva nicaeensis	Moderately sensitive	Israel
M. parviflora	Sensitive	California (USA)
Melilotus sulcatus	Resistant	Israel
Orobanche crenata	Sensitive	Israel, Egypt
O. aegyptiaca	Sensitive	Israel
Portulaca oleracea	Moderately sensitive–sensitive	Egypt, Israel, California (USA), Australia
Digitaria sanguinalis	Sensitive	Israel, California (USA)
Sida spinosa	Sensitive	Mississippi (USA)
Tribulus terrestris	Sensitive	Israel
Xanthium spinosum	Sensitive	Israel
Astragalus boeticus	Moderately sensitive	Israel
Solanum nigrum	Sensitive	Egypt, California (USA), Israel, Australia

Table 11.11 Susceptibility of perennial weed species to soil solarisation

Weed species	Reported	Location
Convolvulus arvensis	Moderately resistant–resistant	Israel, California (USA)
Cyperus esculentus	Resistant	Florida (USA), California (USA)
C. rotundus	Resistant	Egypt, Israel, Mississippi (USA)
Cynodon dactylon	Moderately sensitive	Israel, California (USA), Texas (USA)
Sorghum halepense	Moderately sensitive	Israel, California (USA)

(Table 11.11). In tests at California, both species have been controlled. *Convolvulus arvensis* (field bindweed) has been controlled during the period of solarisation and suppressed for 2–3 weeks afterwards before regrowth occurred. *Cyperus esculentus* (yellow nutsedge) was only reduced by approximately 40%, and *C. rotundus* (purple nutsedge) was partially controlled or increased following solarisation.

The control of seed or seedlings must take into account the seed germination depth and both the temperature and duration of temperature. Control of many small-seeded weeds such as annual bluegrass and common groundsel is more effective than large-seeded weeds such as wild oat since these large seeds can germinate from depths below which thermal killing occurs.

Seed size and germination depth, however, is probably not the primary protection mechanism for weed seeds. Seeds from various weed species vary in their sensitivity to temperature (thermal death point). Common groundsel (*Senecio vulgaris*), for example, is so temperature sensitive that even 4 h at 40°C will reduce germination and 8 h will kill all imbibed seeds. Common purslane (*Portulaca oleracea*), however, can be controlled with either high temperature for a short duration or lower temperature for a longer duration. Treating moist *P. oleracea* seeds at 45°C for up to 8 h did not reduce germination. At 50°C, germination was reduced by 16% at 4 h of treatment or 43% at 8 h. At 60°C, all germination was stopped when treated for 4 and 8 h with 90% reduction at 2 h. Barnyard grass and lambs quarters were more temperature sensitive than *P. oleracea* but less sensitive than *S. vulgaris*. Other weeds such as *Melilotus* sp. and *Medicago* sp. seem to be very heat tolerant.

Soil solarisation has been effective for the control of many weed species in various parts of the world. The treatment must be applied within the correct weather, timing and application perimeters to make the process work. However, when these factors are observed, then control is achieved. Solarisation will not be an effective method of weed control in all crops or in some microclimates where it would normally be effective. This lack of control is most frequently observed under less than optimum application and weather conditions; however, it can be affected by weed species.

Treatment during periods of high radiation followed by good crop management practices can make the treatment effective and preserve the control for longer periods. Cool season crops will be most impacted; however, some summer crops can also be improved through solarisation because of the decreased weed pressure in the crop seedling or new transplant stage.

Soil solarisation has proved to be very effective in controlling several genera of weeds, except *Cyperus* (50–93%). However, the level of killing of this weed depends on soil moisture; the higher the moisture content prior to mulching, the higher the level of killing. Broomrape was controlled by solarisation. It could be controlled for two successive years, in broad bean or in tomato fields, by one treatment.

In general, various weeds were also controlled, completely or partially (75–100%), by soil solarisation; these included *Amaranthus* sp., *Portulaca* spp., *Plantago* spp., *Chenopodium murale, Vicia* sp., *Lactuca scariola, Beta vulgaris, Rumex dentatus, Cynodon dactylon, Coronopus squamatus* and *Sisymbrium irio.*

In contrast, *Malva* spp., *Melilotus indica* and *Convolvulus arvensis* were less affected by solarisation with less than 75% killing.

Soil solarisation controls many annual and perennial weeds. Whilst some weed species are very sensitive to soil solarisation, others are moderately resistant and require optimum conditions (good soil moisture, tight-fitting plastic and high radiation) for control. Winter annual weeds seem to be especially sensitive to solarisation, and control of winter annuals is often evident for more than 1 year following treatment. Soil solarisation is especially effective in controlling weeds in fall-seeded crops such as onions, garlic, carrots, broccoli and other Brassica crops and lettuce. White sweet clover (*Melilotus alba*) is one of the few winter annuals that are poorly controlled. Although summer annual weeds are less temperature sensitive than winter annuals, most summer annuals are relatively easily controlled by soil solarisation. Control of purslane (*Portulaca oleracea*) and crabgrass (*Digitaria sanguinalis*) may be more difficult to achieve. If purslane is controlled, it is a good indicator that the soil has been adequately heated.

Solarisation generally does not control perennial weeds as well as it controls annual weeds because perennials often have deeply buried underground vegetative structures such as roots and rhizomes that may resprout. Seeds of Bermuda grass (*Cynodon dactylon*), Johnson grass (*Sorghum halepense*) and field bindweed (*Convolvulus arvensis*) are controlled by solarisation. Rhizomes of Bermuda grass and Johnson grass may be controlled by solarisation if they are not deeply buried. Solarisation alone is not effective for the control of the rhizomes of field bindweed. Yellow nutsedge (*Cyperus esculentus*) is only partially controlled by soil solarisation. Purple nutsedge (*Cyperus rotundus*) is not significantly affected; marginal solarisation has actually induced purple nutsedge to grow.

Solarisation can minimise weed emergence by causing thermal death to weed seeds. Solarisation performed for 1–4 weeks reduced emergence of prickly sida (*Sida spinosa*), pigweed (*Amaranthus* spp.), morning glories (*Ipomoea* spp.), horse purslane (*Trianthema portulacastrum*) and several grass species (Egley 1983). Solarisation is also an important tool to manage nutsedge, which is often hard to control with regular mulches, because it grows as a rhizome (not requiring sunlight) until it encounters light, then pierces mulches with its sharp growing point and thereafter expands its leaves above the plastic film. Given sufficiently high solarisation temperatures, purple and yellow nutsedges either do not emerge at all, or if they do emerge, then leaf expansion, which is stimulated by sunlight, will occur under the clear plastic, which will mostly prevent puncturing of the plastic. The plant will be exposed to the heat underneath the plastic and thermal death occurs.

However, prolonged use of solarisation can lead to a shift in weed populations when used as a long-term, single solution to weed management. In a study by Ozores-Hampton et al. (2004), beds that were treated with methyl bromide were dominated by pigweed (*Amaranthus retroflexus*), whereas solarised beds were primarily colonised by Bermuda grass (*Cynodon dactylon*), a perennial that is harder to control by solarisation than the annual pigweed.

11.9 Insect Management

Stapleton and DeVay (1995) reported that the greatest use of solarisation as a pest management tool is in greenhouses, organic farms and backyard gardens. It may be possible to combine solarisation with reduced dosages of chemical fumigants to obtain good insect control.

11.9.1 Melon Thrips, *Thrips palmi*

In parts of Japan, hundreds of hectares of greenhouses that produce cucumber or sweet pepper are treated by solarisation for 7 days following crop removal (Horiuchi 1991). The main target pest for this treatment is the melon thrips, *Thrips palmi*. Heating the soil is not the objective in this case, but rather raising the air temperature to heat the crop residue and the thrips that remain on the plants. All plants are pulled from the soil and left in the greenhouse.

11.9.2 Pepper Weevil, *Anthonomus eugenii*

Costello and Gillespie (1993) found that maintaining greenhouse temperatures at 25°C for 10 days after removing all plant residues killed adults of the pepper weevil, *Anthonomus eugenii*. A combination of maintaining 20°C and deploying large numbers of yellow sticky traps to capture adults was also used successfully.

11.10 Integration of Solarisation with Other Management Methods

Under conducive conditions and proper use, solarisation can provide excellent control of soil-borne pathogens in the field, greenhouse, nursery and home garden. However, under marginal environmental conditions, with thermotolerant pest organisms or those distributed deeply in soil, or to minimise treatment duration, it is often desirable to combine solarisation with other appropriate pest management techniques in an integrated pest management approach to improve the overall efficacy of treatment (Stapleton 1997). Solarisation is compatible with numerous other methods of physical, chemical and biological pest management. This is not to say that solarisation is always improved by combining with other methods.

Combining soil solarisation with pesticides, organic fertilisers and biological control agents has led to improved control of pathogens, nematodes and weeds and may be especially useful in cooler areas, against heat-tolerant organisms, or to increase the long-term benefits of solarisation.

11.10.1 Solarisation and Biofumigation

The primary focus of solarisation research will be towards the indirect effects of solarisation and the effectiveness of combining it with other pest management strategies such as biofumigation. Katan and Devay (1991) states that microbial processes, induced by solarisation, may contribute to disease control beyond the physical effect of heating. Both the vulnerability of soil-borne pathogens to mycoparasitic attack and the activity of beneficial microorganisms could be enhanced by solarisation. In particular, a study by Lifshitz et al. (1983) showed that sclerotia of *S. rolfsii* became heavily colonised by bacteria under the influence of the sublethal temperatures generated by solarisation. Cracking of the rind and increased leakage of organic compounds from the sclerotia was reported.

Hence, the germinability of these propagules was compromised.

Organic amendments may increase the effectiveness of solarisation against pests (Gamliel and Stapleton 1993a; Ozores-Hampton et al. 2004; Wang et al. 2008). Wang et al. (2008) showed that following a cowpea cover crop with solarisation was more effective than solarisation alone. In fact, the effectiveness of this combined treatment was comparable to methyl bromide fumigation.

Cabbage residues may or may not improve solarisation success. When cabbage residues were incorporated as large undecomposed pieces, solarisation was not improved and the cabbage may have served as a reservoir for diseases (Chellemi et al. 1997). However, grinding cabbage into powder and mixing it uniformly into the soil may improve solarisation efforts (Gamliel and Stapleton 1993b).

11.10.2 Solarisation and Chemical Controls

Many field trials have shown that, under the prevailing conditions, pesticidal efficacy of solarisation or another management strategy alone could not be improved upon by combining the treatments (Stapleton and DeVay 1995). However, even in such cases, combination of solarisation with a low dose of an appropriate pesticide may provide the benefit of a more predictable treatment which is sought by commercial users. For example, although combining solarisation with a partial dose of 1, 3-dichloropropene did not statistically improve control of northern root-knot nematode (*Meloidogyne hapla*) over either treatment alone; it did reduce recoverable numbers of the pest to near undetectable levels to a soil depth of 46 cm.

Low application rates of fungicides, fumigants or herbicides have been successfully combined with soil solarisation to achieve better pest control. The elevated temperatures seem to increase the activity of fungicides such as metham sodium, so lower rates may be applied. Solarisation speeds up the disappearance of EPTC (Eptam) and vernolate, either by increasing their volatility or their degradation. Other chemicals, such as terbutryn or carbendazim, have slower degradation rates after solarisation, possibly because of changes in the populations of soil microorganisms after solarisation. Although such pesticides may be effective for longer periods than normal, care must be taken that they do not harm the next crop. Chemical controls may be applied either before or after solarisation. A possible disadvantage of combining soil solarisation with chemical control is that the chemical control may reduce the long-term benefits of solarisation.

11.10.3 Solarisation, Amendments and Fertilisers

Solarisation can also be combined with a wide range of organic amendments, such as composts, crop residues, green manures and animal manures and inorganic fertilisers to increase the pesticidal effect of the combined treatments (Ramirez-Villapudua and Munnecke 1987; Gamliel and Stapleton 1993a, b; Chellemi et al. 1997). Incorporation of these organic materials by themselves may act to reduce number of soil-borne pests in soil by altering the composition of the resident microbiota or of the soil physical environment (biofumigation). Combining these materials with solarisation can sometimes greatly increase the biocidal activity of the amendments. However, this appears to be an inconsistent phenomenon, and such effects should not be generalised without first conducting confirmatory research. The concentrations of many volatile compounds emanating from decomposing organic materials into the soil atmosphere have been shown to be significantly higher when solarised (Gamliel and Stapleton 1993b).

Many commercial users of solarisation in California apply manures or other amendments to soil before laying the plastic. There is evidence that these materials release volatile compounds in the soil that kill pests and help stimulate the growth of beneficial soil organisms. For example, the southern root-knot nematode, which was incompletely controlled in lettuce by either solarisation or application of composted chicken manure, was completely controlled by combining the two, resulting in a large yield increase (Gamliel and Stapleton 1993a).

11.10.4 Solarisation and Biological Controls

The successful addition of biological control agents to soil before, during or after the solarisation process in order to obtain increased and persistent pesticidal efficacy has long been sought after by researchers. There have been great hopes of adding specific antagonistic and/or plant growth-promoting microorganisms to solarised soil, either by inundative release or with transplants or other propagative material, to establish a long-term disease-suppressive effect to subsequently planted crops (Katan 1987; Stapleton and DeVay 1995).

It is considered that by combining solarisation with biofumigation and composted amendments, the reintroduction of biocontrol agents such as *Trichoderma* spp. and *Bacillus* spp. may be more effective than either treatment alone in controlling soil-borne diseases. Populations of these two microbial antagonists increase relative to other microorganisms in solarised soil.

Soil solarisation has also been successfully combined with the fungal biological control agents *Trichoderma harzianum* and *Talaromyces flavus*, which were added to the soil or planting material. In another integrated pest management strategy, clear plastic applied to planting rows in summer for solarisation may be left in place and painted silver to control aphid-borne viral diseases in fall vegetable crops.

11.11 Mode of Action

Although the execution of solarisation is simple, the overall mode of action can be complex, involving a combination of several interrelated processes which occur in treated soil and result in increased health, growth, yield and quality of crop plants (Katan 1987; Stapleton and DeVay 1995; Stapleton 1997). The pesticidal activity of solarisation was found to stem from a combination of physical, chemical and biological effects, as described in several comprehensive reviews (Katan 1987; Chen et al. 1991; DeVay et al. 1991; Stapleton 1998, 2000).

11.11.1 Physical Mechanisms

Direct thermal inactivation of soil-borne pathogens and pests is the most obvious and important mechanism of the solarisation process. Under suitable conditions, soil undergoing solarisation is heated to temperatures which are lethal to many plant pathogens and pests. Thermal inactivation requirements have been experimentally calculated for a number of important plant pathogens and pests (Katan 1987; Stapleton and DeVay 1995). Although most mesophilic organisms in soil have thermal damage thresholds beginning around 39–40°C, some thermophilic and thermotolerant organisms can survive temperatures achieved in most types of solarisation treatment (Stapleton and DeVay 1995).

Because solarisation is a passive solar process, soil is heated to maximal levels during the daytime and then cooled at night. The highest temperatures during solarisation are achieved at or near the soil surface, and soil temperature decreases with increasing depth. Typical, diurnal maximum/minimum soil temperatures during summer solarisation of open field soils in the inland valleys of California might be 50/37°C at 10 cm, and 43/38°C at 20 cm with 35/20°C air temperature flux. Solarising soil in closed greenhouses or in containers with a limited volume of soil may produce considerably higher soil temperatures. For example, solarising soil in 3.8-l plastic containers resting on steel pallets under low plastic tunnels constructed using two layers of transparent l m separated by ca 23-cm air space ('double tent') gave maximum/minimum soil temperatures of 75/16°C with corresponding air temperatures of 38/17°C. During solarisation of open fields or greenhouse floors, destruction

Table 11.12 Characteristics of polyethylene types used for soil solarisation

Type of polyethylene	Total luminous transmittance	Greenhouse effect of the medium–long IR radiations
Low-density polyethylene (LDPE)	88–90%	24–38%
Polyvinyl chloride (PVC)	90%	<15
Copolymer of ethylene vinyl acetate (EVA)	80–90%	<25

of soil-borne pest inoculum usually is greatest near the surface and efficacy decreases with increasing depth (Stapleton 1997). There are a number of physical factors which influence the extent of soil heating during solarisation. First, solarisation is dependent on high levels of solar energy, as influenced by both climate and weather. Cloud cover, cool air temperatures and precipitation events during the treatment period will reduce solarisation efficiency (Chellemi et al. 1997). Solarisation is commercially practised mainly in areas with Mediterranean, desert and tropical climates which are characterised by high summer air temperatures. In order to maximise solar heating of soil, transparent plastic film is most commonly used for solarisation. Transparent film allows passage of solar energy into the soil, where it is converted into longer wavelength infrared energy. This long wave energy is trapped beneath the film, creating a 'greenhouse effect'. Opaque black plastic, on the other hand, does not permit passage of most solar radiation. Rather, it acts as a 'black body' which absorbs incoming solar energy. A small portion of the energy is conducted into soil, but most of the solar energy is lost by reradiation into the atmosphere. Nevertheless, solarisation with black or other colours of plastic mulch is sometimes practised under special conditions (Stapleton 1997).

The transparent film used for mulching transmits the visible spectrum of short wavelength of the solar radiation but prevents the long wave infrared radiation from the soil. The passage of light through a semi-transparent body depends on the characteristics of the material that constitutes it and on the angle of the ray, that is, the angle formed by the luminous ray and the perpendicular to the surface of the point of penetration. When both direct and diffuse solar radiation strike the plastic tarp on the soil, they are in part reflected back towards the atmosphere, in part adsorbed by the plastic material and in part penetrate the underlying soil. The short wave incident solar radiation penetrates the polyethylene sheet, but the long wave radiation is prevented from soil, thus trapping the heat resulting in thermal inactivation and production of heat shock proteins and irreversible heat injury. The material used for trapping is therefore, the main element in the phase of capturing and storing the solar energy during the day, but it also acts as a barrier to dispersion of the heat energy during the night when the flux of solar energy has ceased. The transfer of energy from the soil depends on its internal conductivity which is a function of the soil type and its humidity content. The choice of the optimal characteristics of thermic films aims to have a transmittance of ≥80% in the visible spectrum and ≥25% in the long IR. To protect the quality of plastics for agricultural use, the Italian Institute of Plastics has introduced 'marchio di qualita' which ensures the incorporation of characteristics mentioned in Table 11.12. Hence, the PVC films are most efficient in solarisation, but its greenhouse effect depends on the percentage of vinyl acetate which varies between 12 and 14% in the films available in the market.

Apart from solar irradiation intensity, air temperature, plastic film colour and other factors play roles in determining the extent of soil heating via solarisation. These include soil moisture and humidity at the soil/tarp interface, properties of the plastic, soil properties, colour and tilth and wind conditions. The procedure of covering of very moist soil with plastic film to produce microaerobic or anaerobic soil conditions, but without lethal solar heating, can by itself produce varying degrees of soil disinfestation (Katan 1987; Stapleton and DeVay 1995).

11.11.2 Chemical Mechanisms

In addition to direct physical destruction of soil-borne pest inoculum, other changes to the physical soil environment occur during solarisation. Amongst the most striking of these is the increase in concentration of soluble mineral nutrients commonly observed following treatment. For example, the concentrations of ammonium- and nitrate-nitrogen are consistently increased across a range of soil types after solarisation. Results of a study in California showed that in soil types ranging from loamy sand to silty clay, NH_4-N and NO_3-N concentration in the top 15-cm soil depth increased 26–177 kg/ha (Katan 1987; Stapleton and DeVay 1995). Concentrations of other soluble mineral nutrients, including calcium, magnesium, phosphorus, potassium and others, also sometimes increased but less consistently. Increases in available mineral nutrients in soil can play a major role in the effect of solarisation, leading to increased plant health and growth and reduced fertilisation requirements. Increases in some of the mineral nutrient concentrations can be attributed to decomposition of organic components of soil during treatment, whilst other minerals, such as potassium, may be virtually cooked off the mineral soil particles undergoing solarisation. Improved mineral nutrition is also often associated with chemical soil fumigation (Chen et al. 1991).

11.11.3 Biological Mechanisms

In addition to direct physical and chemical effects, solarisation causes important biological changes in treated soils. The destruction of many mesophilic microorganisms during solarisation creates a partial 'biological vacuum' in which substrate and nutrients in soil are made available for recolonisation following treatment (Katan 1987; Stapleton and DeVay 1995). Many soil-borne plant parasites and pathogens are not able to compete as successfully for those resources as other microorganisms which are adapted to surviving in the soil environment. This latter group, which includes many antagonists of plant pests, is more likely to survive solarisation or to rapidly colonise the soil substrate made available following treatment. Bacteria including *Bacillus* and *Pseudomonas* spp., fungi such as *Trichoderma* and some free-living nematodes have been shown to be present in higher numbers than pathogens following solarisation. Their enhanced presence may provide a short- or long-term shift in the biological equilibrium in solarised soils which prevents recolonisation by pests and provides a healthier environment for root and overall plant productivity (Katan 1987; Stapleton and DeVay 1995; Gamliel and Stapleton 1993a).

11.12 Strategies to Enhance Efficacy of Soil Solarisation

Strategies are needed for improving the level of pest control and creating a more long-term effect, lasting throughout a season or over successive seasons. Some of the other reasons for improvements are cost reduction; increasing the reliability and reproducibility of the method, which is climate-dependent; shortening the period during which the soil is occupied with mulch; and making solarisation possible for longer periods during the year and more acceptable.

Improvement of solarisation can be achieved by either using improved plastic or modifying the application technology. For example, by solarising shallow layers of growth substrates, the temperature can be increased to very high levels, thus leading to control of even the thermotolerant pathogen *Monosporascus cannonballus* (Pivonia et al. 2002).

11.12.1 Two Transparent Films

One way to increase soil heating is the use of double-layer mulch, which heats the soil to higher levels than a single one. Raymundo and Alcazar (1986) achieved an increase of 12.5°C at a depth of 10 cm using a double-layer film compared to a single-layer one (60 vs. 47°C, respectively). Ben-Yephet et al. (1987) observed a 98% reduction in the viability of *F. oxysporum* f. sp. *vasinfectum* after 30 days under the double mulch compared with 58% reduction under single mulch, at a depth of 30 cm. Double-layer films form a static air space under and between the plastic layers. This construction apparently acts as an insulator for heat loss from the soil to atmosphere, especially during the night. The use of a double-layer film also offers opportunities for applying solarisation in areas and climatic conditions which are not favourable for solarisation when a single layer is used. In Central Italy, in an area which is climatically marginal for solarisation, double-film solarisation was very effective in reducing the viability of *Pythium, Fusarium* and *Rhizoctonia* in a forest nursery (Annesi and Motta 1994). Solarisation in a closed greenhouse is another version of the double-layer film which improved disease control (Garibaldi and Gullino 1991).

11.12.2 Transparent over Black Double Film

Stevens et al. (1999) reported a 5°C increase in soil temperature when applying solarisation in strips in a cloudy climate. Similarly, Arbel et al. (2003) achieved an increase in temperature by mulching transparent polyethylene sheet over a layer of sprayable black mulch. In field experiments, they observed that the mortality of resting structures of *F. oxysporum, S. rolfsii* and *R. solani* was higher than in the plots which were solarised by a single plastic layer. Consequently, in the solarised plots with mulching transparent polyethylene sheet over a layer of sprayable black mulch, effective control of Fusarium crown and root rot of tomatoes and vine collapse of melons (caused by *M. cannonballus*) was achieved, whilst solarisation with regular films was not effective (Arbel et al. 2003).

11.12.3 Improved Films

A polyethylene film which was formulated with the addition of anti-drip (AD) components prevents condensation of water droplets on the film surface, leading to a 30% increase in irradiation transmittance over regular film. Soil temperatures under AD films were 2–7°C higher than under regular film. Solarisation with AD film in field experiments resulted in effective control of sudden wilt of melons, whilst solarisation with common transparent film had no effect on disease level. Virtually impenetrable films were more effective in raising soil temperatures and killing Fusarium than regular polyethylene (Chellemi et al. 1997).

11.12.4 Sprayable Films

Sprayable polymers also offer a feasible and cost-effective alternative to plastic tarps for soil heating. The plastic-based polymers are sprayed on the soil surface in the desired quantity and form a membrane film, which can maintain its integrity in soil and elevate soil temperatures. Nevertheless, the formed membrane is porous and allows overhead irrigation. Stapleton and Gamliel (1993) achieved effective soil heating and a reduction in the viability of *Pythium* propagules. In Israel, a sprayable polymer product, 'Ecotex', was developed together with the technology to apply it economically on soils for various purposes (Skutelsky et al. 2000). Soil coating with this technology with a black polymer formulation resulted in a membrane film that could raise soil temperatures close to solarisation levels. Soil heating with sprayable mulch is faster than that with plastic film, but the soil also cools down to lower temperatures at night. Overall, soil temperatures under sprayable mulch are lower than those obtained under plastic film. The thickness of the sprayed coat is critical to obtaining effective heating (Skutelsky et al. 2000). Soil heating using sprayable mulches was effective in controlling Verticillium wilt and scab in potato (Gamliel et al. 2001), at a level matching that achieved by solarisation using plastic films.

11.13 Conclusions

The biggest challenge in crop protection sciences is to effectively control soil-borne pests whilst avoiding environmental hazards and degradation of natural resources. Soil solarisation is an additional tool for achieving this task, when it is used in appropriate situations. The integration of pest management methods, rather than relying on one powerful control agent, is not only desirable but also the only feasible solution for coping with our need for methods of controlling soil-borne pests. There are many challenges awaiting the further development of soil solarisation: improvements in implementation technology and control effectiveness, thereby shortening the mulching period and extending the period for solarisation, and a better understanding of control mechanisms which can lead to more effective disturbance of pathogens' life cycles.

An interesting approach, suggested by Grinstein and Ausher (1991), is the use of soil solarisation as a 'cleaning tool' for infested soils, in the framework of crop rotation. Since solarisation frequently has a long-term effect, it can be applied in field every 3–5 years, before the field becomes heavily infested and prior to planting a cash crop, to provide the most benefit from solarisation. This procedure also reduces cost of solarisation per crop and maintains low infestations, provided that a proper crop rotation is practised.

References

Annesi T, Motta E (1994) Soil solarization in an Italian forest nursery. Eur J For Pathol 24:203–209

Arbel A, Siti M, Barak M, Katan J, Gamliel A (2003) Innovative plastic films enhance solarization efficacy and pest control. In: Proceedings of the 10th annual international research conference on methyl bromide alternatives and emission reduction, San Diego, California, USA, pp 91.1–91.3

Bell CE, Elmore CL (1983) Soil solarization as a weed control method in fall planted cantaloupes. Proc West Soc Weed Sci Granada Spain 36:174–177

Ben-Yephet Y, Stapleton JJ, Wakeman RJ, De Vay JE (1987) Comparative effects of soil solarization with single and double layers of polyethylene on survival of *Fusarium oxysporum* f sp *vasinfectum*. Phytoparasitica 15:181–185

Chellemi DO, Olson SM, Mitchell DJ (1994) Effects of soil solarization and fumigation on survival of soilborne pathogens of tomato in northern Florida. Plant Dis 78:1167–1172

Chellemi DO, Olson SM, Mitchel DJ, Secker I, McSorly R (1997) Adaptation of soil solarization to the integrated management of soilborne pests of tomato under humid conditions. Phytopathology 87:250–258

Chen Y, Gamliel A, Stapleton JJ, Aviad T (1991) Chemical, physical and microbial changes related to plant growth in disinfested soils. In: Katan J, DeVay JE (eds) Soil solarization. CRC Press, Boca Raton, pp 103–129

Costello RA, Gillespie DR (1993) The pepper weevil, *Anthonomus eugenii* Cano as a greenhouse pest in Canada. IOBC Bull Work Group IPM Glasshouses 16(2):31–34

DeVay JE, Stapleton JJ, Elmore CL (eds) (1991) Soil solarization. Plant production and protection paper 109. Rome: FAO/UN, 396 pp

Egley GH (1983) Weed seed and seedling reductions by soil solarization with transparent polyethylene sheets. Weed Sci 31:404–409

Elmore CL, Stapleton JJ, Bell CE, DeVay JE (1997) Soil solarization: a nonpesticidal method for controlling diseases, nematodes and weeds. UC DANR Pub 21377, Oakland, CA, 14 pp

Gamliel A, Stapleton JJ (1993a) Effect of soil amendment with chicken compost or ammonium phosphate and solarization on pathogen control, rhizosphere microorganisms and lettuce growth. Plant Dis 77:886–891

Gamliel A, Stapleton JJ (1993b) Characterization of antifungal volatile compounds evolved from solarized soil amended with cabbage residues. Phytopathology 83:899–905

Gamliel A, Skutelski Y, Peretz-Alon Y, Becker E (2001) Soil solarization using sprayable plastic polymers to control soil-borne pathogens in field crops. In: Proceedings of the 8th annual international research conference on methyl bromide alternatives and emission reduction, San Diego, CA, pp 10.1–10.3

Garibaldi A, Gullino ML (1991) Soil solarisation in southern European countries with emphasis in soilborne disease control of protected crops. In: Katan J, De Vay JE (eds) Soil Solarization. CRC Press, Bocan Raton, pp 227–235

Grinstein A, Ausher R (1991) Soil solarization in Israel. In: Katan J, De Vay JE (eds) Soil solarization. CRC Press, Bocan Raton, pp 193–204

Hartz TK, DeVay JE, Elmore CL (1993) Solarization is an effective soil disinfestation technique for strawberry production. Hortic Sci 28:104–106

Himelrick DG, Dozier WA (1991) Soil fumigation and soil solarization in strawberry production. Adv Strawb Prod 10:12–27

Horiuchi S (1991) Solarization for greenhouse crops in Japan. In: De Vay JE, Stapleton JJ, Elmore CL (eds) Soil solarization. Plant production and protection paper 109. Food and Agriculture Organization of the United Nations, Rome

Horowitz M, Roger Y, Herlinger G (1983) Solarization for weed control. Weed Sci 31:170–179

Inada K (1973) Photo-selective plastic film for mulch. J Agric Res Q 7:252–256

Katan J (1987) Soil solarization. In: Chet I (ed) Innovative approaches to plant disease control. Wiley, New York

Katan J, DeVay J (1991) Soil solarization. CRC Press, Boca Raton

Katan J, Greenburger A, Alon H, Grinstein A (1976) Solar heating by polyethylene mulching for control of diseases caused by soil-borne pathogens. Phytopathology 66:683–688

Lifshitz R, Tabachnik M, Katan J, Chet I (1983) The effect of sublethal heating on sclerotia of *Sclerotium rolfsii*. Can J Microbiol 29:1607–1610

McSorley R, Parrado JL (1986) Application of soil solarization to Rockdale soils in a subtropical environment. Nematropica 16:125–140

Ozores-Hampton M, McSorley R, Stansly PA, Roe NE, Chellemi DO (2004) Long term large scale soil solarization as a low-input production system for Florida vegetables. Acta Hortic 638:177–188

Pivonia S, Cohen R, Levita R, Katan J (2002) Improved solarization of containerized medium for the control of *Monosporascus* collapse in melon. Crop Prot 21:907–912

Porter IJ, Merriman PR (1983) Effects of solarization of soil on nematode and fungal pathogens at two sites in Victoria. Soil Biol 15:39–44

Ramirez-Villapudua J, Munnecke DM (1987) Control of cabbage yellows (*Fusarium oxysporum* f sp *conglutinans*) by solar heating of fields amended with dry cabbage residues. Plant Dis 71:217–221

Raymundo SA, Alcazar J (1986) Increasing efficiency of soil solarization in controlling root-knot nematode by using two pastiche layers of plastic mulch. Nematology 18:626

Rubin B, Benjamin A (1984) Solar heating of the soil: involvement of environmental factors in the weed control process. Weed Sci 32:138–142

Seman-Varner R, McSorley R, Gallaher RN (2008) Soil nutrient and plant responses to solarization in an agroecosystem utilizing an organic nutrient source. Renew Agric Food Syst 23:149–154

Skutelsky Y, Gamliel A, Kritzman G, Peretz-Alon Y, Becker E, Katan J (2000) Soil solarization using sprayable plastic polymers to control soil-borne pathogens in field crops. Phytoparasitica 28:269–270

Standifer LC (1984) Effects of solarization on soil weed seed populations. Weed Sci 32:569–573

Stapleton JJ (1997) Solarization: an implementable alternative for soil disinfestation. In: Canaday C (ed) Biological and cultural control of plant diseases, vol 12. APS Press, St. Paul, pp 1–6

Stapleton JJ (1998) Modes of action of solarization and biofumigation. In: Stapleton JJ, DeVay JE, Elmore CL (eds) Soil solarization and integrated management of soilborne pests. Plant production and protection paper 147. FAOIUN, Rome, pp 78–88

Stapleton JJ (2000) Soil solarization in various agricultural production systems. Crop Prot 19:837–841

Stapleton JJ, DeVay JE (1984) Thermal components of soil solarization as related to changes in soil and root microflora and increased plant growth response. Phytopathology 74:255–259

Stapleton JJ, DeVay JE (1995) Soil solarization: a natural mechanism of integrated pest management. In: Reuveni R (ed) Novel approaches to integrated pest management. CRC Press, Boca Raton, pp 309–322

Stapleton JJ, Gamliel A (1993) Feasibility of soil fumigation by sealing soil amended with fertilizers and crop residues containing biotoxic volatiles. Plant Prot Q 3:10–13

Stapleton JJ, McKenry MV, Ferguson L (1999) Methyl bromide alternatives: CDFA approves a solarization technique to ensure against nematode pest infestation of containerized nursery stock. UC Plant Prot Q 9(2):14

Stevens C, Khan VA, Wilson MA, Brown JE, Collins DJ (1999) Use of thermo film-IR single layer and double layer soil solarization to improve solar heating in a cloudy climate. Plasticulture 118:20–34

Wang JF, Luther K, Ho FI, Lin CH, Kirkegaard J (2008) Evaluation of Brassica accessions as potential biocidal green manure to control tomato bacterial wilt. Third international biofumigation symposium, Canberra, Australia

Weingartner DP, Shumaker JR, Smart GC Jr (1975) Incidence of corky ringspot disease on potato as affected by different fumigation rates, cultivars, and harvest dates. Proc Soil Crop Sci Soc Fla 34:194–196

Strobilurin Fungicides

12

Abstract

Natural strobilurins are produced by the forest mushroom fungus *Strobilurus tenacellus*, which grows on fallen pine cones, and secretes into the decaying wood on which they grow. The powerful fungicidal activity of this secretion prevents invasion by other fungi, so protecting the nutrient source of the original mushroom. This fungal antibiotic fights infections of the plants. It is remarkable for its fungitoxic activity. All fungi need to produce their own energy supply in order to grow and produce new spores. This supply is especially important during the early establishment phase of the disease life cycle. It is produced by a complex series of chemical processes in the mitochondria that are part of every living fungal cell.

They form part of the group of quinone outside inhibitor (QoI) fungicides and work by inhibiting mitochondrial respiration that prevents spore germination and mycelial growth in plant pathogens. Their strength was they were highly active on most cereal fungal diseases, including septoria, rusts, mildew and a host of others. They were also credited with physiological benefits, maintaining green leaf area for longer and delaying crop senescence. But they have a very specific mode of action, which made the chemistry vulnerable to resistance. Guidelines for reducing the risk of development of resistance against fungicides are discussed.

After German scientists first discovered strobilurins in 1977, it did not take long for people to realise its potential for use as a fungicide. Thus, the development of what would become one of the most important classes of fungicides began. Strobilurins are produced by fungi to inhibit other organisms (other fungi, in this case) from competing for resources.

Strobilurins are one of the most important classes of agricultural fungicides. Their invention was inspired by a group of fungicidally active natural products. The outstanding benefits they deliver are currently being utilised in a wide range of crops throughout the world. First launched in 1996, the strobilurins now include the world's biggest selling fungicide, azoxystrobin. By 2002, there will be six strobilurin active ingredients commercially available for agricultural use.

The fungus *Strobilurus tenacellus*, which grows on fallen pine cones, produces strobilurin A. This rather insignificant grey to yellowish-brown mushroom grows to a height of 5–7 cm and is

P.P. Reddy, *Recent Advances in Crop Protection*,
DOI 10.1007/978-81-322-0723-8_12, © Springer India 2013

edible, with a mild, slightly bitter taste, but it is remarkable for its fungitoxic activity. Through the production of strobilurin, it is capable of keeping other fungi at bay that might otherwise compete for nutrients. All fungi need to produce their own energy supply in order to grow and produce new spores. This supply is especially important during the early establishment phase of the disease life cycle. It is produced by a complex series of chemical processes in the mitochondria that are part of every living fungal cell. Strobilurins work by blocking electron transfer within this chemical process, thus denying the fungus energy and preventing development, even at the earliest stages of the life cycle, the spore germination stage.

There are two strobilurin active ingredient fungicides available at present under the brand names of Amistar, Landmark, Ensign and Stroby WG. Landmark is a mixture of kresoxim-methyl, strobilurin and epoxiconazole. Epoxiconazole can have an herbicidal activity on some broad-leaved plants so, at present, Landmark should be considered with care. The other products, however, appear to be generally less toxic to nursery stock.

Azoxystrobin has an excellent toxicological profile, and the product label carries no user risk warning phrases and requires only minimum protective clothing for sprayer operators. Both active ingredients, when applied to crops, are non-volatile and therefore cannot enter the environment as a vapour. They also breakdown quickly in the soil to carbon dioxide and other simple molecules. There is no risk of leaching into groundwater, and it has no effect on the germination and development of crops. Tests show no detectable residues in fruit or even wine made from treated crops.

The effects on a wide range of non-target and beneficial organisms are minimal or non-existent, and tests are under way to monitor the effect on protected crop IPM products. Kresoxim-methyl is known to be damaging to fish and some aquatic life; however, good spraying practice should ensure this does not occur. Tests have been carried out showing there is no effect on:

- Carabid beetle, parasitic wasps, adult hoverflies, lacewings and spiders
- Earthworms, honeybees and breeding birds

When sprayed on a crop leaf, the products are quickly bound to the leaf surface providing a protective barrier against incoming spores. Amistar moves slowly through the surface of the leaf tissue, where, unlike other strobilurin-based fungicides, it remains highly active. It diffuses across the leaf to give translaminar protection of the untreated, underside of the leaf. Kresoxim-methyl is locally systemic, and the surface deposits ensure a slow release into the plant over a period of time and rain washing off is minimal. The spray residue on the leaf surface is reactivated by rainfall or dew wetting, enabling repeat uptake over a long period of time. Both products are redistributed steadily in the leaf xylem flow to give complete systemic protection of the new leaf as it grows and curative activity against new establishing mycelia of certain diseases. Spray deposits in leaf axils act as a reservoir for material that is absorbed through the outer leaf sheath and into new leaves not present at the time of spraying. Through this action, they give persistent protection to new growth.

For optimal results, these fungicides should be used early in the disease epidemic because of their powerful effects against spore germination and early pathogen development. Applications should begin at the first signs of disease as they will start reducing active fungi straightaway.

Trials using mist blowers, knapsack sprayers and tractor-mounted sprayers have shown that robust disease control can be achieved using low-volume (50–100 l/ha) and high-volume (over 3,000 l/ha) application. It is important to produce a high-quality spray, using well-maintained equipment that is suitable for the crop. Sufficient water volume must be used to ensure adequate coverage, particularly later in the growing season when foliage density increases.

Application rates are still undergoing extensive trials on ornamentals, and no clear picture has yet emerged on suitable rates and frequency of application. These products should not be used alone. What is becoming clear is that single applications at extended frequencies may well be providing good protection from a range of common diseases. Putting strobilurins into a standard programme and alternating them with existing products is the best way to incorporate them. This method will ensure a minimal risk to resistance. As these products are emulsifiable concentrates, care should be taken on ornamentals with spraying in adverse conditions to avoid scorch.

Many of the newest and most important disease-control chemicals are in the quinone outside inhibitor (Q_0I) family of fungicides. The first fungicides in this family were isolated from wood-rotting mushroom fungi, including one called *Strobilurus tenacellus*. The name *strobilurin* was coined for this chemical family of fungicides in recognition of the source of the first compounds of this type. These fungicides are now more properly referred to as Q_0I fungicides, which are explained in the section on fungicide resistance.

Industry chemists improved on these natural fungicides by making chemical modifications that resulted in compounds which were less subject to breakdown on the leaf surface by sunlight. Several of the Q_0I fungicides currently registered in the United States are considered by the Environmental Protection Agency to be *reduced-risk pesticides*. This means these compounds pose less risk to human health and/or the environment than alternative pesticides available at the time of their commercial introduction.

12.1 Spectrum of Activity

With important exceptions, the Q_0I fungicides control an unusually wide array of fungal diseases, including diseases caused by water moulds, downy mildews, powdery mildews, leaf spotting and blighting fungi, fruit rotters and rusts. They are used on a wide variety of crops, including cereals, field crops, fruits, tree nuts, vegetables, turf grasses and ornamentals.

Whilst Q_0I fungicides provide important benefits, like all fungicides, their use as replacements for older fungicides can sometimes lead to unexpected changes in disease activity. For example, in turf grasses, azoxystrobin provides excellent control of a number of important diseases. However, it can sometimes enhance the severity of certain diseases, such as dollar spot of creeping bent grass and Pythium blight of tall fescue. Mechanisms of disease enhancement are not understood, but one possibility is that use of azoxystrobin at labelled field rates may suppress certain naturally occurring microorganisms that are antagonistic to the pathogen.

Strobilurin fungicides have become a very valuable tool for managing diseases. They are effective against several different plant pathogenic fungi. Also, they have translaminar activity, which means they can move through treated leaves thereby providing control on both leaf surfaces. This group is unique in that these fungicides are the first synthetic, site-specific compounds to provide significant control of plant diseases caused by pathogens from all three major groups of fungi: Oomycota, Ascomycota and Basidiomycota. Azoxystrobin, formulated as Quadris and Abound, was the first fungicide in this group registered by EPA. Trifloxystrobin, formulated as Flint, Stratego and Compass, was the next. Pyraclostrobin, formulated as Cabrio EG and Headline, has also been registered. Quadris is now registered for use on almost all vegetable crops. Quadris is labelled for managing *Cladosporium* leaf blotch, purple blotch, rust, downy mildew and *Rhizoctonia* damping off in bulb crops (onion, garlic, etc.); early blight, late blight and *Rhizoctonia* damping off in carrot and in celery; rust, grey leaf spot, northern corn leaf blight, northern corn leaf spot and *Rhizoctonia* root and stalk rot in corn; anthracnose, belly rot, downy mildew, gummy stem blight/black rot, *Alternaria* and *Cercospora* leaf spots, *Myrothecium* canker, powdery mildew and *Rhizoctonia* root rot in cucurbits; *Alternaria*, *Cercospora* and *Septoria* leaf spots, anthracnose, downy mildew, powdery mildew, white rust and *Rhizoctonia* diseases (e.g. bottom rot) in leafy vegetables; early blight, late blight, black dot, powdery mildew, black scurf and silver scurf in potato; anthracnose, early blight, late blight, black mould, powdery mildew, buckeye rot, *Septoria* leaf spot and target spot in tomato; and *Alternaria*, *Cercospora* and *Ascochyta* leaf spots, powdery mildew, rust, white rust, *Aphanomyces* root rot, southern blight, *Pythium* root rot and *Rhizoctonia* crown rot in the root and tuber subgroup (beets, carrots, ginseng, radishes, etc.) and in the tuberous and corm subgroup (artichokes, sweet potatoes, etc.).

12.2 Diseases Controlled

Target diseases for nearly all strobilurin fungicides include downy mildew, rust, powdery mildew and many leaf spots (*Alternaria, Cercospora, Myrothecium* and *Sphaceloma*, which causes scab on poinsettia) (Fig. 12.1). Heritage and Insignia also provide good to very good control of soil-borne diseases such as *Fusarium, Cylindrocladium* and *Rhizoctonia* stem and root rots but not black root rot (*Thielaviopsis*).

The most consistent and effective control of rust is seen with Heritage. Be sure to use a wetting agent if pustules have formed, or the fungicide may not penetrate the pustule and control will be minimal. This may be because Heritage has a better chance to penetrate the pustule. Rust infections, as well as downy mildew, start inside the leaf and may be hard to reach with a surface-active product without systemic action.

Compass O has been somewhat more effective against *Botrytis* than other strobilurins. This may be due to Compass O redistributing across the leaf and *Botrytis* spores that constantly fall onto the leaves can be killed even when they land on a part of the leaf that was not originally sprayed.

Across the group of five strobilurins, the most variable results were observed when they were tested against *Colletotrichum, Coniothyrium, Phyllosticta, Phytophthora* and *Pythium*. *Colletotrichum, Coniothyrium* and *Phyllosticta* fall into the anthracnose-dieback group of diseases, which can be very hard to control.

It can be difficult to obtain a good fungicide trial for *Phytophthora* and *Pythium* diseases. With *Phytophthora*, results can vary widely within the same trial with some plants dying whilst others that have received the same treatment being unaffected. This makes statistical analysis difficult. *Pythium* is especially hard to evaluate. Sometimes, it can be difficult to determine if the root loss was caused by *Pythium* or by fungicide phytotoxicity.

There is no complete information for each strobilurin on each disease. The strobilurins should be used in a rotational programme for any of the highlighted diseases in Table 12.1. These products have been tested over the past 10–12 years on a wide range of plants under greenhouse and field conditions. Overall, these products have been very safe. Their labels do prohibit use on specific plants, so be sure to read the labels carefully.

Under greenhouse conditions, strobilurin fungicides, namely, azoxystrobin, kresoxim-methyl and trifloxystrobin (at concentrations of 5, 10 and 10 μg ml^{-1}, respectively) showed varying degrees of protection against downy mildew

Fig. 12.1 Powdery mildew on sunflower and rust on dianthus

Table 12.1 Results of strobilurin fungicide efficacy trials at Chase Horticultural Research Inc. (1998 to present)

Disease	Compass O	Cygnus	Disarm O	Heritage	Insignia
Alternaria	Good/excellent	Very good	–	Very good/excellent	Very good/excellent
Botrytis	Very good	None/some	None/some	Fair/good	Poor/very good
Cercospora	Some/excellent	Excellent	Very good/excellent	Very good/excellent	Very good/excellent
Colletotrichum	–	–	Excellent	None/very good	None/some
Coniothyrium	–	–	–	Some	None/good
Cylindrocladium	Some/good	–	–	Very good	Good/excellent
Downy mildew	Good/very good	Some/excellent	–	Very good/excellent	Good/excellent
Fusarium	Some/good	–	None/some	Good/very good	Good/very good
Gliocladium	Very good	–	–	Good/excellent	Very good
Myrothecium	Some/good	Good	Very good/excellent	Very good/excellent	Very good/excellent
Phyllosticta	None	–	–	None/good	Some/excellent
Phytophthora	None/very good	–	–	Very good	None/excellent
Powdery mildew	Very good/excellent	Very good/excellent	–	Very good/excellent	Good/excellent
Pythium	Poor/good	–	–	Some/very good	Good/excellent
Rhizoctonia	Good	None/very good	Good	Very good/excellent	Very good/excellent
Rust	Fair/excellent	Some/excellent	–	Excellent	Very good
Scab	Excellent	Excellent	–	Very good/excellent	Excellent
Sclerotinia	Very good/excellent	None	–	Very good	Very good
Thielaviopsis	None	–	–	None	–

disease (*Sclerospora graminicola*) of pearl millet (*Pennisetum glaucum*). Amongst the three fungicides, azoxystrobin proved to be the best by offering disease protection of 66%. Further, seed treatment along with foliar application of these fungicides to diseased plants showed enhanced protection against the disease to 93, 82 and 62% in treatments of azoxystrobin, kresoxim-methyl and trifloxystrobin, respectively. Foliar spray alone provided significant increase in disease protection levels of 91, 79 and 59% in treatments of azoxystrobin, kresoxim-methyl and trifloxystrobin, respectively. Disease curative activity of azoxystrobin was higher compared to trifloxystrobin and kresoxim-methyl. Tested fungicides showed weaker translaminar activity, as the disease inhibition was marginal when applied on adaxial leaf surface. Partial systemic activity of azoxystrobin was evident by root uptake, whilst trifloxystrobin and kresoxim-methyl showed lack of systemic action in pearl millet. A trend in protection against downy mildew disease similar to greenhouse results was evident in the field trials. Grain yield was significantly increased in all strobilurin fungicide treatments over control, and maximum increase in yield of 1,673 kg ha^{-1} was observed in combination treatments of seed treatment and foliar spray with azoxystrobin (Sudisha et al. 2005).

A strobilurin fungicide, F 500 (Pyraclostrobin), enhances the resistance of tobacco (*Nicotiana tabacum* cv *xanthi*) against infection by either tobacco mosaic virus (TMV) or the wildfire pathogen, *Pseudomonas syringae* pv *tabaci*. F 500 was also active at enhancing TMV resistance in *NahG* transgenic tobacco plants unable to accumulate significant amounts of the endogenous inducer of enhanced disease resistance, salicylic acid (SA). This finding suggests that F 500 enhances TMV resistance in tobacco either by acting downstream of SA in the SA signalling mechanism or by functioning independently of SA. The latter assumption is the more likely because in infiltrated leaves, F 500 did not cause the accumulation of SA-inducible pathogenesis-related PR-1 proteins that often are used as conventional molecular markers for SA-induced disease resistance. However, accumulation of PR-1 proteins and the associated activation of the PR-1 genes were elicited upon TMV infection of tobacco leaves, and both these responses were induced more rapidly in F 500-pretreated plants than in the water-pretreated controls. Taken together, these results suggest that F 500, in addition to exerting direct antifungal activity, may also protect plants by priming them for potentiated activation of subsequently pathogen-induced cellular defence responses (Stefan et al. 2002).

The crop diseases controlled by strobilurin fungicides are presented in Tables 12.2, 12.3, and 12.4.

12.3 Mobility

All of the Q_oI fungicides listed in Table 12.4 exhibit *translaminar movement* (which means 'across the lamina', or leaf blade). When these fungicides are applied, most of the active ingredient is initially held on or within the waxy cuticle of plant surfaces (Fig. 12.2). Some of the active ingredient 'leaks' into the underlying plant cells. For those fungicides with an affinity for the waxy cuticle (such as trifloxystrobin and kresoxim-methyl), active ingredient that 'leaks' all the way through the lamina quickly rebinds to the cuticle on the far side of the leaf blade. Thus, the fungicide can be found on both leaf surfaces even if only one leaf surface was treated. Translaminar movement can take one to several days to be fully effective.

The fungicide azoxystrobin moves translaminarly as well as *systemically* (in the plant's vascular system or 'plumbing'). The fungicides kresoxim-methyl and trifloxystrobin move translaminarly but not systemically. These latter fungicides, however, appear to move as a gas in the layer of still air adjacent to the leaf surface called the *boundary layer*. As they move in the vapour phase, they readily rebind to the cuticle. Fungicides such as kresoxim-methyl and trifloxystrobin – which are not true systemics but which redistribute by these other mechanisms – have been referred to as 'mesostemics', 'quasi-systemics' or 'surface systemics'.

In terms of practical significance, systemic movement (when it occurs) and translaminar

Table 12.2 Efficacy of kresoxim-methyl (Landmark, Stroby WG) against diseases of crop plants

Crop	Fungal pathogen
Cereals	*Pseudocercosporella herpotrichoides*
	Erysiphe graminis
	Puccinia hordei
	P. recondita
	P. striiformis
	Pyrenophora teres
	Rhynchosporium secalis
	Septoria nodorum
	S. tritici
Pome fruits	*Podosphaera leucotricha*
	Alternaria kikuchiana
	A. mali
	Gloeodes pomigena
	Schizothyrium pomi
	Stemphylium vesicarium
	Venturia spp.
	Glomerella cingulata
Strawberries	*Mycosphaerella fragariae*
	Botrytis cinerea
	Sphaerotheca humuli
Beans	*Uromyces appendiculatus*
Peas	*Alternaria alternata*
	Ascochyta pisi
	Erysiphe pisi
	Mycosphaerella pinodes
	Phoma medicaginis
Paprika	*Leveillula taurica*
Tomatoes	*Alternaria solani*
	Phytophthora infestans
Cucurbits	*Pseudoperonospora cubensis*
	Erysiphe cichoracearum
	Sphaerotheca fuliginea
	Alternaria cucumerina
	Mycosphaerella melonis
	Colletotrichum lagenarium
Roses	*Sphaerotheca pannosa*
	Phragmidium spp.
	Diplocarpon rosae
Potatoes	*Alternaria solani*
	Phytophthora infestans
Sunflowers	*Phomopsis helianthi*
Sugar beets	*Cercospora beticola*
	Erysiphe betae
Stone fruit	*Cladosporium carpophilum*
	Sphaerotheca pannosa var. *persica*
	Stigmina carpophila
Black currants/ gooseberries	*Sphaerotheca mors-uvae*
Vines	*Uncinula necator*
	Guignardia bidwellii
	Phomopsis viticola
	Pseudopezicula tracheiphila
	Plasmopara viticola
	Botrytis cinerea
	Elsinoe ampelina
Cabbage	*Albugo candida*
	Mycosphaerella brassicicola
Carrots	*Alternaria dauci*
	Erysiphe heraclei
Celery	*Septoria apiicola*
Asparagus	*Puccinia asparagi*
	Stemphylium sp.
Onions, etc.	*Peronospora destructor*
	Alternaria porri
	Stemphylium botryosum
Tobacco	*Alternaria alternata*
	Erysiphe cichoracearum
Lilies, tulips	*Botrytis* spp.
Chrysanthemums	*Puccinia* spp.

movement help to compensate for incomplete spray coverage. Redistribution in the vapour phase can also help compensate for poor crop coverage but only to a limited extent. These processes may be especially important in crops with dense or difficult-to-spray canopies (e.g. cucurbits). Several days may be required for adequate protection to be achieved via translaminar movement. Thus, growers may not achieve optimum disease control if a Q_oI fungicide is applied with incomplete coverage within 24 h of an infection period.

Another practical consequence of the dynamics of translaminar movement concerns curative disease control. Q_oI fungicides are excellent as preventive fungicides because they all effectively kill germinating spores. However, several of them provide poor performance against certain diseases when used curatively – that is, after infection has taken place. Some Q_oI fungicides bind tightly to the cuticle, where most of the active ingredient can be found. Even though the

Table 12.3 Efficacy of azoxystrobin (Amistar) against diseases of crop plants

Crop	Rate	Diseases controlled
Wheat	250 g a.i./ha	*Erysiphe graminis* f. sp. *tritici*
		Septoria tritici
		S. nodorum
		Puccinia recondita
		P. striiformis
		Pyrenophora tritici-repentis
		Cladosporium spp.
		Alternaria spp.
Barley	250 g a.i./ha	*Erysiphe graminis* f. sp. *hordei*
		Puccinia hordei
		Pyrenophora teres
		Rhynchosporium secalis
Grapevine	125–250 ppm a.i.	*Plasmopara viticola*
		Uncinula necator
		Phomopsis viticola
		Guignardia bidwellii
		Pseudopeziza tracheiphila
Melon/cucumber	100–200 ppm a.i.	*Sphaerotheca fuliginea*
		Erysiphe cichoracearum
		Pseudoperonospora cubensis
		Didymella bryoniae
		Colletotrichum lagenarium
		Alternaria cucumerina
Tomato	25–100 ppm a.i.	*Alternaria solani*
		Colletotrichum coccodes
		C. gloeosporioides
		Septoria lycopersici
		Leveillula taurica
	100–250 ppm a.i.	*Phytophthora infestans*
Rice	100–300 g a.i./ha (foliar application)	*Rhizoctonia solani*
		Pyricularia oryzae
	600 g a.i./ha (paddy granule)	*Cochliobolus miyabeanus*
	3 g ai/nursery box	*Sclerotium oryzae*
Turf	250–600 g a.i./ha	*Rhizoctonia solani*
		R. cerealis
		Pythium spp.
		Colletotrichum graminicola
		Microdochium nivale
		Typhula spp.
		Drechslera poae
		Leptosphaeria korrae
		Laetisaria fuciformis
		Magnaporthe poae
		Gaeumannomyces graminis
Peanut	200–400 g a.i./ha	*Rhizoctonia solani*
		Sclerotium rolfsii
	100–200 g a.i./ha	*Cercospora arachidicola*
		Cercosporidium personatum
		Puccinia arachis

(continued)

Table 12.3 (continued)

Crop	Rate	Diseases controlled
Banana	100–150 g a.i./ha	*Mycosphaerella fijiensis*
		M. musicola
Pecan	100–200 g a.i./ha	*Cladosporium caryigenum*
Peach	75–200 g a.i./ha	*Monilinia fructicola/fructigena*
		Cladosporium carpophilum
Coffee	100–200 ppm a.i.	*Colletotrichum coffeanum*
		Hemileia vastatrix
Citrus	100–200 ppm a.i.	*Elsinoe fawcettii*
		Guignardia citricarpa
Apple	100–200 ppm a.i.	*Venturia inaequalis*
		Alternaria mali
		Podosphaera leucotricha
		Botryosphaeria berengeriana
		Zygophiala jamaicensis
		Gloeodes pomigena
		Mycosphaerella pomi
Pear	100–200 ppm a.i.	*Stemphylium vesicarium*
Japanese pear	100–200 ppm a.i.	*Venturia nashicola*
		Alternaria kikuchiana
		Physalospora piricola

Table 12.4 Efficacy of strobilurin fungicides against diseases of crop plants

Product (company)	Typical use rate (lb a.i./acre)	Strobilurin fungicide	Crops registered for use on
Quilt (Syngenta)	0.07 lb	Azoxystrobin	Corn, soybean, wheat
Quilt Xcel (Syngenta)	0.10 lb	Azoxystrobin	Corn, soybean, wheat
Stratego (Bayer Crop Science)	0.08 lb	Trifloxystrobin	Corn, soybean, wheat
Stratego YLD (Bayer Crop Science)	0.10 lb	Trifloxystrobin	Corn, soybean
Headline AMP (BASF)	0.10 lb	Pyraclostrobin	Corn
Evito T (Arysta Life Science)	0.05 lb	Fluoxastrobin	Corn, soybean
TwinLine (BASF)	0.06 lb	Pyraclostrobin	Wheat
Absolute (Bayer Crop Science)	0.09 lb	Trifloxystrobin	Wheat
Headline (BASF)	0.10 lb	Pyraclostrobin	Corn, soybean, wheat
Quadris (Syngenta)	0.10 lb	Azoxystrobin	Corn, soybean, wheat
Evito (Arysta Life Science)	0.06 lb	Fluoxastrobin	Corn, soybean, wheat
Gem (Bayer Crop Science)	0.10 lb	Trifloxystrobin	Soybean

active ingredient 'leaks' into the leaf blade, it has such a strong affinity for the cuticle that it quickly rebinds with it when the chemical reaches the other side of the leaf. Consequently, at any one time, the dose of active ingredient actually present inside the leaf blade may be low, sometimes too low to suppress the growth of fungi within the leaf. Furthermore, for a number of fungal pathogens, the germinating spore (which starts the infection process on the outside of the plant) is more sensitive to Q_oI fungicides than is the mycelium (the fungal life stage found inside the plant). Thus, the best use of Q_oI fungicides is to apply them before infection takes place (Table 12.5).

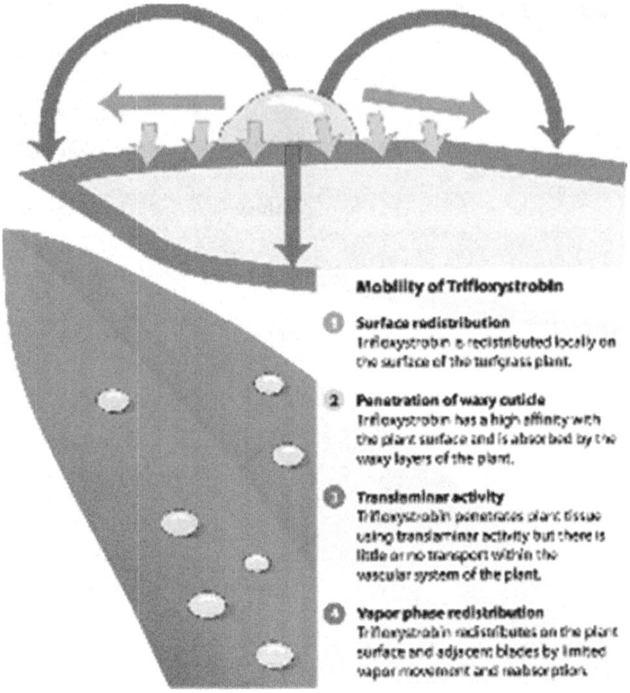

Fig. 12.2 Mobility of trifloxystrobin, an example of a Q_oI fungicide

Table 12.5 Attributes of strobilurin fungicides

Category	Strobilurin fungicides
Mode of action	Inhibition of energy production by blocking electron transport within the mitochondria
Curative	Weak or none
Preventive	Very strong
Residual	Strong
Effect on life cycle	Primarily inhibits spore germination, lesser effect on the rest of the life cycle. Prevents infection
Resistance development	Sudden, qualitative: single gene mutation can result in complete loss of disease control
Systemicity	Mostly locally systemic, translaminar, can have redistribution through vapour phase
Rain fastness	Good rain fastness

12.4 Effects on Plant Health Independent of Disease Control

12.4.1 Growth Enhancement

Several Q_oI fungicides are known to cause growth-promoting effects in certain plants. For example, kresoxim-methyl has been shown to cause changes in the hormonal balance of wheat which results in increased grain yield, apparently from delayed leaf senescence and water-conserving effects. Growth-enhancing effects independent of disease control have been observed in Q_oI-treated plants of several species, although these effects are very much dependent on the crop, fungicide used and environmental conditions.

12.4.2 Phytotoxicity

Whilst the Q_oI fungicides are very valuable for disease control, several are known to cause phytotoxicity in certain, limited circumstances; these are described in product labels. For example, apple cultivars with a genetic background which includes Macintosh are extremely sensitive to azoxystrobin. Indeed, these varieties are so sensitive that they can be injured when a sprayer is used to apply azoxystrobin to another crop (e.g. grapes), rinsed and then used to apply another fungicide to the apple crop! Another example is whilst trifloxystrobin may be used safely on most grapes, it can cause injury to Concord grapes. Kresoxim-methyl is phytotoxic to certain sweet cherry varieties but not others. Producers should be aware of phytotoxicity concerns both for the treated crop and because of the possibility of injury via spray drift.

Another aspect of the phytotoxicity risk is the possibility that tank mixes of Q_0I fungicides with materials that solubilise the cuticle – oils, surfactants and certain liquid formulations of insecticides – could increase their phytotoxicity potential. Although some of the active ingredient can be found within the host tissue, much of the dose of Q_0I fungicides remains on or within the plant cuticle. Application of a spray material that causes abnormally high levels of these fungicides to penetrate into the host tissue could potentially cause phytotoxicity on certain crops or varieties where none is normally expected. Obviously, when applying a previously unused tank mix on a particular crop variety, a wise practice is to test-apply to small areas before treating large acreages.

12.5 Mode of Action

The mode of action of strobilurin fungicides is by inhibition of mitochondrial respiration by blocking electron transfer in complex III (*bc 1* complex) of the mitochondrial electron transport chain (Ammermann et al. 2000). Since *bc 1* complex is conserved in all eukaryotes, at least a partial inhibition of mitochondrial electron transport should be expected also in plant cells following uptake of strobilurin fungicides. The selectivity of strobilurin-type inhibitors of respiration depends less on differential sensitivity of mitochondrial complexes from different sources but strongly on terms of bioavailability or abundance at the target site, which is modified by uptake, partitioning and metabolic degradation as dynamic processes. So, within the class of strobilurin chemistry, compounds vary in their effects on plants according to their biokinetic properties (Koehle et al. 1994). A transient influence on plant mitochondria does not necessarily result in phytotoxicity because the toxicity at the level of organism is determined by the importance of mitochondrial respiration for energy supply which varies with environmental conditions and the life stage of the organism (Sauter et al. 1995). For example, strobilurins cause a much higher rate of cell damage measured by leakage of electrolytes when leaf discs are incubated for some hours in the dark compared to illuminated incubation. Untreated leaf discs did not show significant leakage under the same conditions.

12.6 Resistance

All Q_0I fungicides share a common biochemical mode of action: They all interfere with energy production in the fungal cell. To be precise, they block electron transfer at the site of quinol oxidation (the Qo site) in the cytochrome *bc 1* complex, thus preventing ATP formation. The mode of action of the Q_0I fungicides is highly specific. Of the millions of biochemical reactions that occur in the fungal cell, these fungicides interfere with just one, very specific biochemical site for the fungus. Thus, these are called *site-specific fungicides*. This is important because, commonly, just one mutation at that biochemical site (the target site of the fungicide) can result in a fungicide-resistant strain. If such a fungicide-resistant strain occurs, repeated application of Q_0I fungicides can lead to build-up of a fungicide-resistant pathogen subpopulation.

Experience with the Q_0I fungicides worldwide indicates there is a high risk of development of resistant pathogen subpopulations. Worldwide, resistance has been reported in an increasing number of pathogens of field crops, fruit, vegetable, nut crops, ornamentals and turf grass.

There are two general types of fungicide resistance: quantitative and qualitative. With *quantitative resistance*, resistant strains are somewhat less sensitive to the fungicide as compared to the wild type, but they often can still be controlled with higher rates and/or more frequent applications (within labelled limits, of course). A good example of this type of resistance is that observed with strains resistant to the DMI (demethylation-inhibitor) fungicides, such as propiconazole or triadimefon. With *qualitative resistance*, the resistant strain is vastly less sensitive to the active ingredient and is no longer controlled at labelled field rates. The effect on disease control is the same as if one were spraying water on the crop instead of a fungicide. A good example of this type of resistance is that observed with the benzimidazole fungicides, such as benomyl or thiophanate methyl. Natural occurrences of resistance to the Q_0I fungicides indicate that most cases of control failure are due to resistance of the qualitative type but that instances of quantitative resistance to certain Q_0I fungicides have also been recorded.

Fungicides that share a common biochemical mode of action for poisoning the fungus are thought to be in the same 'fungicide family'. When different fungicidal products share a common mode of action, the fungus does not distinguish between the fungicides, even if the chemical structure of the active ingredients is different and the fungicides are produced by different manufacturers. Biochemically, the fungus sees them all as the same active ingredient. When a fungus is resistant to one fungicide in a chemical family, it is usually resistant to all fungicides in that family. This is called *cross-resistance*. In many situations, fungal strains resistant to Q_0I fungicides exhibit cross-resistance to other Q_0I fungicides. In such cases, efficacy of all Q_0I fungicides may be compromised, even if some of them have never been used on that farm. Cross-resistance only applies within a given chemical family. Therefore, Q_0I-resistant subpopulations can be controlled with other fungicides not in the Q_0I family.

Considering the value of this group of fungicides for managing diseases in vegetable crops, it is essential to use them in a manner that manages resistance. Resistance risk with strobilurins has proved to be higher than expected and more difficult to predict. They are site-specific compounds, which have often indicated a high resistance risk; however, they have a new mode of action (inhibition of mitochondrial respiration) which was thought to be difficult for fungi to overcome. Resistance developed much more quickly than expected, but resistance did develop first in a pathogen with a past history of developing resistance to site-specific compounds. Resistant strains of the cucurbit powdery mildew fungus were found in 1999 after just 2 years of commercial use in four countries. Unexplainably, resistance has not developed yet in North America despite the size of the pathogen population. Just as unexpectedly, resistance has developed in the United States in the fungus causing gummy stem blight and black rot in cucurbits. This was not expected based on past history with benomyl resistance in these pathogens. In the United States, control failure associated with resistance to benomyl was observed in the cucurbit powdery mildew fungus just 1 year after its registration for use on cucurbits, but it was not observed in the gummy stem blight fungus until 23 years after its registration. The cucurbit downy mildew fungus has also developed resistance to strobilurins but not yet in North America.

12.7 Guidelines for Reducing Resistance Risk

Resistance to strobilurins can be managed by limiting their use and by using them as a component of an integrated programme with other fungicides as well as with non-chemical management practices, such as resistant varieties and rotation. This is the standard approach for managing resistance with all fungicides that have potential for resistance development. It is critical to manage resistance before it develops. The goal is not to manage resistance once it has developed but rather to prevent or delay development of resistance. Apply strobilurins in alternation with other systemic fungicides that have a different mode of action when these are registered for the target disease. Do not alternate amongst strobilurins because they all have the same mode of action. The fungicide programme should also include multi-site contact fungicides that have a low risk of resistance, such as chlorothalonil and copper hydroxide. For example, the recommended programme for cucurbit powdery mildew is a strobilurin applied in alternation with myclobutanil tank-mixed with a contact fungicide. No more than half of the applications in a season should include strobilurins. In other areas of the world, a limit of one-third is recommended. Strobilurins have outstanding ability to inhibit spore germination; thus, they should be most useful early in disease development. Use a disease threshold or a disease forecasting system such as TOMcast, when available for the target disease, so that the first application is made at the most critical time. Using contact fungicides alone at the end of the season should be considered. Resistance in the gummy stem blight fungus developed where strobilurins are believed to have been used exclusively. Using a resistance management programme will also minimise yield loss if resistance develops. Resistant strains have exhibited a high degree of resistance. In fungicide evaluation experiments conducted in areas where strobilurin resistance developed, gummy stem blight was more severe on plants sprayed with just strobilurins than on plants that were not sprayed with any fungicide! It is extremely important to evaluate disease control when using any fungicide that is at risk for resistance development and to contact extension staff or the fungicide manufacturer promptly, whilst the crop is still there, if control is inadequate and there are no other feasible explanations, such as poor application timing.

Start by understanding that there is nothing a grower can do that will *eliminate* the risk of fungicide resistance, except to never use the at-risk fungicide. One can *reduce* the risk of its development by following practices that delay development of a resistant subpopulation.

Guidelines for reducing the risk of resistance against fungicides are issued by the Fungicide Resistance Action Committee (FRAC). FRAC is an organisation composed of scientists from manufacturers of the various at-risk fungicides. The guidelines issued by FRAC for each fungicide family are based on scientific principles and up-to-date research. Thus, FRAC guidelines provide a basis for understanding how to reduce the risk in the cropping situations where you work.

In addition to the key guidelines described below, growers should understand that reducing the risk of fungicide resistance begins by using non-fungicidal means for disease control: crop rotation, selection of varieties with reduced susceptibility, sanitation, pathogen-free seed, etc. These practices help reduce overall disease pressure. The occurrence of an adapted mutant with resistance to a fungicide is a matter of chance, like a 'roll of the dice'. The larger the pathogen population, the greater the chance that such a mutant will arise. Reducing disease pressure through non-chemical practices helps lower the chance that a fungicide-resistant mutant will occur; it does this by keeping the overall size of the pathogen population small.

12.7.1 Limit the Number of Applications of Q_oI Fungicides in a Given Season

The more often a Q_oI fungicide is used, the higher is the selection pressure towards the development of a resistant subpopulation. Limiting the number of applications reduces the opportunity for selection pressure, potentially extending the useful life of the Q_oI family of fungicides on a particular crop. This guideline is indicated on product labels, and label restrictions on the seasonal total number of sprays apply to all Q_oI fungicides, not just to the product being used. For example, as of the date of publication, the label for Flint 50W instructed the apple grower not to apply more than four sprays of Flint 50W or other strobilurin (=Q_oI fungicides) during the same season. If a grower makes two applications of Flint 50WG in a given season, only two more applications of any Q_oI fungicide could be made that season, whether it is Flint 50W or Sovran 50WG.

The seasonal limit is higher on crops that receive 8–12 sprays per year than those that receive less. Obviously, if the seasonal limit for

Q_oI fungicides is reached and the crop still needs protection against disease for a longer period of time, alternative fungicides must be used. These alternatives must, of course, not be in the Q_oI group and must be from other fungicide families. Under the current FRAC guidelines, seed treatments and in-furrow treatments do not count towards the seasonal limit, because of the limited mobility of these fungicides within plants.

12.7.2 Limit the Number of Consecutive Applications of a Q_oI Fungicide

The product labels indicate the number of consecutive applications of Q_oI fungicides that are allowed on each crop, before the user must switch to an equal number of applications of non-Q_oI fungicides. For most crops, the number of consecutive applications will be limited to two before the grower must switch to a fungicide with a different mode of action. FRAC guidelines on certain crops are even stricter; for example, on cucurbits, it is advised never to apply Q_oI fungicides consecutively. Like the seasonal limit described above, this guideline is designed to reduce the opportunity for selection pressure towards resistance.

12.7.3 Mixing Q_oI Fungicides with Other Fungicides Can Reduce Selection Pressure Towards Resistance

Mixtures do not prevent resistant mutants from arising on a farm. They can, however, slow the rate of spread of these mutants. A proper mixing partner is one that provides satisfactory disease control when used alone on the target disease. Also, the mixing partner must be from some fungicide family other than the QoI group. Tank-mixing fungicides from the same chemical family do nothing to reduce the risk of fungicide resistance. The application rates of the components should not be reduced below the minimal labelled rate.

12.7.4 From the Standpoint of Fungicide Resistance, Wisest Use of Q_oI Fungicides May Be to Use Them at the Early Stages of Disease Development

Some researchers believe that curative use of a fungicide increases the risk of resistance because the producer is treating a much larger population of spores and mycelium (the body of the fungus) than would be treated preventively. Allowing a build-up of a large population of spores before treatment increases the chances that a resistant mutant will be present when the chemical is applied. Strobilurins have outstanding ability to inhibit spore germination; thus, they should be most useful early in disease development.

12.7.5 Additional Guidelines

Whilst translaminar movement is a wonderful feature of the Q_oI fungicides, potential problems arise when a Q_oI fungicide is tank-mixed with a contact (=protectant) fungicide for resistance management purposes. Since contact fungicides remain on the treated leaf surface, poor coverage of the underside of crop foliage could result in biologically active levels of Q_oI fungicides there through translaminar movement without the presence of the mixing partner. On such leaf surfaces, a Q_oI-resistant strain could take hold and flourish, should it arise. Whenever tank-mixing a Q_oI fungicide with a contact fungicide, always strive for complete coverage of all plant surfaces.

As noted above, growth-enhancing effects independent of disease control have been observed in Q_oI-treated plants of several species. Where these cases occur, this could pose an incentive to use a product inappropriately or excessively. Whilst optimising plant health is always an important objective, inappropriate use or overuse of a Q_oI fungicide for its growth-promoting qualities could be a violation of the product label, and it may also increase selection pressure towards fungicide resistance. Users should be very mindful not to overuse any at-risk fungicide. Once resistance to Q_0I fungicides develops on a farm, there is a very good chance that efficacy of these products against that disease will be compromised for quite some time.

12.8 Commercialisation

Several synthetic strobilurins are now available commercially, and the bulk are used agriculturally (Tables 12.6 and 12.7). The first turf registration was obtained by Zeneca in 1997 for azoxystrobin, better known as 'Heritage' fungicide. Then came trifloxystrobin, Novartis 'Compass', in 1998. (If this doesn't sound quite up to date, it's because Zeneca merged with Novartis to form Syngenta, which then had to divest one of its strobilurins. Thus, Compass ended up in Bayer's hands, whilst Syngenta retained Heritage.) Pyraclostrobin, BASF's 'Insignia', is expected to gain EPA registration soon. For the next few years, at least, this trio will constitute the strobilurin presence in the turf market.

Table 12.6 Strobilurin fungicides (fluoxastrobin)

Strobilurin fungicide group	Strobilurin fungicides
Methoxyacrylate strobilurin fungicidesaa	Azoxystrobin
	Bifujunzhi
	Coumoxystrobin
	Enestroburin
	Jiaxiangjunzhi
	Picoxystrobin
	Pyraoxystrobin
Methoxycarbanilate strobilurin fungicides	Lvdingjunzhi
	Pyraclostrobin
	Pyrametostrobin
Methoxyiminoacetamide strobilurin fungicides	Dimoxystrobin
	Metominostrobin
	Orysastrobin
	Xiwojunan
Methoxyiminoacetate strobilurin fungicides	Kresoxim-methyl
	Trifloxystrobin

Table 12.7 Q_0I fungicides commercially available or expected to be available soon in the United States (updated 2006)

Trade name	Active ingredient	Manufacturer/marketer
Solo products		
Abound 2.08F	Azoxystrobin	Syngenta
Amistar 80WG	Azoxystrobin	Syngenta
Heritage 50WG	Azoxystrobin	Syngenta
Quadris 2.08SC	Azoxystrobin	Syngenta
Reason 500SC	Fenamidone	Bayer
Compass 50WG	Trifloxystrobin	Bayer
Flint 50WG	Trifloxystrobin	Bayer
Gem 500SC	Trifloxystrobin	Bayer
Disarm 480SC	Fluoxastrobin	Arysta
Evito 480SC	Fluoxastrobin	Arysta
Cygnus 50WG	Kresoxim methyl	BASF
Sovran 50WG	Kresoxim methyl	BASF
Cabrio 20EG	Pyraclostrobin	BASF
Headline 2.08EC	Pyraclostrobin	BASF
Insignia 20WG	Pyraclostrobin	BASF
Premixes		
Tanos 50DF	Famoxadone (Q_0I) + cymoxanil	Dupont
Pristine 38WDG	Pyraclostrobin (Q_0I) + boscalid	BASF
Stratego 2.08EC	Trifloxystrobin (Q_0I) + propiconazole	Bayer
Uniform 2.09EC	Azoxystrobin (Q_0I) + mefanoxam	Syngenta
Quilt 1.67SC	Azoxystrobin (Q_0I) + propiconazole	Syngenta
Quadris Opti	Azoxystrobin (Q_0I) + chlorothalonil	Syngenta

12.9 Conclusions

Strobilurins are originally derived from naturally occurring chemicals found in mushrooms. They form part of the group of quinone outside inhibitor (QoI) fungicides and work by inhibiting mitochondrial respiration that prevents spore germination and mycelial growth in plant pathogens. Their strength was they were highly active on most cereal fungal diseases, including septoria, rusts, mildew and a host of others. They were also credited with physiological benefits, maintaining green leaf area for longer and delaying crop senescence. But they have a very specific mode of action, which made the chemistry vulnerable to resistance. Resistance to strobilurins can be managed by limiting their use and by using them as a component of an integrated programme with other fungicides as well as with non-chemical management practices, such as resistant varieties and rotation.

References

Ammermann E, Lorenz G, Schelberger K, Mueller B, Kristen R, Sauter H (2000) BAS 500F - the new broad-spectrum strobilurin fungicide. Proc of the brighton crop protection conference, Alton, Hampshire, England 2:541–548

Koehle H, Gold RE, Ammermann E, Sauter H, Roehl F (1994) Biokinetic properties of BAS 490F and some related compounds. Biochem Soc Trans 22:65

Sauter H, Ammermann E, Benoit R, Brand S, Gold RE, Grammenos W, Koehle H, Lorenz G, Mueller B, Roehl F, Schirmer U, Speakman JB, Wenderoth B, Wingert H (1995) In: Dixon GK, Copping LG, Hollomon DW (eds) Antifungal agents, discovery and mode of action. BIOS Scientific Publishers, Oxford, pp 173–191

Stefan H, Kai S, Harald KU, We C (2002) A strobilurin fungicide enhances the resistance of tobacco against tobacco mosaic virus and *Pseudomonas syringae* pv *tabaci*. Plant Physiol 130:120–127

Sudisha J, Amruthesh KN, Deepak SA, Shetty NP, Sarosh BR, Shekar Shetty H (2005) Comparative efficacy of strobilurin fungicides against downy mildew disease of pearl millet. Pestic Biochem Physiol 81:188–197

13 Variety Mixtures/Cultivar Mixtures/Multilines

Abstract

Cultivar mixtures are defined as 'mixtures of cultivars that vary for many characters including disease resistance, but have sufficient similarity to be grown together'. Cultivar mixtures do not cause major changes to the agricultural system, generally increase yield stability, and in some cases can reduce pesticide use.

Cultivars used in the mixture must possess good agronomic characteristics and may be phenotypically similar for important traits including maturity, height, quality and grain type, depending on the agronomic practices and intended use. Cultivar mixtures in barley for the control of powdery mildew are an example of phenotypically similar mixtures, whereas red- and white-grained sorghum mixtures used in Africa are an example of phenotypically different mixtures.

The principles driving use of variety mixtures for disease control are soundly based in ecology. Epidemics are the exception in natural and semi-natural ecosystems, reflecting the balance derived from the co-evolution of hosts and pathogens. However, in modern agriculture in particular, this balance is far from equilibrium, and epidemics would be frequent were it not for highly effective pesticides and a plant breeding industry which introduces new cultivars to the market with new or different resistance genes. Such a situation is generally profitable when commodity prices are high, but it is costly and rates very poorly on sustainability and ecological or environmental parameter scales. This chapter describes theoretical, experimental and practical results obtained using mixtures of crop cultivars for disease suppression.

High crop yields are a principal objective of modern agriculture. Plant breeding has achieved high crop yields through hybridisation and selection of superior plants. These superior types are often grown in monocultures where each plant is genetically identical to its neighbour. The genetic uniformity for plant height, maturity and quality characteristics also facilitates harvesting, marketing and

P.P. Reddy, *Recent Advances in Crop Protection*,
DOI 10.1007/978-81-322-0723-8_13, © Springer India 2013

Fig. 13.1 Mixture of club wheat cultivars Jackmar and Tyee (Courtesy of C. Mundt)

processing of the crop. Wheat, maize, rice, potatoes, soybeans and banana are some crops that are typically grown in monocultures.

A negative consequence of genetic uniformity is an increase in genetic vulnerability to disease caused by microbial pathogens. Plant diseases can prevent a crop from achieving its yield potential, and the cost of disease and its prevention can dramatically affect the economics of crop production.

If genetic uniformity makes a crop more vulnerable to disease, then one potential, low-cost method of suppressing disease is to increase the genetic diversity of the crop. A simple way to enhance genetic diversity is to mix the seed of cultivars (i.e. plant genotypes) that vary in their susceptibility to specific pathogens. This method ensures genetic diversification with the advantage that it can be used in addition to any other form of disease control (Fig. 13.1) (Wolfe 1988).

13.1 Crop Monoculture and Diversity

It is only in the last 100 years or so that crop monoculture has become predominant in industrialised agriculture for field and plantation crops. The reasons were for simplicity of planting, harvesting and other operations, which could all be mechanised, and for uniform quality of the crop product. However, monoculture produced severe disadvantages, such as vulnerability to diseases, pests and weeds, and yield instability, which necessitated, for example, the large-scale use of pesticides, fertilisers and growth regulators.

Breeders, of course, have tried to breed for disease and pest resistance, but success has often been short-lived, because of the scale of monoculture and the poor management of resistant varieties after their release into agricultural production. Poor management in this case means the use of individual resistant varieties on a large scale which usually leads rapidly to selection of new pathogen races able to overcome the resistance.

13.1.1 Levels of Monoculture

From the point of view of disease, there is a need to think of monoculture at three levels:

- *Species monoculture*: The production of, say, wheat, as a single species on large areas, often in continuous cultivation or in wheat dominated rotations.
- *Variety monoculture*: Within the species monoculture, single varieties are often used continuously on large areas, providing maximum opportunity for selection of pathogens and pests that are well adapted to growing on the particular variety.
- *Resistance monoculture*: Even though different varieties may be used simultaneously, they may have the same disease resistance so that they appear identical to a particular pathogen.

To avoid or reduce some of the problems of monoculture, we need to introduce and manage diversity in better ways. At the highest level, species monoculture is difficult to change, at least in the short term. At the variety level, diversification is easy to manage, in the form of variety mixtures within the field. Variety mixtures can be produced commercially or by the farmer at low cost, to produce good disease control and yield stability. Composition of mixtures can be changed to delay selection of pathogen races able to overcome more than one component of each mixture. The main disadvantage is that the quality of the mixture may not be acceptable to the end user of the crop product.

Fig. 13.2 Mixtures of red-grained and white-grained sorghum (Courtesy of Dr. Henry Ngugi, Pennsylvania State University)

Breeders can diversify at the resistance level to produce lines of a single variety that possess different resistance genes (multiline varieties), but this is difficult, time-consuming and often not legally acceptable. Also, the differences in disease resistance amongst the component lines may be small relative to differences amongst varieties. On the other hand, the lines can be selected to vary in disease resistance but to be uniform for good quality.

13.2 What Is a Cultivar Mixture?

Wolfe (1988) defined cultivar mixtures as 'mixtures of cultivars that vary for many characters including disease resistance, but have sufficient similarity to be grown together'. Cultivar mixtures do not cause major changes to the agricultural system, generally increase yield stability, and in some cases can reduce pesticide use. They are also quicker and cheaper to formulate and modify than 'multilines', which are defined as mixtures of genetically uniform lines of a crop species (near-isogenic lines) that differ only in a specific disease or pest resistance (Browning and Frey 1981).

Cultivars used in the mixture must possess good agronomic characteristics and may be phenotypically similar for important traits including maturity, height, quality and grain type, depending on the agronomic practices and intended use. Cultivar mixtures in barley for the control of powdery mildew are an example of phenotypically similar mixtures, whereas red- and white-grained sorghum mixtures used in Africa are an example of phenotypically different mixtures (Fig. 13.2).

The principles driving use of variety mixtures for disease control are soundly based in ecology. Epidemics are the exception in natural and semi-natural ecosystems, reflecting the balance derived from the co-evolution of hosts and pathogens. However, in modern agriculture in particular, this balance is far from equilibrium, and epidemics would be frequent were it not for highly effective pesticides and a plant breeding industry which introduces new cultivars to the market with new or different resistance genes. Such a situation is generally profitable when commodity prices are high, but it is costly and rates very poorly on sustainability and ecological or environmental parameter scales.

Finckh and Mundt (1992) clearly illustrated why genetically heterogeneous populations should be treated in a holistic way, as in their wheat experiments the yield variation was between 52 and 58% which was attributable to disease in monocultures, whereas in mixtures, this dropped to between 10 and 31%, illustrating some of the potential different plant–plant interaction effects. Day (1984) recorded powdery mildew reduction of around 35% unrelated to growth stage or absolute level of infection.

There is a clear relationship between increased number of components in a mixture and increased disease control, but even two-component mixtures can achieve disease control. Cox et al. (2004) found a proportion range of two-component wheat mixtures to be more effective at reducing leaf rust of wheat.

Mundt and Leonard (1985) found that mixtures based on clumps of 200 oat seeds were not effective in reducing crown rust but random mixtures were, and with bean rust, reducing genotype unit areas (GUA) from 0.84 to 0.023 m^2 resulted in progressive disease reduction (Mundt and Leonard 1986). Using much more contrasting GUAs of 0.003 m^2 with wheat yellow rust and leaf rust, Brophy and Mundt (1991) again determined that the smaller GUA was more effective at reducing disease in mixtures. For *Rhynchosporium secalis* on winter barley, the optimum GUA or patch size was determined at about 4 m^2, which led to experiments to determine whether a patchy arrangement of component cultivars might be better than homogeneous mixing even within fields. Patchy sowing was found to both reduce disease and increase yield more than homogeneous sowings compared with the mean of monocultures sown alongside, even in the absence of disease (Newton and Guy 2008).

Row mixtures of susceptible and resistant cultivars achieved 94% less severe rice blast than when grown as monocultures and increased the yield by 89% in 3,342 ha in Yunnan Province (Zhu et al. 2000). In the former German Democratic Republic, where up to 92% of the spring barley crop was grown as mixtures, powdery mildew declined from over 50% to less than 10%, thereby reducing the fungicide requirement substantially (Wolfe 1997).

Garrett and Mundt (1999) achieved 36–37% in late blight of potato (*Phytophthora infestans*) in resistant and susceptible mixtures compared with the mean of the components grown in monoculture. Susceptible cultivars benefit most in mixtures, whilst partial and fully resistant cultivars are little affected. In bean crop, mixtures have been used to reduce anthracnose (*Colletotrichum lindemuthianum*).

Coffee is grown in mixtures to control its rusts, where mixtures have been planted on a large scale in Colombia (Moreno Ruiz and Castillo Zapata 1990). Mixtures of coffee genotypes have been planted on a large scale in an effort to proactively reduce anticipated infection by *Hemileia vastatrix* in Colombia (Moreno Ruiz and Castillo Zapata 1990).

Another situation in which mixtures may be of economic interest is for the protection of susceptible host genotypes with superior agronomic characteristics. In that case, the deployment of the susceptible host in combination with an agronomically inferior but disease-resistant genotype may be a solution (Garrett and Mundt 1999).

Variety mixtures or multilines can improve significantly the control of any disease that has an airborne dispersal phase (rusts, mildews, *septorioses, helminthosporioses, Rhynchosporium* and even *Pseudocercosporella herpotrichoides*), often to the extent that the use of fungicide becomes uneconomic. Because of this and other interactions amongst the components, mixtures provide a buffer against environmental variation so that yield is stable amongst environments.

Stability of yield is extremely important for the farmer. Because of environmental variation amongst mixture components, it is not possible to forecast which component will give the best yield in the next season. The safest gamble, therefore, is always to grow the mixture.

13.3 Variety and Species Mixtures in Practice

Variety and species mixtures are not only being used extensively in small-scale subsistence agriculture worldwide but also in large-scale systems. Cultivar mixtures are grown to the extent of 20,000 ha of coffee in Colombia, 14,000 ha of barley in Poland and Denmark and 7,000 ha of wheat in Pacific Northwest USA. Areas grown are substantial.

13.3.1 Reasons for Growing Mixtures

Farmers grow mixtures for the following reasons:

- Protection from airborne diseases such as rusts and powdery mildews, *Rhynchosporium, Septoria* and *Pseudocercosporella* (cultivar and species mixtures)
- Protection from cold injury (in the USA, Pakistan and Poland)
- To achieve better quality (Switzerland, Coffee in Colombia)
- To achieve higher yield stability

13.3.2 Uses of Growing Mixtures

Part of mixture production is for animal feed; however, cereal cultivar mixtures in Switzerland, Poland and the USA are used for bread and beer production. Most interesting is the fact that the highest quality coffee of Colombia is almost all produced in cultivar mixtures to protect the coffee from the coffee rust disease. These mixtures are perennial and have been successful since 1982 on a large scale.

13.4 Crops and Diseases Suited to Cultivar Mixtures

The effectiveness of cultivar mixtures has been demonstrated most commonly for foliar diseases of small grains in which host plants are small and for which there is frequent inoculum exchange amongst host genotypes. The effectiveness of cultivar mixtures in the control of foliar diseases of plants is related to the probability that pathogen propagules will fail to encounter susceptible tissues. Following Garrett and Mundt (1999), there are several inherent characteristics of each specific plant disease that affect this probability.

13.4.1 Size of Genotype Unit Area

Genotype unit area (GUA) is defined as the area occupied by an independent unit of host tissue of the same genotype (Mundt and Browning 1985). Normally, GUA is the size of an individual plant, but not always. For example, if individual plants are intertwined in the field, then the effective GUA may be smaller. And in the case of clonally propagated plants, the GUA could be larger than the individual plant (Garrett and Mundt 1999). On an individual plant basis, crop species with a small GUA include wheat, barley, oats and rice. Those with a moderate-sized GUA include beans, potatoes and corn, whereas fruit trees have a large GUA.

A genotype unit area that is very small is in most cases the ideal for suppression of disease. But specific combinations of genotype patterning and dispersal gradients may result in other GUA optima (Garrett and Mundt 1999). In general, as GUA increases, the effectiveness of mixtures for disease suppression decreases (Mundt 1989). GUA interacts with the dispersal gradient of the pathogen, as shown in Fig. 13.3.

Mundt and Leonard (1986), working with bean rust, demonstrated that the least amount of disease was associated with the smallest genotype unit area (Table 13.1). Mundt and Browning (1985) reported similar results with oat crown rust and maize common rust.

- The area under the disease progress curve (ADPC) for each treatment divided by the ADPC for the pure-line susceptible treatment.
- Mixture treatments consisted of 50% susceptible plants and 50% resistant plants.
- Mixture treatments consisted of 25% susceptible and 75% resistant plants.

13.4.2 Steepness of Dispersal Gradient

The shape of a pathogen's dispersal gradient and its interaction with genotype unit area influences the effectiveness of cultivar mixtures for disease suppression. Steeper dispersal gradients result in more inoculum loading on the plant (genotype) on which it was produced, with fewer spores lost as a result of the dilution or barrier effects. Consequently, splash-dispersed pathogens are less suited to control by cultivar mixtures than are wind-dispersed pathogens (Fig. 13.4).

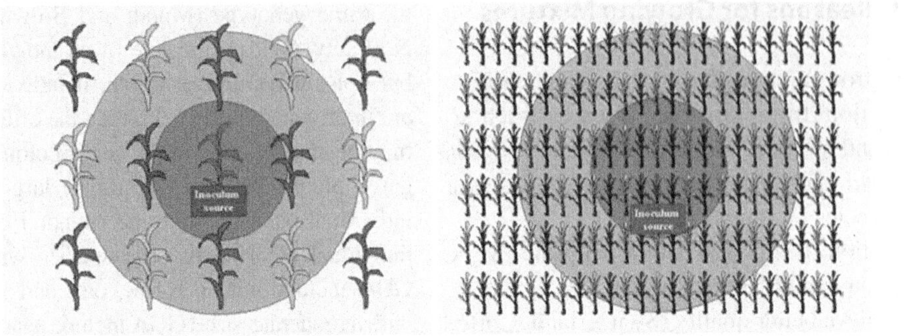

Fig. 13.3 Each figure depicts two pathogen propagule dispersal gradients (*circles*) originating from the centre host plant superimposed over a mixture of two plant genotypes (as indicated by colour). Within each dispersal gradient, propagule loads decrease with distance from the source. *Left* – when the size of individual plants is large relative to the dispersal gradient or the gradient is steep, inoculum tends to land on and reinfect the host plant on which it originated. *Right* – when the size of individual plant is small relative to the dispersal gradient or the gradient is shallow, a greater share of inoculum falls on plants with a genotype different from the source host plant

Table 13.1 Effect of spatial arrangement of susceptible and resistant snap beans on the increase of bean rust in mixtures of susceptible and resistant plants and in pure stands of susceptible plants in field plots

	Relative area under the disease progress curve[a]			
	Genotype unit area (m²)	1982[b]	1983[c]	1984[c]
Pure-line susceptible	13.40	1.00	1.00	1.00
Mixture	0.023	0.73	0.42	0.57
	0.093	0.76	0.29	0.58
	0.372	0.99	0.71	0.56
	0.836	0.90	1.31	0.72

[a]The area under the disease progress curve (ADPC) for each treatment divided by the ADPC for the pure-line susceptible treatment

[b]Mixture treatments consisted of 50% susceptible and 50% resistant plants

[c]Mixture treatments consisted of 25% susceptible and 75% resistant plants

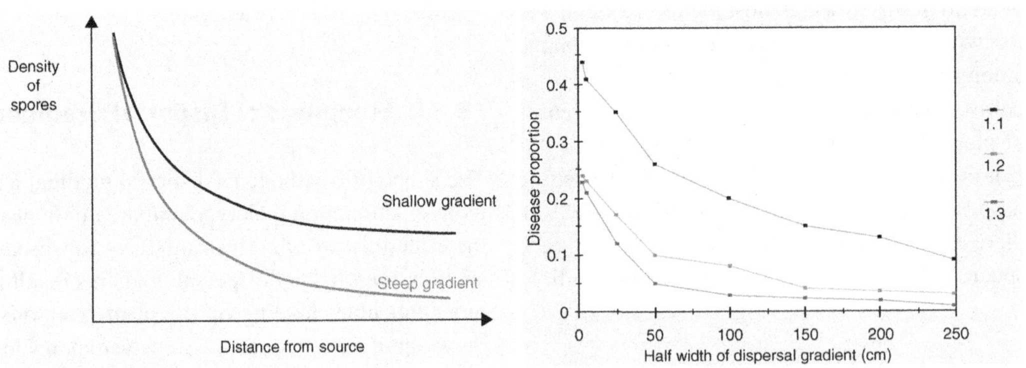

Fig. 13.4 Computer simulations have illustrated that cultivar mixtures are most effective in reducing spread of diseases caused by pathogens with shallow spore dispersal gradients. This figure shows the predicted effect of the relative steepness of the spore dispersal gradient on the amount of disease in three mixtures of susceptible and resistant cultivars (ratio of susceptible to resistant cultivars = 1:1, 1:2 and 1:3) after six pathogen generations in a model simulating barley powdery mildew (Fitt and McCartney 1986) (*Note*: Steep gradients have small half widths)

13.4.3 Ultimate Lesion Size

Continuous expansion of individual lesions, in the cases that it is possible, decreases the degree of disease suppression achieved by mixing host genotypes because it increases the rate of new infection of the susceptible genotype (Garrett and Mundt 1999). Thus, analogous to genotype unit area, cultivar mixtures for disease suppression are most effective when the size of individual lesions remains small. Non-expanding lesions require a proportionally greater number of infections for the disease epidemic to progress. The dilution and barrier effects influence the number of new infections but not the rate of lesion expansion. Leaf rust of wheat (*Puccinia recondita*) is an example of a disease with a small determinant lesion size. Stripe rust of wheat (*Puccinia striiformis*) shows substantial lesion expansion, and, thus, cultivar mixtures may be expected to have somewhat less impact on suppression of this disease (Lannou et al. 1994).

13.4.4 Degree of Host Specialisation

Most research on use of cultivar mixtures for disease suppression has focused on diseases caused by biotrophic, obligate pathogens that interact with their hosts on a gene-for-gene basis. The presence of differential, qualitative resistance to races of the pathogen in different host genotypes is commonly one of the most important criteria for selecting the cultivars that make up the mixture. With differential qualitative resistance, each host genotype potentially benefits from being in a mixed population; that is, relative to a pure-line stand of a genotype, the cultivar mixture will have a reduced proportion of tissue susceptible to the races that can infect it (Table 13.2) (Jeger et al. 1981a, b; Garrett and Mundt 1999).

13.5 Use of Cultivar Mixtures to Manage Multiple Diseases

In field conditions, crops can be attacked by multiple diseases, and cultivar mixtures can be used to manage and control those multiple diseases by combining cultivars with differential disease resistance to different diseases. The basic principles are the same as those used for controlling a single disease, with the addition of the possible interactions between the diseases present in the crop.

Experimental results show that cultivar mixtures can indeed control multiple diseases, and that the level of control is associated with the already mentioned characteristics of the specific plant disease (size of genotype unit area, dispersal gradient, lesion

Table 13.2 Inherent characteristics of plant pathosystems that predict whether or not mixtures of cultivars will provide disease suppression (From Garrett and Mundt 1999)

Host	Pathogen	Small host genotype unit area	Shallow dispersal gradient	Small lesion size	Short pathogen generation time	Strong host specialisation
Coffee (Fig. 13.5a)	*Hemileia vastatrix*	–	+	+	–	+
Pepper	*Xanthomonas campestris*	–	–	–	+	+
Potato (Fig. 13.5b)	*Phytophthora infestans*	–	+	–	+	+
Rice (Fig. 13.5c)	*Magnaporthe oryzae*	+	+	+	+	+
Wheat	*Blumeria graminis*	+	+	+	+	+
Wheat (Fig. 13.5d)	*Puccinia triticina*	+	+	+	–	+
Wheat	*Puccinia striiformis*	+	+	–	–	+
Wheat	*Mycosphaerella graminicola*	+	–	–	–	–
Wheat	*Rhizoctonia cerealis*	+	–	–	–	–

+ Characteristic favours disease suppression by cultivar mixtures
– Characteristic does not favour disease suppression by cultivar mixtures

Fig. 13.5 (**a**) Uredinia of coffee rust *Hemileia vastatrix*. (**b**) White sporangia and sporangiophores of late blight pathogen *Phytophthora infestans*, at the margins of necrotic potato leaf lesions. (**c**) Mature rice blast lesions showing necrotic borders caused by *Magnaporthe grisea*. (**d**) Uredinia of rust on wheat *Puccinia* sp.

size, pathogen generation time and degree of host specialisation).

Cox et al. (2004) studied the effect of wheat mixtures for the control of tan spot (caused by *Pyrenophora tritici-repentis*) and leaf rust (caused by *Puccinia triticina*). Mixtures consisted of two varieties, one resistant to tan spot and susceptible to leaf rust and the other vice versa. For both diseases, severity was lower on the susceptible cultivar in the mixture as compared with monoculture. The decrease was higher in leaf rust, which is a typical windborne pathogen, highly specialised, polycyclic and with a shallow dispersal gradient. Tan spot, by contrast, is a residue-borne disease with low number of generations per growing season, capable of infecting several species, and with a steep dispersal gradient.

In another example, Ngugi et al. (2001) used mixtures of sorghum to control anthracnose (caused by *Colletotrichum sublineolum*) and leaf blight (caused by *Exserohilum turcicum*), considered the most destructive diseases of high yielding cultivars in Eastern Africa. In this case, they used one sorghum cultivar susceptible to both diseases and another with good resistance to both. Mixtures were effective in controlling both diseases, delaying the time when the disease was first observed and lowering the rate of disease progress. The effect was more pronounced with leaf blight, which is a wind-dispersed disease with shallow dispersal gradient and uniformly distributed incidence. Anthracnose is splash dispersed, which is normally associated with steep gradients.

One important practical advantage of using cultivar mixtures to control multiple diseases is the relative ease of combining varieties with resistance to different diseases compared with the development of a single variety resistant to all diseases considered. Mixtures can therefore extend the options in deployment of host resistance.

13.6 How Many Cultivars Make a Good Mixture?

The number of cultivars in the mixture can influence the disease-control benefit achieved from it. Mundt (1994) showed that increasing the number of cultivars up to five gave a trend towards decreasing the severity of stripe rust on wheat

(Table 13.3) but with potentially diminishing returns beyond three or four cultivar components. Newton et al. (1997), as shown in the figure below (Fig. 13.6), obtained similar results in the control of scald on winter barley.

13.7 Effect of Cultivar Mixtures on Epidemic Development

The development of a plant disease epidemic is a function of the initial inoculum, the rate of disease development and the duration of crop growth.

Table 13.3 Percent reduction of stripe rust severity (relative to mean of component cultivars in pure stand) for wheat cultivar mixtures tested in field plots at three locations (From Mundt 1994)

Components in mixture	Number of mixtures	Mean[a]
2	10	31
3	10	42
4	5	49
5	1	48
Mean of all mixtures		39

[a]Mean over three locations and 3–4 replications per location

Cultivar mixtures affect the epidemic factors as follows:

- The initial inoculum level can be reduced. Compared to a pure-line susceptible population, the initial inoculum is lowered when a given race or strain is not completely virulent on all genotypes in the mixture.
- The rate of secondary infection can be reduced. As the proportion of susceptible tissue available for a given race of the pathogen is lowered, the number of new, secondary infections by this race is reduced, which results in a decrease in the observed 'apparent infection rate'.

Leonard's (1969) classic formula describes this second effect based on a single pathogen genotype in a simple mixture of one susceptible and one immune plant genotype:

$$\frac{X'}{X_0} = m^n \frac{X}{X_0}$$

X = Proportion of infected tissue in a population composed only of the susceptible genotype

X' = Proportion of infected host tissue for the susceptible genotype in the mixture

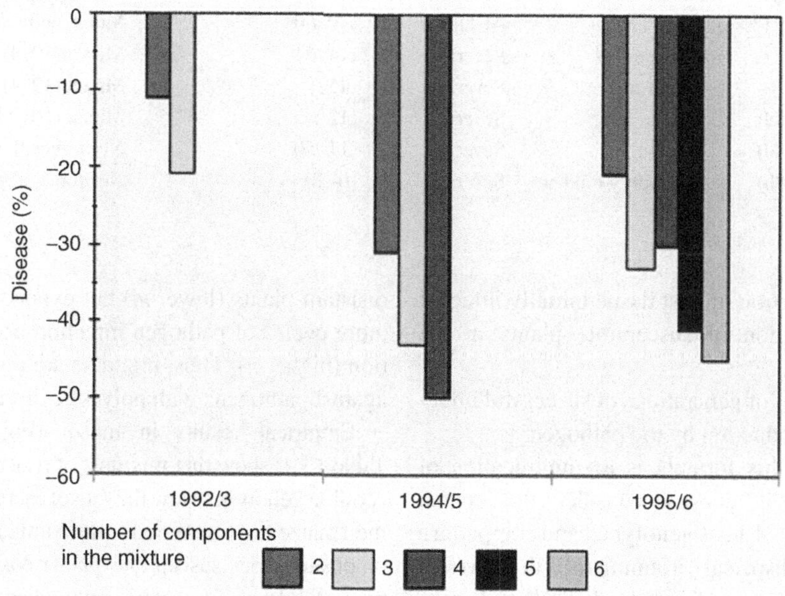

Fig. 13.6 Change in scald severity in mixtures of winter barley cultivars compared as influenced by the number of component cultivars (Adapted from Newton et al. 1997)

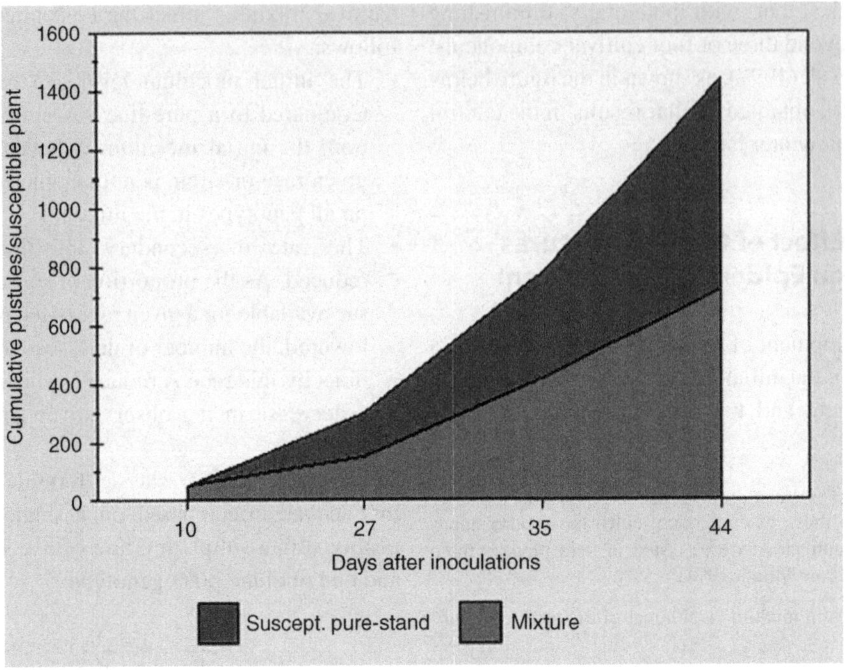

Fig. 13.7 Cumulative numbers of common maize rust pustules on susceptible plants in pure stands and in mixtures of 25% susceptible and 75% resistant plants (Adapted from Mundt and Leonard 1986)

Table 13.4 Examples of degree of disease suppression achieved through the use of cultivar mixtures

Plant	Pathogen	Trait	Disease reduction (%)	Reference
Maize (Fig. 13.8c)	Rust	Pustules/plant	50	Mundt and Leonard (1986)
Snap beans (Fig. 13.8a)	Bean rust	AUDPC	30–60	Mundt and Leonard (1986)
Wheat	Stripe rust	Severity	14–64	Mundt (1994)
Wheat	Leaf rust	Severity	45	Mundt (1994)
Barley (Fig. 13.8d)	Scald	Severity	12	Mundt et al. (1994)
Barley (Fig. 13.8d)	Scald	Severity	11–50	Newton et al. (1997)
Barley (Fig. 13.8b)	Powdery mildew	Severity	0–20	Newton et al. (1997)

X_0 = Proportion of host tissue initially infected

m = Proportion of susceptible plants in the mixture

n = Number of generations of successful infection and reproduction by the pathogen

Although this formula is a simplification of reality (e.g. it doesn't consider the spatial configuration of host genotypes and the pattern of inoculum dispersal), it summarises the 'mixture effect' as a function of m and n. It predicts that the disease suppressing effects of the mixture will increase linearly with an increasing proportion of resistant plants (lower m) but exponentially with more cycles of pathogen infection and reproduction (higher n). Thus, mixtures are most effective against pathogens with polycylic disease cycles.

Empirical results in maize (Fig. 13.7 and Table 13.4) show that mixtures of resistant and susceptible genotypes slow the rate of increase of common maize rust (measured as the cumulative number of pustules per susceptible plant) compared with pure stands of susceptible genotypes. Table 13.4 shows other examples of cultivar mixtures in which some degree of disease suppression was achieved.

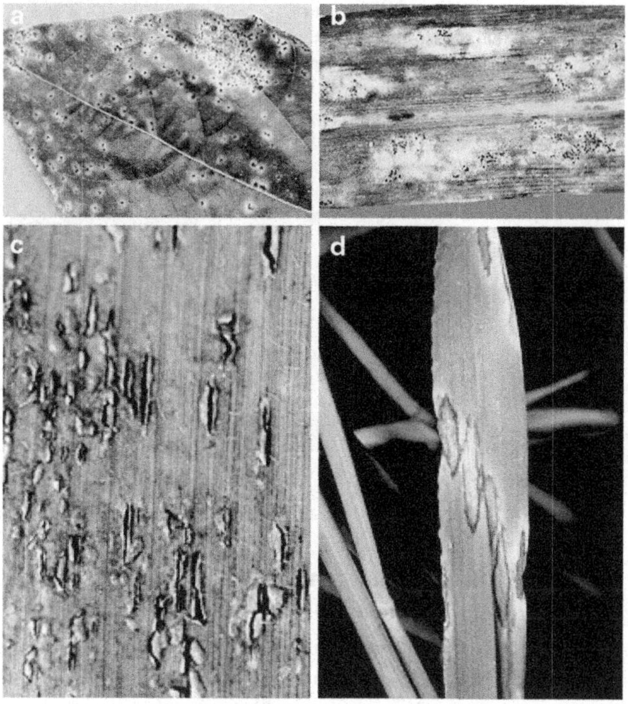

Fig. 13.8 (**a**) Bean rust, *Uromyces appendiculatus,* on the upper leaf surface. (**b**) Cottony mycelial growth with dark, spherical cleistothecia of powdery mildew fungus *Blumeria graminis* on barley; (**c**) Uredinia of corn rust pathogen, *Puccinia sorghi*. (**d**) Typical leaf symptoms of barley scald

13.8 Mechanisms of Variety Mixtures for Reducing Epidemics

The use of varietal mixtures is an epidemic control strategy that has been shown to be effective against airborne pathogens of crops developing polycyclic epidemics. Basically, host mixtures may restrict the spread of diseases relative to the mean of their components, provided that the components differ in their susceptibility. Most studies, however, have been developed for specialised pathogens and specific resistance genes; when a pathogen develops epidemic cycles in a mixture, the number of new lesions generated for each cycle is considerably reduced compared to what would happen in a pure stand. This operates in three principal ways that have been identified through both experimental and theoretical work.

The first mechanism of disease reduction is the decrease in the spatial density of susceptible plants. In a mixture, the probability of a spore released from a lesion to be deposited on susceptible tissue is reduced in relation to the density of susceptible plants. When the distance between susceptible plants increases, it becomes increasingly unlikely for a spore to land on a suitable host. In addition to host density, the presence of resistant plants in the canopy provides a physical barrier against spore dispersal (Fig. 13.9).

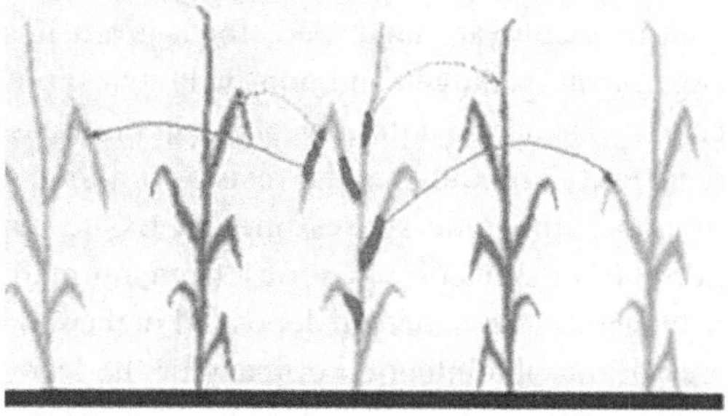

Fig. 13.9 In a mixture, a part of the spores released from an infected plant is lost on resistant plants (in *green*). The number of spores producing new infections on susceptible plants (in *yellow*) is considerably reduced due to the lower density of susceptible plants and barrier effect of resistant plants

These two mechanisms appear to be mechanical effects related to the way the pathogen spores are spread and to the distribution of resistant

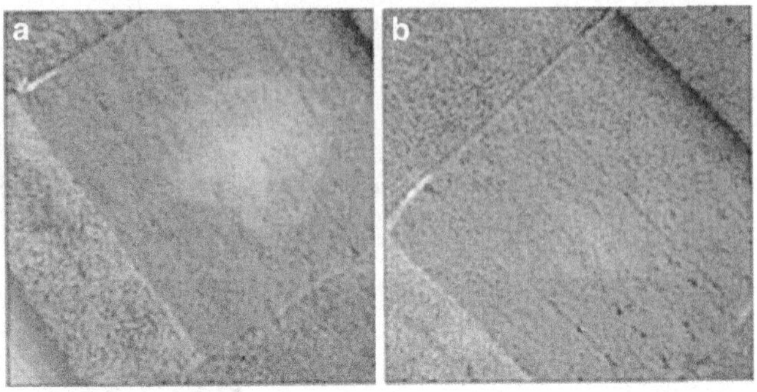

Fig. 13.10 (**a**) Focal development of disease in a field of pure line of wheat inoculated with yellow rust at centre. (**b**) Disease development of yellow rust in field plot of mixture of wheat varieties with different resistance genes

and susceptible hosts. Therefore, the magnitude of disease reduction that can be expected depends on parameters such as the spore dispersal gradient, the lesion growth rate, plant size, distribution of the plant genotypes (groups of plants or random distribution), etc. Pathosystems with the best characteristics for effective use of mixtures include particularly the airborne pathogens of small-grain cereals (e.g. barley powdery mildew). In host mixtures, the genetic diversity of the pathogen population is greater than in a cultivar stand, and, for a given host component, pathogen and non-pathogen spores coexist. Therefore, a third mechanism of disease reduction in mixtures is the resistance induced by non-pathogenic spores on host tissue that prevents or reduces infections from normally pathogenic spores that are deposited in the same area. Either the infection efficacy or the lesion productivity can be reduced. Induced resistance is a general mechanism in pathosystems, and its characteristics may vary from case to case. It has been suggested, however, that even much localised effects in terms of susceptible tissue area protected by a non-pathogenic spore may result in significant disease reduction at the epidemic level. Experimental studies have shown that induced resistance could account for 20–40% of the disease reduction in mixtures.

Variety mixtures do not eliminate the pathogen as a fungicide might. Rather, they reduce the rate of disease progress by eliminating large numbers of spores at each cycle of pathogen multiplication. Spores are lost on resistant plants or because of the larger distance between susceptible plants and the infection processes are perturbed by induced resistance. The result is a high level of partial resistance (Fig. 13.10).

13.9 Effect of Cultivar Mixtures on the Evolution of Pathogen Races or Pathotypes

The effect of the cultivar mixtures on pathogen evolution can be analysed based on two key questions:

Will a given resistance be more durable when deployed in a mixture than deployed in a monoculture?
The lower level of exposure to the pathogen population of the resistance gene in a mixture compared with monoculture will reduce selection pressure on the pathogen population and therefore increase the gene's durability (Mundt 2002).

Considering a given number of resistance genes, will they be more durable if deployed in a mixture than deployed sequentially in monoculture or combined into a single host genotype?
Mixtures support more diverse pathogen populations than do pure stands, and that diversity is positively related to the degree of disease control provided by the mixture (Mundt 2002). The mixtures lead, generally, to the evolution of complex pathogen races potentially capable of overcoming many resistance genes and therefore reduce the efficacy of the crop mixture as a measure to reduce disease. However, the appearance of complex races does not mean that those races will be

selected for within the mixtures. There are several factors that can prevent the prevalence of complex races. The relative efficacy and durability of the resistance genes deployed in crop mixtures compared with the other two options depends on the rate of progress of those races in the mixture, which can be affected by several factors:

13.9.1 Diversity Within Pathotypes

Increasing diversity reduces the rate of increase of complex pathotypes in crop mixtures (Lannou 2001).

13.9.2 Fitness Cost Associated with Virulence

The ability of complex races to attack multiple host genotypes is countered by a reduction in fitness associated with the lack of avirulence genes (Lannou 2001; Mundt 2002).

13.9.3 Differential Adaptation

Complex races can infect different host genotypes but have reduced infection efficiency compared with the specific simple races corresponding to each component of the mixture, reducing therefore their rate of progress (Lannou 2001; Mundt 2002).

13.9.4 Density Dependence

It is the decrease in pathogen multiplication rate associated with an increase in lesion density. It could affect simple and complex races differentially during an epidemic and reduce selection for complex races (Lannou and Mundt 1997).

The progress towards complex pathogen races may be relatively slow. Single races or pathotypes capable of overcoming a single resistance gene should progress faster than complex races capable of overcoming multiple resistance genes, precluding the former of becoming dominant within the pathogen population.

13.10 Mechanisms by Which Cultivar Mixtures Suppress Disease

Cultivar mixtures do not completely suppress or eliminate the disease. Rather, mixtures reduce the rate of disease progress by eliminating large numbers of spores at each cycle of pathogen multiplication. Spores are eliminated from the epidemic process by deposition on resistant plants and by dilution because of the greater distance between plants of the same genotype. Moreover, the infection process may be slowed by the induction of defence responses in susceptible plants by strains of the pathogen that are avirulent on specific host genotypes. The result is a level of disease suppression owing to multiple epidemiological and physiological mechanisms. The four mechanisms by which cultivar mixtures suppress disease are summarised as follows.

13.10.1 Dilution and Barrier Effect

Increasing the distance between susceptible plants reduces/slows the rate of plant to plant spread. The presence of resistant plants in the canopy provides a physical barrier against spore dispersal, interrupting spore movement. The number and size of the resistant plants and the physics of spore dispersal influence the strength of the barrier effect

A decrease in the density of susceptible plants slows disease development. If plants within the population possess different race-specific genes, the relative ability of any given pathogen race to spread from plant to plant is reduced because distance between plants of the same genotype is increased. For barley powdery mildew, Chin and Wolfe (1984) demonstrated that the increased distance between plants of the same genotype in cultivar mixtures was the most important mechanism of control, especially early in the epidemic. The ideal spatial arrangement of host genotypes is one in which plants susceptible to the same pathogen race do not occur as neighbours.

Similarly, the barrier effect is caused by the presence of a resistant plant that acts as a barrier to the spread of pathogen propagules (e.g. spores).

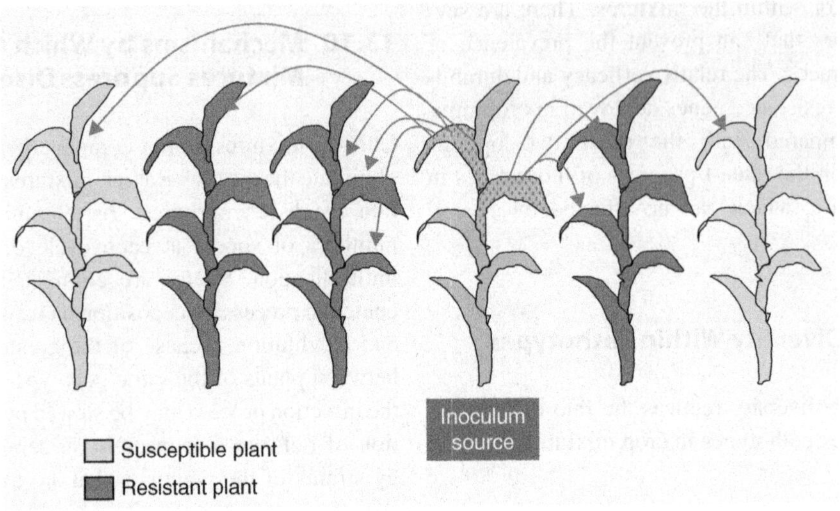

Fig. 13.11 Induced resistance

For both of these mechanisms, the size of the host plant influences the effectiveness of the cultivar mixture. In general, mixture effectiveness decreases with increasing size of the host individuals (Garrett and Mundt 1999). The expected mixture effect for cereals, for example, is stronger than that expected for apples.

In a mixture, the number of released spores creating new infections can be reduced considerably by lowering the density of susceptible plants (dilution effect). Moreover, some of the spores released from an infected plant are captured by a resistant plant and consequently, removed from the epidemic process (barrier effect).

13.10.2 Induced Resistance

Induced resistance occurs when biochemical host defences are triggered by inoculation with an avirulent race. Triggering these defences slows the infection processes of virulent pathogen races to which the host is normally susceptible (Fig. 13.11).

Induced resistance occurs when spores of an avirulent strain or race land on and trigger a biochemical defence response on an incompatible host. This induction of defence responses reduces partially the susceptibility of the host plant to infection by spores of a virulent strain or race (Lannou et al. 1995). Either the infection efficacy or the number of new spores produced as a result of infection can be reduced (Martinelli et al. 1993) (Fig. 13.12). Induced resistance is a nonspecific, general mechanism in many pathosystems, and its characteristics vary from disease to disease. Some mechanisms of induced resistance are localised to tissues in the vicinity of an infection, but other mechanisms may affect a larger part of the plant. It has been suggested, however, that even much localised induction of resistance by an avirulent spore may result in significant disease reduction at the epidemic level (Lannou et al. 1995).

Experimental studies indicate that induced resistance may account for 20–40% of the disease reduction in mixtures when two or more pathogen races are active in the crop (Lannou and de Vallavieille-Pope 1997). According to Calonnec et al. (1996), up to one-third of the reduction in infection by *Puccinia striiformis* in wheat mixtures was due to induced resistance. In this system, induced resistance is particularly important because of the indeterminate (continually expanding) nature of stripe rust lesions. In barley powdery mildew, it is believed that induced resistance plays an important role during the latter stages of an epidemic (Chin and Wolfe 1984).

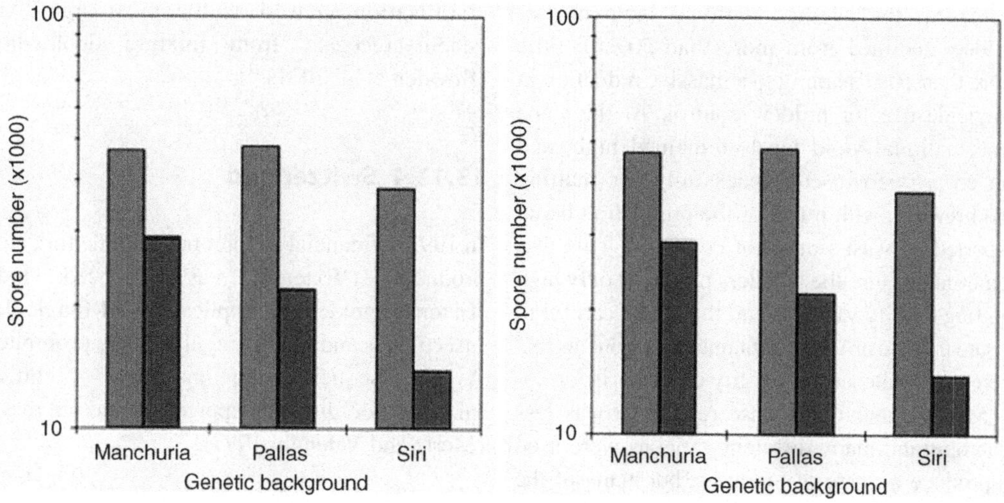

Fig. 13.12 Numbers of colonies and spores of powdery mildew formed by *Blumeria graminis* f. sp. *hordei* on 2×5 cm barley leaf segments. Seedlings were inoculated either with an avirulent inducer isolate and then a virulent isolate or with the virulent isolate only. Three sets of four near-isogenic lines, with different race-specific genes, were used. Values are the number of colonies per leaf segment (*left*) or number of spores per lesion (*right*) (*red* virulent, *green* avirulent and then virulent) (From Martinelli et al. 1993)

13.10.3 Modification of the Microclimate

The presence in the component cultivar of plant attributes (i.e. plant height, canopy traits, etc.) that modify the microclimate towards less favourable conditions for the disease can help suppression of the disease. In one pathogen generation, the combined effect of the four mechanisms in slowing the pathogen spread may be small. It is the multiplicative effect over several pathogen generations that lead to the greatest suppression of the disease (Wolfe 1988).

In a mixture of glutinous (35–40 cm taller and much more susceptible to rice blast) and non-glutinous rice, Zhu et al. (2005) have found that the interplanting of both types reduced the number of days with 100% humidity at 0800 h (from 20 for the pure stands to 2.2) and the mean percentage of glutinous leaf area covered by dew (from 84 to 36%). Both variables are critical for the development of the disease. This change in the environmental conditions was a substantial contributor to panicle blast control (over 90% reduction in incidence on the glutinous cultivar and 30–40% on the non-glutinous one), regardless of the effects of other mechanisms.

13.11 Reported Successes with Cultivar Mixtures

Crop cultivar mixtures have been sown commercially in numerous countries with encouraging results.

13.11.1 Former GDR

From 1984 to 1990, cultivar mixtures comprised a substantial percentage (up to 92%) of the barley acreage of the former German Democratic Republic (East Germany). This country is the most remarkable example of large-scale use of this disease-control strategy in industrialised agriculture.

One of the most remarkable examples of the large-scale use of variety mixtures in industrialised agriculture was the development during the 1980s of the use of spring barley mixtures in the former German Democratic Republic. Following the recognition of the problems caused by the powdery mildew pathogen in monoculture of barley varieties, and of the high cost of western fungicides, the Government implemented the use of barley mixtures nationwide. As the acreage increased, the average national incidence of mildew declined from more than 50% to little more than 10%, leading to a massive reduction in fungicide use for mildew control. At the same time, national yield levels remained high, and the crops were used successfully for malting and brewing, with much of the production being exported to west European countries. This was achieved because the breeders produced only high malting quality varieties and they were careful to ensure that the mixtures contained components that were well matched for quality characteristics.

Management for disease resistance was less than optimal; many different varieties were used to produce a range of mixtures, but many of the varieties contained the same resistance genes, that is, variety diversity was far greater than resistance diversity. Despite this disadvantage, the mixtures were effective until the time of political reunification in Germany when the whole project was stopped. Variety and resistance monoculture has now been re-established at the cost of a large-scale expansion of fungicide use combined with overdependence on the single Mlo resistance gene.

13.11.2 China

A varietal diversification programme was tested in the Yunnan province to control rice blast. Involving 812 ha in the first year and 3,342 ha in the second year, mixtures of susceptible and resistant cultivars reduced the average rice blast severity on the susceptible varieties from 1 to 20% (Zhu et al. 2000). This experiment was unique in both its scale and experimental design.

13.11.3 USA

Wheat cultivar mixtures are increasingly popular in northeastern Oregon and southeastern Washington (Mundt 1994) for the objectives of stripe rust suppression and stabilisation of yields. In 1998, 10% of soft white winter wheat and 76% of club wheat fields in this region were sown to cultivar mixtures. In Kansas, wheat variety blends occupied 7% of fields in 2000, with yield stabilisation viewed as the most significant benefit received from mixture deployment (Bowden et al. 2001).

13.11.4 Switzerland

In 1992, a financial support programme for cereal production (‘Extenso’) was introduced. Under ‘Extenso’ guidelines, applications of fungicides, insecticides and growth regulators are prohibited. As a consequence, the importance of cultivar mixtures for disease suppression has increased (Mertz and Valenghi 1997).

13.11.5 Denmark

In 1979, seed companies were allowed for the first time to produce and sell variety mixtures of spring barley in Denmark. Winter barley mixtures with powdery mildew resistance were released in the mid-1980s. In 1996, 62,000 ha of barley (9.7% of the total) were sown to variety mixtures (Munck 1997). For the 1997 growing season, 49 different mixtures were marketed, involving 20 different varieties from six resistance groups to powdery mildew.

A committee appointed by the Danish State Seed Testing Station (now Danish Plant Directorate) approved the mixtures according to the following criteria:

- Only varieties from the Danish National List of Varieties could be used as components.
- Mixtures should be composed so as to reduce harmful organisms (mainly powdery mildew, rust and nematodes).
- A mixture should be composed of at least four varieties representing at least three different sources of resistance to powdery mildew.
- The components should be uniform with respect to maturity.
- A mixture should be composed of equal amounts of the component varieties.

A major revision of the criteria is under preparation in 1997. In the first years, only few mixtures were registered. For instance in 1981, six mixtures were approved, all of which included

Table 13.5 Grain yield, yield increase in sprayed plots, % powdery mildew and scald in spring barley mixtures and their components (Data from the Danish Agricultural Advisory Centre)

		Yield in unsprayed plots (hkg/ha)		Yield increase in sprayed plots (hkg/ha)		% Powdery mildew		% Scald	
Barley varieties	Powdery mildew	1995	1996	1995	1996	1995	1996	1995	1996
No. of trials	resistance	5	4	5	4	4	4	4	4
Mixture	–	62.5	69.0	2.3	0.4	1.0	0.0	3.4	0.1
Alexis	Mlo	61.5	69.4	3.0	1.1	0.0	0.0	5.0	0.2
Meltan	Ru, Im 9, Hu 4	63.9	64.9	0.1	1.2	0.0	0.0	1.9	0.1
Goldie	Ar, Ia, U	63.1	69.7	1.6	1.5	0.0	0.0	4.3	0.1
Canut	Ly, La	60.1	–	4.7	–	1.6	–	10.6	–
Lambda	Ri, Tu 2	–	70.2	–	1.5	–	0.0	–	0.1

at least one variety with *Laevigatum* resistance. For the growing season 1997, 49 different mixtures have been approved, involving 20 different varieties from six resistance groups (Algerian, Arabische, Monte Cristo, Mlo, Ricardo and Rupee). The most frequently occurring varieties are Lamba (Ri, Tu2) in 71% of the mixtures and Meltan (Ru, lm9, Hu4) and Goldie (Ar, La, U), in 51 and 57%, respectively. Three-quarters of the mixtures include a variety possessing Mlo resistance.

In the mid-1980s, winter barley varieties with powdery mildew resistance were released, and an increasing interest for winter barley mixtures led to a law which allowed mixtures from 1987. The criteria for winter barley mixtures were similar to those for spring barley: Only varieties from the Danish National or EU List of Varieties can be mixed and only with equal amounts of the components. The mixtures may be composed of three or four varieties, which are uniform with respect to maturity. In addition, information on winter hardiness and resistance to powdery mildew is considered. Eight mixtures were approved for the season 1996–1997.

13.11.5.1 Area of Variety Mixtures

Based on information from second-generation certified seed, mixtures were grown on 6–15% of the spring barley area in the years 1980–1997; from 1986, it was 11% or more. The actual area was probably 2–3% greater because some farmers use their own seed for mixing. During the last 2 years, the area has decreased slightly, possibly due to lesser mixing effects and increasing areas with malting barleys, which are grown only in monoculture. In 1996, 62,000 ha (9.7%) were grown with spring barley mixtures. The area with winter barley mixtures was considerably less, that is, up to 4%, and in 1995–1996, the area was 2,400 ha (1.2%).

13.11.5.2 Grain Yield and Powdery Mildew in Barley Mixtures

From 1979 to 1991, the Danish Agricultural Advisory Centre conducted more than 230 trials with spring barley mixtures in sprayed and unsprayed trials. In summary, the mixtures yielded from 1 to 7% (average 3%), more than the mean of the varieties in pure stand in unsprayed trials. In the same period, the powdery mildew score (0–10 with 0 = no powdery mildew) varied between 0, 1 and 7 in the mixtures. The mean powdery mildew scores of the component varieties were equal to the mixture scores in years with low levels of powdery mildew or higher in years with high levels. Since 1983, a mixture has been used as a standard in the national trials, the mixture changed over the years as single component varieties were replaced with more up-to-date varieties. Using this scheme, greater stability over years is expected. When comparing the variation in yield of variety mixtures with that of the highest yielding variety, it is clear that the stability of mixtures is greater over years than that of varieties in pure stand. In the last 2–3 years, the yields of mixtures compared to the means of components have declined, probably because component resistances such as Mlo, Tu2 and lm9 are still effective (Table 13.5).

Fig. 13.13 (a) Interspecies mixture – wheat, oat and barley. (b) Wheat varietal mixture

From 1983 to 1986, trials with winter barley mixtures were conducted at the Danish Government Research Stations. In unsprayed plots, the levels of leaf blotch and scald were reduced significantly, and small but non-significant yield increases were obtained commonly in the mixtures compared with the mean yields of the component varieties grown alone. It was concluded that the winter barley mixtures had a greater ability to utilise growth factors than did pure varieties.

13.11.5.3 Other Mixture Trials

During the last 5 years, the Danish Agricultural Advisory Centre has tested four variety mixtures of winter wheat (Fig. 13.13). The results obtained on disease reduction and yield increase are similar to those from trials with barley mixtures. Mixtures of pea varieties have been tested for 3 years, and in each year, the mixtures yielded more than the means of the four components (2–4%). However, sales of winter wheat or pea mixtures are not yet allowed in Denmark.

13.11.6 Poland

Use of barley cultivar mixtures for disease suppression was initiated in the early 1990s and has now reached about 90,000 ha/year (Gacek 1997). The main emphasis has been on spring-sown feed barley (eight mixtures recommended), but two spring-sown malting mixtures and three winter barley mixtures also have been recommended. The mixtures are designed for control of powdery mildew, but more general recommendations for their use are given to farmers. The following two paragraphs and table summarise the information given to farmers on these mixtures.

The mixtures are designed particularly for control of powdery mildew, but more general recommendations for their use are given to farmers:

- Because of their broad genetic variation, variety mixtures have greater environmental plasticity than pure varieties and are therefore recommended for use in the country as a whole.
- Yields of the mixtures are greater and more stable than those of pure varieties.
- Crops are healthier, mainly through reduced mildew infection, and less likely to lodge.
- The reduced need for fungicides lowers production costs and reduces contamination of the environment.
- Variety mixtures should be cultivated in the same way as pure varieties.

Before introduction into commercial production, candidate mixtures and their variety components are grown in replicated, small plot (10 m^2) field trials for 3 years at three or four sites. The best performing mixtures are selected for multiplication (initially as pure varieties with the final year as a mixture). Examples of trial performance of four recently selected barley mixtures are presented in Table 13.6.

Future research on variety mixtures will include practical aspects such as control of pests and weeds

Table 13.6 Trial performance of four recently selected barley mixtures

JP 5	JP 6	JP 7	JP 8
Boss – 63.6	Boss – 56.4	Boss – 64.7	Boss – 64.7
Edgar – 62.2	Ekol – 55.2	Rambo – 64.1	Rambo – 64.1
Ekol – 61.6	Rabel – 57.1	Rabel – 65.4	Bryl – 64.7
Mean – 62.5	Mean – 56.2	Mean – 64.7	Mean – 64.5
Mix – 64.5	Mix – 57.0	Mix – 67.0	Mix – 67.0
% – 103	% –	% – 104	% – 67.0

and pre-breeding for mixing ability together with long-term investigation of plant development and interplant interactions. There will also be further investigation of cereal species mixtures, which are even more widely used by Polish farmers.

13.11.7 Colombia

Mixtures of coffee genotypes have been planted on a large scale in an effort to proactively reduce anticipated infection by *Hemileia vastatrix* in Colombia (Moreno Ruiz and Castillo Zapata 1990).

13.11.8 Switzerland

Intensification of agriculture in Switzerland after World War II increased production too far and led to problems such as high financial input and negative impact on the environment. The government therefore introduced new agricultural policies. An essential part was a shift from price support to direct payments, linked increasingly to IP (integrated production). In 1992, financial support for extensive cereal production ('Extenso') was introduced (800 SFr/ha). 'Extenso' production means no application of fungicides, insecticides or growth regulators. Many farmers realise that 'Extenso' is an economical and ecological alternative, particularly where mixtures are used to provide disease control in place of fungicides. As a consequence, the importance of mixtures has increased since 1992.

The list of recommended barley varieties, which all originate from outside Switzerland, changes every year. The seed suppliers offer pre-made mixtures, mostly with two components only, which are tested and recommended by the research stations, advisers or farm schools. Other variety mixtures are made up by the farmers themselves.

In contrast, the wheat seed market is more stable; for example, today's most popular variety has covered 60–70% of the wheat area for more than 10 years. Most wheat varieties used are bred in Switzerland in programmes in which disease resistance is the first criterion for selection, followed by baking quality, lodging resistance and yield. Fungicides, insecticides and growth regulators are never used during the breeding and recommendation process. As wheat price is based on quality and there are only few varieties within the highest class, there is currently only one favourable combination available for use as a variety mixture.

Legislation to allow for the legal maintenance and sale of 'population varieties' was initiated through the revision of the 'Ordinance on Production and Circulation of Cereal Seed'. Since 1995, registration of breeding lines for the sole purpose of use as mixture components has become possible. Such breeding lines are subject to the same protection rules as varieties. So far, however, no breeding line mixture has been submitted for homologation.

13.12 Agronomic Considerations

With regard to cultivar mixtures, a basic question concerns whether increased genetic diversity amongst the individual crop plants is compatible with the production and marketing goals of the production system. Genotype and species mixtures are common in traditional agriculture.

Current evidence also suggests that mixtures can work in commercial and modern agriculture (Mundt 1994; Bowden et al. 2001).

Mixtures have been frequently used for objectives other than disease control. Bowden et al. (2001) listed three advantages cultivar mixtures can provide: stabilisation of yield (particularly when GxE, i.e. genotype by environment interaction effects, account for a significant variation in yield), compensation effects (a strong variety compensates for a weak or injured variety) and disease control. Disease control may help to achieve the other two goals, but there also can be a direct effect of the mixture on yield stabilisation and compensation.

Potential disadvantages of mixing cultivars also need to be considered. One practical disadvantage is the added time and cost involved in mixing. For example, some farmers lack the equipment to adequately mix seed. Incompatibility of the varietal components is another potential concern, especially with regard to plant height and maturity. This problem restricts the options for mixtures to components with similar heights and maturation times (Bowden et al. 2001). Another potentially important agronomic disadvantage of mixtures is the loss of the opportunity to adjust management practices to the specific requirements of each variety (e.g. plant density, fertilisation, planting date).

Marketing restrictions and processing quality are often cited as major limitations to the use of mixtures. However, cultivars of the same market class are often bulked during handling and shipping. The German experience with barley demonstrated that adequate malting quality could be maintained in the face of widespread deployment of cultivars mixtures (Wolfe 1992).

13.13 Conclusions

For a particular host–pathogen system, it is very important to thoroughly test if mixtures can be useful for suppression of disease and stabilisation of yield. Experimentation with simple mixtures of susceptible and resistant genotypes can be very helpful for testing whether the three basic mechanisms (dilution, barrier and induced resistance) are active for that system (Garrett and Mundt 1999). The benefits achieved by cultivation of variety mixtures need to be carefully weighed against the agronomic considerations and requirements of the production system.

From this and other examples, it is clear that variety mixtures can be used successfully on a large scale, but to do so requires more publicity and information on the potential value and advantages of mixtures for individual farmers, together with incentives related to the benefits for the environment as a whole.

References

Bowden R, Shoyer J, Roozeboom K, Claasen M, Evans P, Gordon B, Heer B, Janssen K, Long J, Martin J, Schlegel A, Sears R, Witt M (2001) Performance of wheat variety blends in Kansas. Extension Bulletin No. 128, Kansas State University: College of Agriculture (www.oznet.ksu.edu/library/crpsl2/SRL128.pdf)

Brophy LS, Mundt CC (1991) Influence of plant spatial patterns on disease dynamics, plant competition and grain yield in genetically diverse wheat populations. Agric Ecosyst Environ 35:1–12

Browning JA, Frey KJ (1981) The multiline concept in theory and practice. In: Jenkyn JF, Plumb RT (eds) Strategies for the control of cereal disease. Blackwell Scientific, London, pp 37–39

Calonnec A, Goyeau H, de Vallavieille-Pope C (1996) Effects of induced resistance on infection efficiency and sporulation of *Puccinia striiformis* on seedlings in varietal mixtures and on field epidemics in pure stands. Euro J Plant Pathol 102:733–741

Chin KM, Wolfe MS (1984) The spread of *Erysiphe graminis* f. sp. *hordei* in mixtures of barley varieties. Plant Pathol 33:89–100

Cox CM, Garret KA, Bowden RL, Fritz AK, Dendy SP, Heer WF (2004) Cultivar mixtures for the simultaneous management of multiple diseases: tan spot and leaf rust of wheat. Phytopathology 94:961–969

Day KL (1984) The effect of cultivar mixtures on foliar disease and yield in barley and wheat. MSc thesis, University of New Castle, Upon, Tyne

Finckh MR, Mundt CC (1992) Plant competition and disease in genetically diverse wheat populations. Oecologia 91:82–92

Fitt BDL, McCartney HA (1986) Spore dispersal in relation to epidemic models. In: Leonard KJ, Fry WE (eds) Plant disease epidemiology, vol 1. McMillan, New York, pp 311–345

Gacek E (1997) Summarized variety mixture information given to polish farmers. In: Wolfe MS (ed) Variety mixtures in theory and practice. European Union Variety and

Species Mixtures Working Group of COST Action 817. Online at: http://www.scri.ac.uk/research/pp/pestanddisease/rhynchosporiumonbarley/otherwork/cropmixtures/varietymixtures

Garrett KA, Mundt CC (1999) Epidemiology in mixed host populations. Phytopathology 89:984–990

Jeger MJ, Griffiths E, Jones DG (1981a) Disease progress of non-specialized fungal pathogens in intraspecific mixed stands of cereal cultivars. I. Models. Annu Appl Biol 98:197–198

Jeger MJ, Griffiths E, Jones DG (1981b) Disease progress of non-specialized fungal pathogens in intraspecific mixed stands of cereal cultivars. II. Field experiments. Annu Appl Biol 98:199–210

Lannou C (2001) Intrapathotype diversity for aggressiveness and pathogen evolution in cultivar mixtures. Phytopathology 91:500–510

Lannou C, de Vallavieille-Pope C (1997) Mechanisms of variety mixtures for reducing epidemics. In: Wolfe MS (ed) Variety mixtures in theory and practice. European Union Variety and Species Mixtures Working Group of COST Action 817. Online at: http://www.scri.ac.uk/research/pp/pestanddisease/rhynchosporiumnbarley/otherwork/cropmixtures/varietymixtures

Lannou C, Mundt CC (1997) Evolution of a pathogen population in host mixtures: rate of emergence of complex races. Theor Appl Genet 94:991–999

Lannou C, de Vallavieille-Pope C, Biass C, Goyeau H (1994) The efficacy of mixtures of susceptible and resistant hosts in two wheat rusts of different lesion size: controlled condition experiments and computerized simulations. J Phytopathol 140:227–237

Lannou C, de Vallavieille-Pope C, Goyeau H (1995) Induced resistance in host-mixtures and its effect on disease control in computer-simulated epidemics. Plant Pathol 44:478–489

Leonard KL (1969) Factors affecting rates of stem rust increase in mixed plantings of susceptible and resistant oat varieties. Phytopathology 59:1845–1850

Martinelli JA, Brown JKM, Wolfe MS (1993) Effects of barley genotype on induced resistance to powdery mildew. Plant Pathol 42:195–202

Mertz U, Valenghi D (1997) "Extenso" production and cereal mixtures in Switzerland. In: Wolfe MS (ed) Variety mixtures in theory and practice. European Union Variety and Species Mixtures Working Group of COST Action 817. Online at: http://www.scri.ac.uk/research/pp/pestanddisease/rhynchosporiumonbarley/otherwork/cropmixtures/varietymixtures

Moreno Ruiz G, Castillo Zapata J (1990) The variety Colombia: a variety of coffee with resistance to rust (*Hemileia vastatrix* Berk and Br). CENICAFE, Chinchina-Caldas

Munck L (1997) Variety mixtures: 19 years of experience in Denmark. In: Wolfe MS (ed) Variety mixtures in theory and practice. European Union Variety and Species Mixtures Working Group of COST Action 817. Online at: http://www.scri.ac.uk/research/pp/pestanddisease/rhynchosporiumonbarley/otherwork/cropmixtures/varietymixtures

Mundt CC (1989) Modeling disease increase in host mixtures. In: Leonard KJ, Fry WE (eds) Plant disease epidemiology, vol 2. McGraw Hill, New York, pp 150–181

Mundt CC (1994) Use of host genetic diversity to control cereal diseases: implications for rice blast. In: Zeigler RS, Leong SA, Teng PS (eds) Rice blast disease. CAB International, London, pp 293–308

Mundt CC (2002) Use of multiline cultivars and cultivar mixtures for disease management. Annu Rev Phytopathol 40:381–410

Mundt CC, Browning JA (1985) Development of crown rust epidemics in genetically diverse oat populations: effect of genotype unit area. Phytopathology 75:607–610

Mundt CC, Leonard KJ (1985) Effect of host genotype unit area on epidemic development of crown rust following focal and general inoculations of mixtures of immune and susceptible oat plants. Phytopathology 75:1141–1145

Mundt CC, Leonard KJ (1986) Effect of host genotype unit area on development of focal epidemics of bean rust and common maize rust in mixtures of resistance and susceptible plants. Phytopathology 76:895–900

Mundt CC, Hayes PM, Schon CC (1994) Influence of barley mixtures on severity of scald and net blotch and on yield. Plant Pathol 43:356–361

Newton AC, Guy DC (2008) The effect of uneven, patchy cultivar mixtures on disease control and yield in winter barley. Field Crop Res 110:225–228

Newton AC, Ellis RP, Hackett CA, Guy DC (1997) The effect of component number on *Rynchosporium secalis* infection and yield in mixtures of winter barley cultivars. Plant Pathol 45:930–938

Ngugi HK, King SB, Holt J, Julian M (2001) Simultaneous temporal progress of sorghum anthracnose and leaf blight in crop mixtures with disparate patterns. Phytopathology 91:720–729

Wolfe MS (1988) The use of variety mixtures to control disease and stabilize yield. In: Simmonds NW, Rajaram S (eds) Breeding strategies for resistance to the rust of wheat. CIMMYT, Mexico, pp 91–100

Wolfe MS (1992) Barley diseases: maintaining the value of our varieties. In: Munk L (ed) Barley genetics, vol VI. Munksgaard International, Copenhagen, pp 1055–1067

Wolfe MS (1997) Variety mixtures: concept and value. In: Wolfe MS (ed) Variety mixtures in theory and practice. European Union Variety and Species Mixtures Working Group of COST Action 817. Online at: http://www.scri.ac.uk/research/pp/pestanddisease/rhynchosporiumonbarley/otherwork/cropmixtures/varietymixtures

Zhu Y, Chen H, Fan J, Wang Y, Li Y, Chen J, Fan JX, Yang S, Hu L, Leung H, Mew TW, Teng P, Wang Z, Mundt CC (2000) Genetic diversity and disease control in rice. Nature 406:718–722

Zhu Y, Fang H, Wang Y, Fang J, Yang S, Mew TW, Mundt CC (2005) Panicle blast and canopy moisture in rice cultivar mixtures. Phytopathology 95:433–438

Biointensive Integrated Pest Management

14

Abstract

Biointensive IPM is defined as 'A systems approach to pest management based on an understanding of pest ecology. It begins with steps to accurately diagnose the nature and source of pest problems, and then relies on a range of preventive tactics and biological controls to keep pest populations within acceptable limits. Reduced-risk pesticides are used if other tactics have not been adequately effective, as a last resort, and with care to minimize risks'.

Biointensive IPM incorporates ecological and economic factors into agricultural system design and decision-making and addresses public concerns about environmental quality and food safety. The benefits of implementing biointensive IPM can include reduced chemical input costs, reduced on-farm and off-farm environmental impacts and more effective and sustainable pest management. An ecology-based IPM has the potential of decreasing inputs of fuel, machinery and synthetic chemicals – all of which are energy intensive and increasingly costly in terms of financial and environmental impact. Such reductions will benefit the grower and society.

BIPM options may be considered as proactive or reactive. Cultural control practices are generally considered to be proactive strategies. Proactive practices include crop rotation; resistant crop cultivars including transgenic plants, disease-free seed and plants; crop sanitation; spacing of plants; altering planting dates; mulches; etc. The reactive options mean that the grower responds to a situation, such as an economically damaging population of pests, with some type of short-term suppressive action. Reactive methods generally include inundative releases of biological control agents, mechanical and physical controls, botanical pesticides and chemical controls.

P.P. Reddy, *Recent Advances in Crop Protection*,
DOI 10.1007/978-81-322-0723-8_14, © Springer India 2013

Through 'green revolution' in late 1960s, India achieved self-sufficiency in food production, which was hailed as a breakthrough on the farm front by international agricultural experts. But still, the country has not achieved self-sufficiency in production of horticultural crops. Most of the growth in food production during the green revolution period is attributed to the use of improved crop varieties and higher levels of inputs of fertilisers and pesticides. The modern agricultural techniques such as use of synthetic fertilisers and pesticides are continuing to destroy stable traditional ecosystems, and the use of high-yielding varieties of crops has resulted in the elimination of thousands of traditional varieties with the concurrent loss of genetic resources. The introduction of high-yielding varieties changed the agricultural environment leading to numerous pest problems of economic importance. In the process of intensive farming, the environment has been treated in an unfriendly manner.

Prof. Swaminathan (2000) emphasised the need for 'ever green revolution' keeping in view the increase in population. The increase in population and diminishing per capita availability of land demand rise in productivity per unit area. In India, annual crop losses due to pests, diseases and weeds have been estimated to be about Rs. 600,000 million in 2005. Increasing yields from existing land requires effective crop protection to prevent losses before and after harvest. The challenge before the plant protection scientist is to do this without harming the environment and resource base.

14.1 Integrated Pest Management (IPM)

Integrated pest management is an important principle on which sustainable crop protection can be based. IPM allows farmers to manage pests in a cost-effective, environmentally sound and socially acceptable way. According to FAO, IPM is defined as 'A pest management system that in the context of the associated environment and the population dynamics of the pest species, utilizes all suitable techniques and methods, in a compatible manner as possible and maintains the pest populations at levels below those causing economic injury'.

14.2 Biointensive Integrated Pest Management (BIPM)

Biointensive IPM incorporates ecological and economic factors into agricultural system design and decision-making and addresses public concerns about environmental quality and food safety. The benefits of implementing biointensive IPM can include reduced chemical input costs, reduced on-farm and off-farm environmental impacts and more effective and sustainable pest management. An ecology-based IPM has the potential of decreasing inputs of fuel, machinery and synthetic chemicals – all of which are energy intensive and increasingly costly in terms of financial and environmental impact. Such reductions will benefit the grower and society.

Over-reliance on the use of synthetic pesticides in crop protection programmes around the world has resulted in disturbances to the environment, pest resurgence, pest resistance to pesticides and lethal and sublethal effects on non-target organisms, including humans. These side effects have raised public concern about the routine use and safety of pesticides. At the same time, population increases are placing ever-greater demands upon the 'ecological services', that is, provision of clean air, water and wildlife habitat – of a landscape dominated by farms. Although some pending legislation has recognised the costs to farmers of providing these ecological services, it is clear that farmers will be required to manage their land with greater attention to direct and indirect off-farm impacts of various farming practices on water, soil, and wildlife resources. With this likely future in mind, reducing dependence on chemical pesticides in favour of ecosystem manipulations is a good strategy for farmers.

Biointensive IPM is defined as 'A systems approach to pest management based on an understanding of pest ecology. It begins with steps to accurately diagnose the nature and source of pest problems, and then relies on a range of preventive tactics and biological controls to keep pest populations within acceptable limits. Reduced-risk pesticides are used if other tactics have not been adequately effective, as a last resort, and with care to minimize risks' (Benbrook 1996).

The primary goal of biointensive IPM is to provide guidelines and options for the effective management of pests and beneficial organisms in an ecological context. The flexibility and environmental compatibility of a biointensive IPM strategy make it useful in all types of cropping systems. Biointensive IPM would likely decrease chemical use and costs even further.

14.2.1 Components of Biointensive IPM

An important difference between conventional and biointensive IPM is that the emphasis of the latter is on proactive measures to redesign the agricultural ecosystem to the disadvantage of a pest and to the advantage of its parasite and predator complex. At the same time, biointensive IPM shares many of the same components as conventional IPM, including monitoring, use of economic thresholds, record keeping and planning.

14.2.1.1 Planning

Good planning must precede implementation of any IPM programme but is particularly important in a biointensive programme. Planning should be done before planting because many pest strategies require steps or inputs, such as beneficial organism habitat management, that must be considered well in advance. Attempting to jump-start an IPM programme in the beginning or middle of a cropping season generally does not work.

When planning a biointensive IPM programme, some considerations include:

- Options for design changes in the agricultural system (beneficial organism habitat, crop rotations)
- Choice of pest-resistant cultivars
- Technical information needs
- Monitoring options, record keeping, equipment, etc.

When making a decision about crop rotation, consider the following questions: Is there an economically sustainable crop that can be rotated into the cropping system? Is it compatible? Important considerations when developing a crop rotation are as follows:

- How might the cropping system be altered to make life more difficult for the pest and easier for its natural controls? What two (or three or several) crops can provide an economic return when considered together as a biological and economic system that includes considerations of sustainable soil management?
- What are the impacts of this season's cropping practices on subsequent crops?
- What specialised equipment is necessary for the crops?
- What markets are available for the rotation crops?

Management factors should also be considered. For example, one crop may provide a lower direct return per acre than the alternate crop, but may also lower management costs for the alternate crop, with a net increase in profit.

14.2.1.2 Pest Identification

A crucial step in any IPM programme is to identify the pest. The effectiveness of both proactive and reactive pest management measures depends on correct identification. Misidentification of the pest may be worse than useless; it may actually be harmful and cost time and money. Help with positive identification of pests may be obtained from university personnel, private consultants, the Cooperative Extension Service, books and websites.

After a pest is identified, appropriate and effective management depends on knowing answers to a number of questions. These may include:

- What plants are hosts and non-hosts of this pest?
- When does the pest emerge or first appear?
- Where does it lay its eggs?
- For plant pathogens, where is the source(s) of inoculum?
- Where, how and in what form does the pest overwinter?

Monitoring (field scouting), economic injury and action levels are used to help answer these and additional questions.

14.2.1.3 Monitoring

Monitoring involves systematically checking crop fields for pests and beneficials, at regular intervals and at critical times, to gather information about the crop, pests and natural enemies. Sweep nets, sticky traps and pheromone traps can be used to collect insects for both identification and population density information. Leaf counts are one method for recording plant growth stages. Records of rainfall and temperature are sometimes used to predict the likelihood of disease infections.

The more often a crop is monitored, the more information the grower has about what is happening in the fields. Monitoring activity should be balanced against its costs. Frequency may vary with temperature, crop, growth phase of the crop and pest populations. If a pest population is approaching economically damaging levels, the grower will want to monitor more frequently.

14.2.1.4 Economic Injury and Action Levels

The economic injury level (EIL) is the pest population that inflicts crop damage greater than the cost of control measures. Because growers will generally want to act before a population reaches EIL, IPM programmes use the concept of an economic threshold level (ETL or ET), also known as an action threshold. The ETL is closely related to the EIL and is the point at which suppression tactics should be applied in order to prevent pest populations from increasing to injurious levels.

ETLs are intimately related to the value of the crop and the part of the crop being attacked. For example, a pest that attacks the fruit or vegetable will have a much lower ETL (i.e. the pest must be controlled at lower populations) than a pest that attacks a non-saleable part of the plant. The exception to this rule is an insect or nematode pest that is also a disease vector. Depending on the severity of the disease, the grower may face a situation where the ETL for a particular pest is zero, that is, the crop cannot tolerate the presence of a single pest of that particular species because the disease it transmits is so destructive.

14.2.2 BIPM Options

BIPM options may be considered as proactive or reactive.

14.2.2.1 Proactive Options

Proactive options, such as crop rotations and creation of habitat for beneficial organisms, permanently lower the carrying capacity of the farm for the pest. The carrying capacity is determined by the factors like food, shelter, natural enemy complex and weather, which affect the reproduction and survival of a pest species. Cultural control practices are generally considered to be proactive strategies. Proactive practices include crop rotation; resistant crop cultivars including transgenic plants, disease-free seed and plants; crop sanitation; spacing of plants; altering planting dates; mulches; etc.

The proactive strategies (cultural controls) include:

- Healthy, biologically active soils (increasing below-ground diversity)
- Habitat for beneficial organisms (increasing above-ground diversity)
- Appropriate plant cultivars

Intercropping

Intercropping is the practice of growing two or more crops in the same, alternate or paired rows in the same area. This technique is particularly appropriate in vegetable production. The advantage of intercropping is that the increased diversity helps 'disguise' crops from insect pests and, if done well, may allow for more efficient utilisation of limited soil and water resources.

Strip Cropping

Strip cropping is the practice of growing two or more crops in different strips across a field wide enough for independent cultivation. It is commonly practised to help reduce soil erosion in hilly areas. Like intercropping, strip cropping increases the diversity of a cropping area, which in turn may help 'disguise' the crops from pests. Another advantage to this system is that one of the crops may act as a reservoir and/or food source for beneficial organisms.

The options described above can be integrated with no-till cultivation schemes and all its variations (strip till, ridge till, etc.) as well as with hedgerows and intercrops designed for beneficial organism habitat. With all the cropping and tillage options available, it is possible, with creative and informed management, to evolve a biologically diverse, pest-suppressive farming system appropriate to the unique environment of each farm.

Disease-Free Seed and Plants

These are available from most commercial sources and are certified as such. Use of disease-free seed and nursery stock is important in preventing the introduction of disease.

Resistant Varieties

These are continually being bred by researchers. Growers can also do their own plant breeding simply by collecting non-hybrid seed from healthy plants in the field. The plants from these seeds will have a good chance of being better suited to the local environment and of being more resistant to insects and diseases. Since natural systems are dynamic rather than static, breeding for resistance must be an ongoing process, especially in the case of plant disease, as the pathogens themselves continue to evolve and become resistant to control measures.

Perhaps the greatest single technological achievement is the advance in breeding crops for resistance to pests. Cultivation of resistant varieties is the cheapest and best method of controlling pests. One of the important components of IPM is the use of resistant cultivars to key pests. Under All India Coordinated Research Projects of Indian Council of Agricultural Research, large number of highly/moderately resistant varieties are released to the farmers (Table 14.1).

Biotech Crops

Gene transfer technology is being used by several companies to develop cultivars resistant to insects, diseases and nematodes. An example is the incorporation of genetic material from *Bacillus thuringiensis* (*Bt*), a naturally occurring bacterium, into brinjal and potatoes, to make the plant tissues toxic to shoot and fruit borer and potato beetle larvae, respectively.

Whether or not this technology should be adopted is the subject of much debate. Opponents are concerned that by introducing *Bt* genes into plants, selection pressure for resistance to the *Bt* toxin will intensify, and a valuable biological control tool will be lost. There are also concerns about possible impacts of genetically modified plant products (i.e. root exudates) on non-target organisms as well as fears of altered genes being transferred to weed relatives of crop plants. Whether there is a market for gene-altered crops is also a consideration for farmers and processors. Proponents of this technology argue that use of such crops decreases the need to use toxic chemical pesticides.

Transgenic crop varieties in horticultural crops (tomato, potato, brinjal, beans, cabbage, cauliflower, musk melon, banana, coffee) have been developed by cloning *Bt* endotoxin genes which are cultivated in large areas. In 2006, India is the fifth largest GM crop-growing country (3.8 million ha) in the world only next to the USA (54.6 million ha), Argentina (18 million ha), Brazil and Canada. Combining a host gene for resistance with pathogen-derived genes or with genes coding for antimicrobial compounds provides for a broad and effective resistance in many host–pathogen combinations (Table 14.2).

Sanitation

It involves removing and destroying the overwintering or breeding sites of the pest as well as preventing a new pest from establishing on the farm (e.g. not allowing off-farm soil from farm equipment to spread nematodes or plant pathogens to your land). This strategy has been particularly useful in horticultural and tree–fruit crop situations involving twig and branch pests. If, however, sanitation involves removal of crop residues from the soil surface, the soil is left exposed to erosion by wind and water. As with so many decisions in farming, both the short- and long-term benefits of each action should be considered when tradeoffs like this are involved.

Table 14.1 Horticultural crop varieties resistant to pests/diseases

Crop	Pest/disease	Resistant varieties
Banana	*Radopholus similis*	Kadali, Pedalimoongil, Ayiramkapoovan, Peykunnan, Kunnan, Pisang Seribu, Tongat, Vennettu Kunnan, Anaikomban
	Panama wilt (*Fusarium oxysporum* f. sp. *cubense*)	Robusta, Dwarf Cavendish
Citrus	*Tylenchulus semipenetrans*	Trifoliate orange, Swingle citrumelo
	Gummosis, leaf fall, fruit rot (*Phytophthora* spp.)	Cleopatra mandarin, Rangpur lime, trifoliate orange rootstocks
Grapevine	Root-knot nematode, *Meloidogyne incognita*	Black Champa, Dogridge, 1613, Salt Creek, Cardinal, Banquabad
Papaya	Ring spot virus	Rainbow
Passion fruit	Root-knot nematode, *Meloidogyne incognita*	Yellow, Kaveri
Potato	Late blight, *Phytophthora infestans*	Kufri Sutlej, Kufri Badshah, Kufri Jawahar (in plains), Kufri Jyoti, Kufri Giriraj, Kufri Kanchan, Kufri Meghad (in hills)
Tomato	Bacterial wilt, *Ralstonia solanacearum*	Arka Abha, Arka Alok, Arka Shreshta, Arka Abhijit, Megha, Shakthi, Sun 7610, Sun 7611
	Powdery mildew (PM)	Arka Ashish
	Fusarium and *Verticillium* wilt	Vaishali, Rupali, Rashmi
	Leaf curl virus	Avinash-2, Hisar Anmol
	Root-knot nematode	Hisar Lalit, Pusa Hybrid-2, Arka Vardan
Brinjal	Bacterial wilt, *Ralstonia solanacearum*	Arka Nidhi, Arka Keshav, Arka Neelkanth, Arka Anand, Swarna Shree, Swarna Shyamali, Surya, Ujjwala
	Phomopsis blight	Pusa Bhairav
	Little leaf	Pusa Purple Long, Pusa Purple Cluster (field resistant)
Chilli	TMV, CMV, leaf curl	Pusa Sadabahar, Punjab Lal, Pusa Jwala
	Thrips	NP 46A (T)
	Powdery mildew	Arka Suphala (T)
	Dieback and powdery mildew	Musalwadi (T)
	Mosaic, leaf curl	Pant C-1
	Leaf curl and fruit rot	Jawahar 218 (T)
	Viruses	Arka Harita, Arka Meghana
French bean	Angular leaf spot, mosaic	Pant Anupama
	Rust, bacterial blight	Arka Anoop
	Rust	Arka Bold, Swarna Priya, Swarna Latha, Arka Anoop
	Rust, *Alternaria* leaf spot	Arka Bold
Pea	Powdery mildew	Pusa Pragati, Jawahar Matar 5, Jawahar Peas 83
	PM, rust	Arka Ajit, Arka Karthik, Arka Sampoorna
	Fusarium wilt	JP Batri Brown 3, JP Batri Brown 4
Cowpea	Bacterial blight	Pusa Komal
Pigeon pea	*Fusarium* wilt	Maruti
Field bean	Viral diseases, jassid, aphid and pod borer	Pusa Sem-2, Pusa Sem-3
Cluster bean	PM, *Alternaria* leaf spot	Gomah Manjari
Okra	YVMV	Pusa Sawani, Arka Abhay, Arka Anamika, Hisar Unnat, DVR-1, DVR-2, IIVR-10, Varsha Upkar, P-7, Pusa A-4, Parbhani Kranti (T), Punjab Kesari, Punjab Padmini, Sun-40, Makhmali
	YVMV, fruit borer	Pusa A-4

(continued)

Table 14.1 (continued)

Crop	Pest/disease	Resistant varieties
Cucumber	PM	Swarna Poorna
	PM, downy mildew (DM), angular leaf spot, anthracnose	Poinsette
Cabbage	Black rot	Pusa Mukta
	Black leg	Pusa Drum Head
Cauliflower	Black rot	Pusa Snowball K-1
	Black rot and curd blight	Pusa Shubhra
	Curd blight	Pusa Synthetic
	DM	Pusa Hybrid-2
Onion	Purple blotch, basal rot, thrips	Arka Pitamber, Arka Kirtiman, Arka Lalima
	Purple blotch, *Alternaria porri*	Arka Kalyan
Garlic	Purple blotch, *Stemphylium* disease	Agrifound White
Muskmelon	PM	Arka Rajhans, Pusa Madhuras (MR)
	PM, DM	Punjab Rasila
	Fusarium wilt	Pusa Madhuras, Durgapura Madhu, Arka Jeet, Punjab Sunehari (MR), Harela
Watermelon	PM, DM, anthracnose	Arka Manik
Pumpkin	Fruit fly	Arka Suryamukhi
Ridge gourd	PM, DM	Swarna Uphaar
Bottle gourd	Blossom end rot	Arka Bahar (T)
	CMV	Punjab Komal
Carrot	PM, root-knot nematode	Arka Suraj
Amaranth	White rust	Arka Arunima, Arka Suguna (MR)
Palak	*Cercospora* leaf spot	Arka Anupama
China aster	Root-knot nematode, *M. incognita*	Shashank, Poornima (MR)
Tuberose	Root-knot nematode, *M. incognita*	Sringar, Suvasini (T)
Mentha	Root-knot nematode, *M. incognita*	Kukrail, Arka Neera
Black pepper	Root-knot nematode, *M. incognita*	IISR Pournami (T)
	Foot rot, *Phytophthora capsici*	IISR Shakthi
Cardamom	Mosaic	IISR Vijetha
	Rhizome rot	IISR Avinash
Ginger	Root-knot nematodes	IISR Mahima
	Soft rot	Maran
Cumin	Fusarium wilt	GC-4

MR Moderately resistant, *T* Tolerant

Spacing of Plants

It heavily influences the development of plant diseases. The distance between plants and rows, the shape of beds and the height of plants influence air flow across the crop, which in turn determines how long the leaves remain damp from rain and morning dew. Generally speaking, better air flow will decrease the incidence of plant disease. However, increased air flow through wider spacing will also allow more sunlight to the ground. This is another instance in which detailed knowledge of the crop ecology is necessary to determine the best pest management strategies. How will the crop react to increased spacing between rows and between plants? Will yields drop because of reduced crop density? Can this be offset by reduced pest management costs or fewer losses from disease?

Altered Planting Dates

This can at times be used to avoid specific insects or diseases. For example, squash bug infestations

Table 14.2 Development of transgenics in vegetable crops

Vegetable crop	Target pathogen	Transgene/s	Institute
Potato	Tuber moth	Bt Cry 1Ab	CPRI, Shimla
	Potato virus Y	Coat protein	CPRI, Shimla
Tomato	Leaf curl virus	Leaf curl virus sequence	IIHR, Bangalore, IAHS, Bangalore
		Replicase gene	IARI, New Delhi
	Fungal diseases	Chitinase, glucanase	IIHR, Bangalore
		Alfalfa glucanase	IAHS, Bangalore
		OXDC	JNU, New Delhi
	Lepidopteron pests	Bt Cry 1Ab	IARI, New Delhi, Proagro PG-S (India) Ltd.
Brinjal	Fungal diseases	Chitinase, glucanase, thaumatin encoding genes	
	Lepidopteron pests	Bt Cry 1Ab	IARI, New Delhi, Proagro PG-S (India) Ltd.
Cabbage	Lepidopteron pests	Bt Cry 1Ab	IARI, New Delhi, Proagro PG-S (India) Ltd.
		Cry 1H/Cry 9C	Proagro PG-S (India) Ltd.
Cauliflower	Lepidopteron pests	Bt Cry 1Ab	IARI, New Delhi, Proagro PG-S (India) Ltd.
		Cry 1H/Cry 9C	Proagro PG-S (India) Ltd.

on cucurbits can be decreased by the delayed planting strategy, that is, waiting to establish the cucurbit crop until overwintering adult squash bugs have died. To assist with disease management decisions, the Cooperative Extension Service (CES) will often issue warnings of 'infection periods' for certain diseases, based upon the weather.

In some cases, the CES also keeps track of 'degree days' needed for certain important insect pests to develop. Insects, being cold-blooded, will not develop below or above certain threshold temperatures. Calculating accumulated degree days, that is, the number of days above the threshold development temperature for an insect pest, makes the prediction of certain events, such as egg hatch, possible. University of California has an excellent website that uses weather station data from around the state to help California growers predict pest emergence.

Some growers gauge the emergence of insect pests by the flowering of certain non-crop plant species native to the farm. This method uses the 'natural degree days' accumulated by plants. For example, a grower might time cabbage planting for 3 weeks after the *Amelanchier* species (also known as saskatoon, shadbush or service berry) on their farm are in bloom. This will enable the grower to avoid peak egg-laying time of the cabbage maggot fly, as the egg hatch occurs about the time *Amelanchier* species are flowering (Couch 1994). Using this information, cabbage maggot management efforts could be concentrated during a known time frame when the early instars (the most easily managed stage) are active.

Optimum Growing Conditions

Plants that grow quickly and are healthy can compete with and resist pests better than slow-growing, weak plants. Too often, plants grown outside their natural ecosystem range must rely on pesticides to overcome conditions and pests to which they are not adapted.

Mulches

Living or nonliving mulches are useful for suppression of insect pests and some plant diseases. Hay and straw, for example, provide habitat for spiders. Research in Tennessee showed a 70% reduction in damage to vegetables by insect pests when hay or straw was used as mulch. The difference was due to spiders, which find mulch more habitable than bare ground (Reichert and Leslie 1989). Other researchers have found that living mulches of various clovers reduce insect pest damage to vegetables and orchard crops. Again, this reduction is due to natural predators and parasites that are provided habitat by the clovers.

Mulching helps to minimise the spread of soil-borne plant pathogens by preventing their transmission through soil splash. Winged aphids are repelled by silver- or aluminium-coloured mulches. Recent springtime field tests at the Agricultural Research Service in Florence, South Carolina, have indicated that red plastic mulch suppresses root-knot nematode damage in tomatoes by diverting resources away from the roots (and nematodes) and into foliage and fruit (Adams 1997).

14.2.2.2 Reactive Options

The reactive options mean that the grower responds to a situation, such as an economically damaging population of pests, with some type of short-term suppressive action. Reactive methods generally include inundative releases of biological control agents, mechanical and physical controls, botanical pesticides and chemical controls.

Biological Controls

Biological control is the use of living organisms – parasites, predators or pathogens – to maintain pest populations below economically damaging levels and may be either natural or applied. A first step in setting up a biointensive IPM programme is to assess the populations of beneficials and their interactions within the local ecosystem. This will help to determine the potential role of natural enemies in the managed horticultural ecosystem. It should be noted that some groups of beneficials (e.g. spiders, ground beetles, bats) may be absent or scarce on some farms because of lack of habitat. These organisms might make significant contributions to pest management if provided with adequate habitat.

(a) *Natural Biological Control*: It results when naturally occurring enemies maintain pests at a lower level than would occur without them and is generally characteristic of biodiverse systems. Mammals, birds, bats, insects, fungi, bacteria and viruses all have a role to play as predators, parasites and pathogens in an agricultural system.

Creation of habitat to enhance the chances for survival and reproduction of beneficial organisms is a concept included in the definition of natural biocontrol. Farmscaping is a term coined to describe such efforts on farms. Habitat enhancement for beneficial insects, for example, focuses on the establishment of flowering annual or perennial plants that provide pollen and nectar needed during certain parts of the insect life cycle. Other habitat features provided by farmscaping include water, alternative prey, perching sites, overwintering sites and wind protection. Beneficial insects and other beneficial organisms should be viewed as mini-livestock, with specific habitat and food needs to be included in farm planning.

The success of such efforts depends on knowledge of the pests and beneficial organisms within the cropping system. Where do the pests and beneficials overwinter? What plants are hosts and non-hosts? When this kind of knowledge informs planning, the ecological balance can be manipulated in favour of beneficials and against the pests.

It should be kept in mind that ecosystem manipulation is a two-edged sword. Some plant pests (such as the tarnished plant bug and lygus bug) are attracted to the same plants that attract beneficials. The development of beneficial habitats with a mix of plants that flower throughout the year can help prevent such pests from migrating *en masse* from farmscaped plants to crop plants.

(b) *Applied Biological Control*: It is also known as augmentative biocontrol and involves supplementation of beneficial organism populations, for example, through periodic releases of parasites, predators or pathogens. This can be effective in many situations – for instance, well-timed inundative releases of *Trichogramma* egg wasps for codling moth control.

Most of the beneficial organisms used in applied biological control today are insect parasitoids and predators. They control a wide range of pests from caterpillars to mites. Some species of biocontrol organisms, such as *Eretmocerus californicus*, a parasitic wasp, are specific to one host – in this case, the sweet potato whitefly. Others, such as green lacewings, are generalists and will attack many species of aphids and whiteflies.

Information about rates and timing of release is available from suppliers of beneficial organisms. It is important to remember that released insects are mobile; they are likely to leave a site if the habitat is not conducive to their survival. Food, nectar and pollen sources can be 'farm-scaped' to provide suitable habitat.

The quality of commercially available applied biocontrols is another important consideration. For example, if the organisms are not properly labelled on the outside packaging, they may be mishandled during transport, resulting in the death of the organisms. A recent study by Rutgers University noted that only two of six suppliers of beneficial nematodes sent the expected numbers of organisms and only one supplier out of the six provided information on how to assess product viability.

Whilst augmentative biocontrols can be applied with relative ease on small farms and in gardens, applying some types of biocontrols evenly over large farms has been problematic. New mechanised methods that may improve the economics and practicality of large-scale augmentative biocontrol include ground application with 'biosprayers' and aerial delivery using small-scale (radio-controlled) or conventional aircraft.

Inundative releases of beneficials into greenhouses can be particularly effective. In the controlled environment of a greenhouse, pest infestations can be devastating; there are no natural controls in place to suppress pest populations once an infestation begins. For this reason, monitoring is very important. If an infestation occurs, it can spread quickly if not detected early and managed. Once introduced, biological control agents cannot escape from a greenhouse and are forced to concentrate predation/parasitism on the pest(s) at hand.

An increasing number of commercially available biocontrol products are made up of microorganisms, including fungi, bacteria, nematodes and viruses.

Of late, biological suppression of pests has become an intensive area of research because of environmental concerns. About 60% of the natural control of insect pests is by the natural enemies of pests such as parasitoids, predators and pathogens. The Australian ladybird beetle, *Cryptolaemus montrouzieri*, has been found very effective against mealy bugs infesting grapes, guava, citrus, mango, pomegranate, ber and custard apple. The encyrtid parasite, *Leptomastix dactylopii,* is effective against mealy bug, *Planococcus citri,* on guava, citrus, pomegranate, ber and custard apple. *Bacillus thuringiensis* (*Bt*) is effective against tomato fruit borer, okra fruit borer and diamondback moth on cabbage and cauliflower.

Several methods of enrichment and conservation of natural enemies include providing nesting boxes for wasps and predatory birds; retaining pollen and nectar-bearing flowering plants like Euphorbia, wild clover on bunds to provide supplementary food for natural enemies; and placing bundles of paddy straw in fields for attracting predatory spiders. In addition, erecting perching sites, placing water pans and retaining bushes (Acalypha, Hibiscus, Crotons) help in retention of predatory birds.

The last decade has witnessed a tremendous breakthrough in biological control of diseases and nematodes like *Rhizoctonia, Pythium, Fusarium, Macrophomina, Ralstonia* and *Meloidogyne* in banana, tomato, eggplant, pea, grapes, cucumber, black pepper, cardamom, ginger and turmeric, especially by using species of *Trichoderma, Pochonia, Pseudomonas* and *Bacillus* (Tables 14.3, 14.4, 14.5, 14.6, 14.7, 14.8, 14.9, 14.10, 14.11, and 14.12).

Mechanical and Physical Controls

Methods included in this category utilise some physical component of the environment, such as temperature, humidity or light, to the detriment of the pest. Common examples are tillage, flaming, flooding, soil solarisation and plastic mulches to kill pests.

Heat or steam sterilisation of soil is commonly used in greenhouse operations for control of soil-borne pests. Floating row covers over vegetable crops exclude flea beetles, cucumber beetles and adults of the onion, carrot, cabbage and seed corn root maggots (Fig. 14.1). Insect screens are used in greenhouses to prevent aphids, thrips, mites and other pests from entering ventilation ducts. Large, multi-row vacuum machines have been used for pest management in strawberries and vegetable crops. Cold storage reduces post-harvest disease problems on produce.

Table 14.3 Biological control of fruit crop pests

Fruit crop	Pest	Biocontrol agent/dosage
Apple	Woolly aphid, *Eriosoma lanigerum*	*Aphelinus mali* – 1,000 adults or mummies/infested tree
	San Jose scale, *Quadraspidiotus perniciosus*	*Encarsia periniciosi* – 2,000 adults/infested tree
	Codling moth, *Cydia pomonella*	*Chilocorus infernalis* – 20 adults or 50 grubs/tree, *Trichogramma embryophagum* – 2,000 adults/tree, *Steinernema carpocapsae*
Citrus	Cottony cushion scale, *Icerya purchasi*	*Rodolia cardinalis* – 10 beetles/infested plant
	Mealy bug, *Planococcus citri*	*Cryptolaemus montrouzieri* – 10 beetles/infested plant, *Leptomastix dactylopii* 3,000 adults/ha
	Red scale, *Aonidiella aurantii*	*Chilocorus nigrita* – 15 adults/infested tree
	Scale insect, *Coccus viridis*	*Verticillium lecanii* – 16×10^4 spores/ml + 0.05% teepol
	Leaf miner, *Phyllocnistis citrella*	*S. carpocapsae*
Grapevine	Mealy bug, *Maconellicoccus hirsutus*	*C. montrouzieri* – 2,500–3,000 beetles/ha or 10 beetles/vine
Guava	Green shield scale, *Chloropulvinaria psidii*	*C. montrouzieri* – 10–20 beetles/infested plant
	Aphid, *Aphis gossypii*	*V. lecanii* – 10^9 spores/ml + 0.1% teepol

Table 14.4 Biological control of fruit crop diseases

Fruit crop	Disease/pathogen/s	Potential biocontrol agent/s
Banana	Panama wilt, *Fusarium oxysporum* f. sp. *cubense*	*T. viride, Aspergillus niger, Pseudomonas fluorescens, T. viride + P. fluorescens* – sucker treat
Citrus	Root rot, *Phytophthora* spp.	*Trichoderma viride/T. harzianum* at 100 kg/ha, *Penicillium funiculosum, Pythium nunn* – soil treat
	Canker, *Xanthomonas campestris* pv. *citri*	*Aspergillus niger* AN 27
Mulberry	Leaf spot, *Cercospora moricola*	*T. viride, T. harzianum, Pseudomonas fluorescens*
	Cutting rot, *F. solani*	*T. virens, T. harzianum, T. pseudokoningii*
Grapevine	Powdery mildew, *Uncinula necator*	*Ampelomyces quisqualis* – dispersal from wick cultures at 15 cm of shoot growth and bloom
	Downy mildew, *Plasmopara viticola*	*Fusarium proliferatum* weekly spray starting from 15 cm of shoot growth – 10^6 spores/ml
Guava	Anthracnose, *Pestalotia psidii, Colletotrichum gloeosporioides*	*T. harzianum*
	Wilt, *Gliocladium roseum* and *F. solani*	*Penicillium citrinum, Aspergillus niger* AN 17, *T. harzianum*
Mango	Anthracnose, *Colletotrichum gloeosporioides*	*T. harzianum, Streptosporangium pseudovulgare*
	Powdery mildew, *Oidium mangiferae*	*S. pseudovulgare*
	Bacterial canker, *Xanthomonas campestris* pv. *mangiferaeindicae*	*Bacillus coagulans*
Apple	Scab, *Venturia inaequalis*	*Chaetomium globosum, A. pullulans, Microsphaeropsis* sp., *Chaetomium globosum, Cladosporium* spp., *Trichothecium roseum* – foliar spray
	Collar rot, *Phytophthora cactorum*	*Enterobacter aerogenes, Bacillus subtilis, T. virens* – soil treat
	White root rot, *Dematophora necatrix*	*T. viride, T. harzianum, T. virens* – soil treat

(continued)

Table 14.4 (continued)

Fruit crop	Disease/pathogen/s	Potential biocontrol agent/s
Pear	Blue mould, *Penicillium expansum*, grey mould, *Botrytis cinerea*	*C. infirmo-miniatus* YY6, *C. laurentii* RR87-108, *R. glutinis* HRB6 – fruit spray 3 weeks or 1 day prior to harvest – 10^8 cfu/ml, *Pantoea agglomerans* CPA-2 – post-harvest fruit dipping in 8×10^8 cfu/ml
	Fire blight, *Erwinia amylovora*	*Pseudomonas fluorescens* – foliar spray
Peach	Brown rot, *Monilia fructicola*	*Bacillus subtilis* (B-3) – post-harvest fruit line spray at 5×10^8 cfu/g, *Pseudomonas syringae* – post-harvest fruit dipping in 10^7 cfu/ml
	Twig blight, *Monilia laxa*	*Penicillium frequentans* – spray shoots in early growing season – 10^{8-9} spores/ml
	Crown gall, *Agrobacterium tumefaciens*	*Agrobacterium radiobacter* K84, K1026 – root dip treat
Strawberry	Grey mould, *Botrytis cinerea*	*Trichoderma* products (BINAB TF and BINAB T), *Bacillus pumilus, Pseudomonas fluorescens, Gliocladium roseum* – spray flowers and fruits – white flower bud to pink fruit – 10^6 spores/ml, *G. roseum* – bee vectoring of flowers – 10^9 cfu/g of powder
Passion fruit	Collar rot, *Rhizoctonia solani*	*T. harzianum, Trichoderma* sp.
Amla	Bark splitting, *Rhizoctonia solani*	*Aspergillus niger* AN 27

Table 14.5 Biological control of vegetable crop pests

Vegetable crop	Pest	Biocontrol agent/dosage
Potato	Cut worm, *Agrotis ipsilon, A. segetum*	*Steinernema carpocapsae, S. bicornutum, Heterorhabditis indica*
Tomato	Fruit borer, *Helicoverpa armigera*	*Trichogramma brasiliensis/T. chilonis/T. pretiosum* – 50,000/ha, *Ha* NPV – 250 LE/ha
Brinjal	Fruit and shoot borer, *Leucinodes orbonalis*	*S. carpocapsae, H. indica*
Chilli	Fruit borer, *H. armigera*	*Ha* NPV-250 LE/ha
Beans	Mite, *Tetranychus* spp.	*Phytoseiulus persimilis* – 10 adults/plant or release 1–6 leaves with predatory mites
Pigeon pea	Pod borer, *Helicoverpa armigera*	*Ha* NPV-250 LE/ha
Cabbage	Diamondback moth, *Plutella xylostella*	*S. carpocapsae, S. glaseri, S. feltiae, S. bicornutum, H. bacteriophora*
Mushroom	*Lycoriella auripila, L. mali, L. solani, Megaselia halterata*	*S. feltiae*

Table 14.6 Biological control of vegetable crop diseases

Crop	Disease/pathogen/s	Biocontrol agent/mode of application
Potato	Black scurf, *Rhizoctonia solani*	*T. harzianum, T. viride* – tuber treat, *Aspergillus niger* AN27, *Verticillium biguttatum* – soil treat, *Laetisaria arvalis* – tuber treat, binucleate *Rhizoctonia*
	Wilt, *Ralstonia solanacearum*	*Bacillus cereus, B. subtilis*
Tomato	Damping off, *Pythium aphanidermatum*	*T. viride, T. harzianum, Pseudomonas aeruginosa* 7NSK2
	Wilt, *Fusarium oxysporum* f. sp. *lycopersici*	*T. viride, T. harzianum, Aspergillus niger,* non-pathogenic *F. oxysporum, F. oxysporum* f. sp. *dianthi, P. fluorescens* strains Pf1, *P. putida, Penicillium oxalicum, Pythium oligandrum, Bacillus subtilis* strains FZB-G, *Streptomyces* spp. – seed treat, seed and soil treat

(continued)

Table 14.6 (continued)

Crop	Disease/pathogen/s	Biocontrol agent/mode of application
Brinjal	Damping off, wilt, *Phytophthora* sp., *Pythium aphanidermatum, F. solani*	*T. viride, T. harzianum, T. koningii* – seed and soil treat
	Collar rot, *Sclerotinia sclerotiorum*	*T. viride, T. virens, Bacillus subtilis* – soil treat
Bell pepper	*Phytophthora capsici*	*T. viride, T. harzianum* – fruit treat
	Damping off, *Pythium aphanidermatum*	*Streptomyces griseoviridis* – seed and soil treat
French bean	Dry root rot, *Macrophomina phaseolina*	*Pseudomonas cepacia* UPR5C – seed treat
	Wilt, *Fusarium oxysporum* f. sp. *phaseoli*	*Streptomyces* spp. – seed treat
Pea	Root rot, *Aphanomyces euteiches*	*P. fluorescens* PRA25 – seed treat, *Pseudomonas cepacia* AMMD – seed treat
	Damping off, *Pythium ultimum*	*Pseudomonas cepacia* AMMD – seed treat, *P. putida* NIR – seed treat
	Wilt, *F. oxysporum* f. sp. *udum*	*T. viride, T. harzianum, T. koningii* – seed treat, *Aspergillus niger* AN27
Cluster bean	Bacterial blight, *Xanthomonas axonopodis* pv. *cyamopsidis*	*Aspergillus niger* AN27
Cabbage	Damping off, *R. solani*	*T. viride, T. harzianum, T. koningii* – seed treat
Cauliflower	Blight, *Alternaria brassicola*	*Streptomyces griseoviridis* – seed treat
Okra	*Rhizoctonia solani*	*Bradyrhizobium japonicum, Rhizobium* spp. – seed treat
Carrot	Soft rot, *Sclerotinia sclerotiorum*	*Coniothyrium minitans* – soil treat
	Root rot, *R. solani*	*T. virens* GL-21
Radish	Wilt, *F. oxysporum* f. sp. *raphani*	*P. fluorescens* strains WCS374, WCS417r – soil treat
	Root rot, *R. solani*	*Laetisaria rosiepellis, Pythium acanthicum* – soil treat
Beet root	Damping off, *Pythium debaryanum, P. ultimum*	*Penicillium* spp. + *P. fluorescens* – seed treat, *Pythium oligandrum*
Cucumber	Wilt, *Fusarium oxysporum* f. sp. *cucumerinum, R. solani*	*Colletotrichum orbiculare, F. oxysporum* f. sp. *niveum, Pseudomonas putida* 89B-27, *Serratia marcescens,* Tobacco necrosis virus
	Powdery mildew	*Ampelomyces quisqualis* – foliar spray
	Cucumber mosaic virus	*P. fluorescens* strain 89B-27
Water melon	Wilt, *Fusarium oxysporum* f. sp. *solani, F. oxysporum f. sp. niveum*	*T. viride, Aspergillus niger* – seed and soil treat, *Penicillium janczewski*
Musk melon	Wilt, *F. oxysporum, F. solani, R. solani*	*T. harzianum, Aspergillus niger* – seed treat
Onion	Soft rot, *Sclerotium cepivorum*	*Chaetomium globosum, Trichoderma* sp. C62 – soil treat

Although generally used in small or localised situations, some methods of mechanical/physical control are finding wider acceptance because they are generally more friendly to the environment.

Chemical Controls (Reduced-Risk Pesticides)

Included in this category are both synthetic pesticides and botanical pesticides.

(a) *Synthetic Pesticides*: They comprise a wide range of man-made chemicals used to control insects, mites, nematodes, plant diseases and vertebrate and invertebrate pests. These powerful chemicals are fast acting and relatively inexpensive to purchase.

Pesticides are the option of last resort in IPM programmes because of their potential negative impacts on the environment, which result from the manufacturing process as well as from their application on the farm. Pesticides should be used only when other

Table 14.7 Biological control of ornamental crop diseases

Crop	Disease/pathogen/s	Biocontrol agent/mode of application
Rose	Grey mould, *Botrytis cinerea*	*T. viride, T. harzianum* – cutting treat
Gladiolus	Yellows and corm rot, *F. oxysporum* f. sp. *gladioli*	*T. virens, T. harzianum* – corm and soil treat
Chrysanthemum	Wilt, *Fusarium oxysporum*	*T. harzianum* – soil treat – 160 kg/ha
	Rhizoctonia solani	*Aspergillus niger* AN27
Carnation	Wilt, *F. oxysporum* f. sp. *dianthi*	*Pseudomonas fluorescens* strain WCS 417r – soil treat, *P. putida* WCS 358r – root dip treat, *Alcaligenes* sp., *Bacillus* sp., *Arthrobacter* sp., *Hafnia* sp., *Serratia liquefaciens*
Gerbera	*Phytophthora cryptogea*	*Trichoderma* spp. – soil treat
Narcissus	Wilt, *F. oxysporum* f. sp. *narcissi*	*Streptomyces griseoviridis* – bulb treat, *Mini medusa polyspora* – bulb treat
Zinnia	*Rhizoctonia solani*	*T. virens* GL-21, *T. virens* GL-20
Marigold	*Pythium ultimum*	*Glomus intraradices, G. mosseae* – soil treat

Table 14.8 Biological control of medicinal and aromatic crop diseases

Medicinal/aromatic crop	Disease/pathogen/s	Biocontrol agent
Opium poppy	Sclerotinia rot and blight, *Sclerotinia sclerotiorum*	*T. harzianum, T. viride, T. koningii, T. virens* – soil treat
	Downy mildew, *Peronospora arborescens*	*Trichoderma* spp. – seed treat
Periwinkle	*Phytophthora parasitica*	*Phytophthora parasitica* var. *nicotianae* – soil treat
Jasmine	Root rot, *Macrophomina phaseolina*	*T. viride, T. harzianum* – cutting treat
Chinese rose	Wilt, *Fusarium oxysporum*	*Aspergillus niger* – soil treat
Menthol mint	Stolon decay, *Sclerotinia sclerotiorum*	*T. harzianum, T. virens* – sucker treat
	Verticillium dahliae	*Verticillium nigrescens*

Table 14.9 Biological control of tuber crop diseases

Tuber crop	Disease/pathogen/s	Biocontrol agent
Yam	*Botrytis theobromae*	*T. viride*
Cassava	*Phytophthora drechsleri*	*T. viride*
Elephant foot yam	*Sclerotium rolfsii*	*T. harzianum, T. pseudokoningii*

Table 14.10 Biological control of plantation crop pests

Plantation crop	Pest	Biocontrol agent/dosage
Coconut	Black headed caterpillar, *Opisina arenosella*	*Goniozus nephantidis* – 3,000 adults/ha
	Rhinoceros beetle, *Oryctes rhinoceros*	Baculovirus – 10 infected beetles/tree
Areca nut	*Ischnaspis longirostris*	*Chilocorus nigrita* – 20–50 adults/plant
Coffee	Mealy bugs, *Planococcus* and *Pseudococcus* spp.	*Cryptolaemus montrouzieri* – 2–10 beetles/infested plant

Table 14.11 Biological control of plantation crop diseases

Crop	Disease/causal agent	Biocontrol agents reported
Coconut	Stem bleeding, *Thielaviopsis paradoxa* (*Ceratostomella paradoxa*)	*Trichoderma virens, T. harzianum* phosphobacteria
	Basal stem rot, *Ganoderma lucidum*	*G. virens, T. harzianum*
Areca nut	Bud rot, *Phytophthora* spp.	*Trichoderma* spp.
	Fruit rot, *Phytophthora arecae, Colletotrichum capsici*	*Trichoderma* spp., *P. fluorescens*
	Foot rot/anabe, *Ganoderma lucidum*	*T. harzianum*
Tea	Red root rot, *Poria hypolateritia*	*T. harzianum*
	Brown root, *Fomes noxius*	*G. virens, T. harzianum*
	Black root rot, *Rosellinia arcuata*	*G. virens, T. harzianum*
Coffee	Black root, *Pellicularia koleroga*	*G. virens, T. harzianum*
	Brown root, *Fomes noxius*	*G. virens, T. harzianum*
	Santhaveri wilt, *F. oxysporum* f. sp. *coffeae*	*G. virens, T. harzianum*
Rubber	Brown rot, *Phellinus noxius*	*T. viride, T. harzianum, T. hamatum*
Betel vine	Foot and root rot, *Phytophthora parasitica* pv. *piperina*	*T. viride, T. harzianum* – soil treat
	Collar rot, *Sclerotium rolfsii*	*T. harzianum, T. viride, T. koningii, T. virens* – soil treat

Table 14.12 Biological control of spice crop diseases

Spice crop	Disease/causal organism	Effective biocontrol agents/mode of application
Black pepper	Foot rot, *Phytophthora capsici*	*Trichoderma harzianum, T. virens, Glomus fasciculatum* – soil treat, *Pseudomonas fluorescens, Bacillus* sp. – foliar spray
	Anthracnose, *Colletotrichum gloeosporioides*	*P. fluorescens* – foliar spray
	Slow decline, *Radopholus similis, Meloidogyne incognita, Phytophthora capsici*	*T. harzianum, T. virens, Paecilomyces lilacinus, Pochonia chlamydosporia* – soil treat
Cardamom	Damping off, *Pythium vexans*	*T. harzianum, T. viride*-soil treat in solarised nursery beds
	Clump rot/rhizome rot, *Pythium vexans, Rhizoctonia solani, Meloidogyne incognita*	*T. harzianum* – soil treat
	Capsule rot, *Phytophthora meadii, P. nicotianae* var. *nicotianae*	*T. harzianum, T. viride, T. virens, T. hamatum* – soil treat
Ginger	Rhizome rot, *Pythium aphanidermatum, P. myriotylum*	*T. harzianum, T. virens* – soil solarisation + soil treat, *.T viride, P. fluorescens* – seed treat, *Aspergillus niger* AN27 – soil treat
	Yellows, *Fusarium oxysporum* f. sp. *zingiberi, M. incognita*	*T. harzianum, T. virens, T. hamatum* – soil solarisation + soil treat, rhizome treat
	Bacterial wilt, *Ralstonia solanacearum*	Avirulent *Ralstonia solanacearum, P. fluorescens*, endophytic bacteria – soil treat
Turmeric	Rhizome rot, *Fusarium* sp., *Pythium graminicola, P. aphanidermatum, R. similis, M. incognita*	*T. harzianum, T. viride, .T virens* – soil treat
Fenugreek	Root rot, *R. solani*	*T. viride, P. fluorescens* – seed treat
Coriander	Root rot/wilt, *Fusarium oxysporum* f. sp. *corianderii*	*T. viride, T. harzianum, Streptomyces* sp. –seed treat
Cumin	Wilt, *F. oxysporum* f. sp. *cumini*	*Trichoderma* spp., *T. virens* – soil treat
Vanilla	Root rot, *Phytophthora meadii, F. oxysporum* f. sp. *vanillae*	*T harzianum, P fluorescens* – soil treat
Mustard	Damping off, *P. aphanidermatum*	*T. viride, T. harzianum* – seed and soil treat

Fig. 14.1 Floating row covers over vegetable crops to exclude insect pests

measures, such as biological or cultural controls, have failed to keep pest populations from approaching economically damaging levels.

If chemical pesticides must be used, it is to the grower's advantage to choose the least-toxic pesticide that will control the pest but not harm non-target organisms such as birds, fish and mammals. Pesticides that are short-lived or act on one or a few specific organisms are in this class. Examples include insecticidal soaps, horticultural oils, copper compounds (e.g. Bordeaux mixture), sulphur, boric acid and sugar esters.

(b) *Biorational Pesticides*: Biorational pesticides are generally considered to be derived from naturally occurring compounds or are formulations of microorganisms. Biorationals have a narrow target range and are environmentally benign. Formulations of *Bacillus thuringiensis*, commonly known as *Bt*, are perhaps the best-known biorational pesticide. Other examples include silica aerogels, insect growth regulators and particle film barriers.

(c) *Sugar Esters*: Sugar esters have performed as well as or better than conventional insecticides against mites and aphids in apple orchards, psylla in pear orchards and whiteflies, thrips and mites on vegetables. However, sugar esters are not effective against insect eggs. Insecticidal properties of sugar esters were first investigated a decade ago when a scientist noticed that tobacco leaf hairs exuded sugar esters for defence against some soft-bodied insect pests. Similar to insecticidal soap in their action, these chemicals act as contact insecticides and degrade into environmentally benign sugars and fatty acids after application.

(d) *Inorganic Chemicals*: Spray application of K_2HPO_4 or KH_2PO_4 at 3.5 g/l of water has been reported to control powdery mildew in rose and carnation. Similarly, the above treatment was also found effective for the management of powdery mildew on mango, grapes and cucurbits.

(e) *Botanical Pesticides*: They can be as simple as pureed plant leaves, extracts of plant parts or chemicals purified from plants. Pyrethrum, neem formulations and rotenone are examples of botanicals. Some botanicals are broad-spectrum pesticides. Others, like ryania, are very specific. Botanicals are generally less harmful in the environment than synthetic pesticides because they degrade quickly, but they can be just as deadly to beneficials as synthetic pesticides. However, they are less hazardous to transport and in some cases can be formulated on-farm. The manufacture of botanicals generally results in fewer toxic by-products.

Neem products such as cake, oil, neem seed kernel extract (NSKE), neem seed powder extract (NSPE), pulverised NSPE and soaps are being used extensively to manage horticultural crop pests (bean fly, *Ophiomyia phaseoli*; serpentine leaf miner, *Liriomyza trifolii*, on several crops; cucurbit fruit fly, *Bactrocera cucurbitae*; tomato fruit borer, *Helicoverpa armigera*; brinjal fruit and shoot borer, *Leucinodes orbonalis*; water melon and chilli thrips, *Thrips* spp.; chilli yellow mite, *Polyphagotarsonemus latus*, and okra leafhopper, *Amrasca biguttula biguttula*) (Krishna Moorthy and Krishna Kumar 2002).

Table 14.13 Insect pests on which soaps were found effective

Crops	Pest on which effective
Cabbage and cauliflower	Diamondback moth, leaf webber, aphids, young *Spodoptera* larva
Tomato	Whitefly, red spider mites, fruit borer (egg-laying stage), leaf miner
Okra	Leaf hopper, whitefly, aphids
Cucurbits	Fruit fly, leaf miner
Mango	Leaf hopper
Ornamental crops	Mites, whitefly

The soap sprays were highly effective on leafhoppers, aphids, red spider mites and white flies in many vegetables, but moderately effective on thrips in water melon and chillies (Table 14.13).

(f) *Compost Teas*: They are most commonly used for foliar disease control and applied as foliar nutrient sprays. The idea underlying the use of compost teas is that a solution of beneficial microbes and some nutrients is created then applied to plants to increase the diversity of organisms on leaf surfaces. This diversity competes with pathogenic organisms, making it more difficult for them to become established and infect the plant.

An important consideration when using compost teas is that high-quality, well-aged compost be used, to avoid contamination of plant parts by animal pathogens found in manures that may be a component of the compost. There are different techniques for creating compost tea. The compost can be immersed in the water, or the water can be circulated through the compost. An effort should be made to maintain an aerobic environment in the compost/water mixture.

14.2.3 Case Studies

14.2.3.1 Mango Pests and Diseases (Uttarakhand)

- Spraying of 0.3% copper oxychloride for control of die-back, anthracnose and red rust diseases wherever they appeared during September–October
- Ploughing of orchard in November–December to expose pupae of fruit flies, midges, leafhoppers and eggs of mealy bugs to natural enemies
- Polythene banding of tree trunk in December–January and application of 5% NSKE and *Beauveria bassiana* in January
- Spraying of 0.2% sulfex for the control of powdery mildew disease
- Spraying of *Verticillium lecanii* in orchards for control of hoppers
- Fixing methyl eugenol traps (wooden blocks impregnated with methyl eugenol) to control fruit flies from April to August
- Mechanical removal of mango leaf webber larvae and webs by leaf web-removing device (developed by the Central Institute of Subtropical Horticulture, Lucknow) from April to September–October

The BIPM package was successfully validated in 16.8 ha of mango orchards in Gulab Khera, Habibpur, Budhadia, Pathak Ganj, Rehman Khera and Kanar villages in Malihabad and Kakori belt of mango near Lucknow on Dashehari variety during 2000–2004. As a result of adoption of IPM, yield of mango increased from 6.0 to 9.0 MT/ha as it was 3.5–7.0 MT/ha earlier in non-IPM orchards. By adopting IPM, the mango growers in that area earned a profit of Rs. 30,000/- to Rs. 55,000/-, whilst the farmers who did not adopt IPM earned a profit of Rs. 17,000/- to 35,000/- per ha only (Table 14.14) (Amerika Singh et al. 2004).

14.2.3.2 Apple Pests and Diseases (Himachal Pradesh)

- Use of 5% urea at leaf-shedding stage for early decomposition of the infested leaves and to encourage the population of antagonists in the plant rhizosphere
- Use of Bordeaux paint during autumn on the naked plant stem to overcome the direct effect of UV rays on the plant skin to reduce sunburn and canker disease complex
- Overwinter spray of Bordeaux mixture as eradicative action to pathogens (root and collar rots) and total disinfection of plant surface
- Use of Neemarin at pink bud stage, that is, pre-bloom stage to manage blossom thrips population
- Use of *Bacillus thuringiensis* at fruit development stage for the management of fruit scrapper insect pests

Table 14.14 Economics of BIPM in mango

Parameters	IPM plots	Non-IPM plots	% Increase
Yield MT/ha	6.0–9.0	3.5–7.0	28.57–71.43
Net profit (Rs./ha)	30,000–55,000	17,000–35,000	57.14–76.47

Table 14.15 Economics of apple BIPM

Parameters	IPM plots	Non-IPM plots	% increase
Yield (MT/ha)	4.99	4.12	21.11
Net profit (Rs/ha)	145,733	110,889	31.42
Benefit to cost ratio	4.07	3.01	35.21

Fig. 14.2 BIPM of tomato fruit borer using African marigold as a trap crop

- Use of *Trichoderma viride* (Bioderma) for the control of root rot fungus

During the period 2001–2004, BIPM for apple crop was validated and promoted in 30 ha of orchards in Kotkhai, Jubbal, Theneder and Rohru villages of Himachal Pradesh. By adopting IPM package, farmers were able to harvest 580 boxes (4.99 MT/ha) of apple in IPM plots as compared to 380 boxes (4.12 MT/ha) in case of non-IPM plots. Apple growers who adopted IPM earned a profit of Rs. 145,733/-, whilst the farmers who did not adopt IPM earned a profit of Rs. 110.889/-. Average benefit to cost ratio of IPM to non-IPM was 4.07–3.01 (Table 14.15) (Amerika Singh et al. 2004).

14.2.3.3 Tomato Fruit Borer

Use of African marigold (*Tagetes erecta*) as a trap crop for the management of fruit borer on tomato involves planting one row of 45-day-old marigold seedlings after every 16 rows of 25-day-old tomato seedlings and spraying of *Ha* NPV at 250 LE/ha or 4% NSKE or 4% pulverised NSPE, 28 and 45 DAP coinciding with peak flowering (Srinivasan et al. 1994) (Fig. 14.2).

During the period 2001–2004, BIPM technology in tomato was validated and promoted in more than 40 ha area in 42 villages covering 88 families located 40 km from Bangalore. Similarly, near Varanasi also IPM technology has been validated in eight villages in about 40 ha area

Table 14.16 Economics of tomato fruit borer BIPM

Centre	Yield (MT/ha)	Net returns (Rs.)	Benefit to cost ratio
Bangalore – IPM	74.03	249,721	4.82
Non-IPM	45.05	69,704	0.61
Varanasi – IPM	14.25	39,917	3.30
Non-IPM	13.00	38,167	2.02
Ranchi – IPM	22.29	56,705	1.87
Non-IPM	18.77	41,776	1.32

Fig. 14.3 BIPM in cabbage using Indian mustard as a trap crop

covering 100 families in tomato. Near Ranchi, IPM technology has been validated and promoted in 20 villages with the support of 100 farming families covering an area of 40 ha together. In IPM validation studies conducted at three locations (Bangalore, Varanasi and Ranchi), IPM fields recorded higher tomato fruit yields of 74.038, 14.250 and 22.293 MT/ha as compared to 45.056, 13.000 and 18.772 MT/ha in non-IPM fields, respectively (Amerika Singh et al. 2004) (Table 14.16).

14.2.3.4 Cabbage Pests

BIPM using Indian mustard as a trap crop involves planting of paired rows of mustard after every 25 rows of cabbage/cauliflower and spraying of 4% NSKE at primordial formation. Two more sprays of 4% NSKE may be given at 10–15-day interval after the first spray. The IPM plots gave 152% more returns than pure cabbage crop. IPM controls diamondback moth (*Plutella xylostella*), leaf webber (*Crocidolomia binotalis*), stem borer (*Hellula undalis*), aphids (*Brevicoryne brassicae, Hyadaphis erysimi*) and bug (*Bagrada cruciferarum*) (Fig. 14.3) (Srinivasan and Krishna Moorthy 1991).

The IPM gave 60% more yield and 152% more returns than pure cabbage crop (Table 14.17) (Khaderkhan et al. 1998).

In another study, the benefit to cost ratio in IPM and non-IPM plots was 2.42 and 0.83, respectively (Table 14.18) (Krishna Moorthy et al. 2003).

14.2.3.5 Okra Pests and Diseases

- Planting of yellow vein mosaic virus resistant (YVMV) hybrids, namely, Sun-40 and Makhmali
- Sowing of sorghum/maize as border crop
- Installation of yellow sticky polythene traps smeared with castor oil and delta traps set up for whitefly and other small sucking pests

Table 14.17 Economics of cabbage BIPM

Practice	Yield (MT/ha)	% increase	Net returns (Rs.)	% increase
Farmers' practice	20	–	19,817	–
IPM practice	32	60	49,251	152

Table 14.18 Economics of cabbage BIPM

Treatment	Yield (MT/ha)	Net returns (Rs/ha)	Benefit to cost ratio
IPM plots	55	30,085	2.42
Non-IPM plots (farmers' practice)	35	5,090	0.83

Table 14.19 Yield and economics of BIPM in okra

Parameter	IPM	Non-IPM	% increase
Yield (MT/ha)	10.30	7.24	42
Net returns (Rs/ha)	64,797	34,678	86
Benefit to cost ratio	1.28	0.72	77

- Erection of bird perches at 25/ha for facilitating predation of borer larvae
- Installation of pheromone traps at 5/ha for monitoring *Earias vitella*
- Three sprays of 5% NSKE for hopper, whitefly and mites starting at 28 DAS
- Five releases of *Trichogramma chilonis* at 1 lakh/ha starting from 42 DAS at weekly interval
- Rouging out YVMV-affected plants from time to time

In okra crop, BIPM technology has been validated in about 3 ha area in Raispur village near Ghaziabad during 2003–2004. IPM fields gave higher yields of 10.305 MT/ha as compared to 7.246 MT/ha in non-IPM fields (Amerika Singh et al. 2004) (Table 14.19).

14.2.3.6 Black Pepper Foot Rot and Nematode Disease Complex

Integrated management of foot rot (*Phytophthora capsici*) and nematodes (*Meloidogyne incognita* and *Radopholus similis*) on black pepper was achieved by:

- Mixing VAM and *Trichoderma harzianum* in solarised nursery mixture to raise healthy and robust seedlings
- Application of *T. harzianum* and FYM in planting pit
- Field application of neem cake at 1 kg/vine mixed with 50 g of *T. harzianum* during August (Sarma 2003)

14.2.3.7 Cardamom Capsule Rot

Management of capsule rot (*Phytophthora* spp.) of cardamom was achieved by two applications of *T. harzianum* at 1 kg/plant (grown on decomposed coffee pulp and FYM in 1:1 ratio) during May and July integrated with foliar spray of Akomin (potassium phosphonate) (Anandraj and Eapen 2003).

14.2.3.8 Banana Burrowing Nematode

Integration of neem cake at 200 g/plant with *Glomus mosseae* at 100 g/plant (containing 25–30 chlamydospores/g of inoculum) was most effective in reducing the *Radopholus similis* population both in soil and roots, whilst karanj cake with *G. mosseae* gave maximum increase in fruit yield of banana. Mycorrhizal root colonisation and number of chlamydospores of *G. mosseae* were maximum in neem cake amended soil (Table 14.20) (Parvatha Reddy et al. 2002).

Table 14.20 Effect of *Glomus mosseae* and oil cakes on population of *Radopholus similis* and yield of banana

Treatment	Dose (g)/plant	Population of *R. similis*		Yield (kg)/plant
		Roots (10 g)	Soil (250 ml)	
G. mosseae	200	112	122	8.64
Castor cake	400	146	132	8.18
Karanj cake	400	118	128	10.34
Neem cake	400	118	112	8.91
G. mosseae + castor cake	100 + 200	90	108	12.68
G. mosseae + karanj cake	100 + 200	76	80	16.61
G. mosseae + neem cake	100 + 200	48	62	14.80
Control	–	218	184	5.45
CD (P=0.05)		11.97	8.31	0.84

Table 14.21 Effect of neem cake, Pasteuria penetrans and Paecilomyces lilacinus on root galling, nematode multiplication rate and yield of tomato

Treatment			
Nursery (m^2)	Main field (per plant)	Root-knot index	Yield (kg)/6 m^2
Neem cake – 1 kg	*P. lilacinus* – 0.5 g	3.4	9.168
Neem cake – 1 kg	*P. penetrans* (28×10^4 spores)	3.2	9.312
P. lilacinus – 20 g	*P. penetrans* (28×10^4 spores)	3.0	9.504
P. penetrans (28×10^7 spores)	*P. lilacinus* – 0.5 g	2.9	9.624
Neem cake – 0.5 kg + *P. lilacinus* – 10 g	*P. penetrans* (28×10^4 spores)	2.5	9.672
Neem cake – 0.5 kg + *P. penetrans* (28×10^4 spores)	*P. lilacinus* – 0.5 g	2.0	9.984
Neem cake – 0.5 kg + *P. lilacinus* – 10 g + *P. penetrans* (28×10^4 spores)	–	2.6	9.600
Control	–	4.6	8.352
CD (P=0.05)		0.14	0.100

14.2.3.9 Tomato Root-Knot Nematode

In nursery, integration of *Pasteuria penetrans* (at 28×10^4 spores/m^2), *Paecilomyces lilacinus* (at 10 g/m^2 with 19×10^9 spores/g) and neem cake (at 0.5 kg/m^2) gave maximum increase in plant growth and number of seedlings/bed. Parasitisation of *M. incognita* females was highest when neem cake was integrated with *P. penetrans*, whilst parasitisation of eggs was highest when neem cake was integrated with *P. lilacinus*. In field, planting of tomato seedlings (raised in nursery beds amended with neem cake + *P. penetrans*) in pits incorporated with *P. lilacinus* (at 0.5 g/plant) gave least root galling and nematode multiplication rate and increased fruit weight and yield of tomato (Table 14.21) (Parvatha Reddy et al. 1997).

14.2.4 Transfer of Technology

The constraint for wider adoption of eco-friendly crop protection technologies is transfer of technology. Crop protection technologies developed have not reached the small and marginal farmers. Unless these technologies are assessed in farmer's fields and refined to suit local conditions, the fruits of research will not benefit farmers. Researchers and extension personnel should work hand in hand for successful transfer of crop protection technologies. Communication media such as radio, TV, audio and video cassettes, agri-portals, farmer's field schools and KVKs should be used for effective transfer of technologies.

14.2.5 Conclusions

Globalisation driven by WTO is opening up fantastic opportunities for export of agricultural products and processed food from India. It is a revolution, which is taking place, and our farmers will miss this golden opportunity if they are not equipped with the right crop protection technologies to produce agricultural products of international standards without pesticide residues. The challenge before the plant protection scientists is to prevent crop losses due to pests before and after harvest without harming the environment. There is a need to develop low input and eco-friendly crop protection technologies so as to be very competitive in the international market.

Padma Vibhushan awardee Prof M.S. Swaminathan, an eminent agricultural scientist of international repute, has said that 'The ever-green revolution will be triggered by farming systems that can help produce more from the available land, water and labour resources without either ecological or social harm' (Swaminathan 2000). Let us rededicate ourselves to achieve the dream of 'Ever-Green Revolution' of Prof Swaminathan.

References

Adams S (1997) Seein' red: colored mulch starves nematodes. Agricultural Research. October, p 18

Anandraj M, Eapen SJ (2003) Achievements in biological control of diseases of spice crops with antagonistic organisms at Indian Institute of Spices Research, Calicut. In: Ramanujam B, Rabindra RJ (eds) Current status of biological control of plant diseases using antagonistic organisms in India. Project Directorate of Biological Control, Bangalore, pp 189–215

Benbrook CM (1996) Pest management at the crossroads. Consumers Union, Yonkers, 272 pp

Couch GJ (1994) The use of growing degree days and plant phenology in scheduling pest management activities. Yankee Nursery, Quarterly Fall, pp 12–17

Khaderkhan H, Nataraju MS, Nagaraja GN (1998) Economics of IPM in tomato. In: Reddy PP, Kumar NKK, Verghese A (eds) Advances in IPM for horticultural crops. Association for Advancement of Pest Management in Horticultural Ecosystems, Division of Entomology and Nematology, Indian Institute of Horticultural Research, Bangalore, pp 151–152

Krishna Moorthy PN, Krishna Kumar NK (2002) Advances in the use of botanicals for the IPM of major vegetable pests. In: Proceedings of the international conference on vegetables, Bangalore. Dr. Prem Nath Agricultural Science Foundation, Bangalore, pp 262–272

Krishna Moorthy PN, Krishna Kumar NK, Girija G, Varalakshmi B, Prabhakar M (2003) Integrated pest management in cabbage cultivation. Extension Bulletin No. 1, Indian Institute of Horticultural Research, Bangalore, 10 pp

Parvatha Reddy P, Nagesh M, Devappa V (1997) Effect of integration of *Pasteuria penetrans*, *Paecilomyces lilacinus* and neem cake for the management of root-knot nematode infecting tomato. Pest Managmt Hortil Ecosystems 3:100–104

Parvatha Reddy P, Rao MS, Nagesh M (2002) Integrated management of burrowing nematode (*Radopholus similis*) using endomycorrhiza (*Glomus mosseae*) and oil cakes. In: Singh HP, Chadha KL (eds) Banana. AIPUB, Trichy, pp 344–348

Reichert SE, Leslie B (1989) Prey control by an assemblage of generalist predators: spiders in garden test systems. Ecology Fall, pp 1441–1450

Sarma YR (2003) Recent trends in the use of antagonistic organisms for the disease management in spice crops. In: Ramanujam B, Rabindra RJ (eds) Current status of biological control of plant diseases using antagonistic organisms in India. Project Directorate of Biological Control, Bangalore, pp 49–73

Singh Amerika, Trivedi TP, Sardana HR, Sabir N, Krishna Moorthy PN, Pandey KK, Sengupta A, Ladu LN, Singh DK (2004) Integrated pest management in horticultural crops – a wide area approach. In: Chadha KL, Ahluwalia BS, Prasad KV, Singh SK (eds) Crop improvement and production technology of horticultural crops. Horticulture Society of India, New Delhi, pp 621–636

Srinivasan K, Krishna Moorthy PN (1991) Indian mustard as a trap crop for management of major lepidopterous pests on cabbage. Trop Pest Mang 37:26–32

Srinivasan K, Krishna Moorthy PN, Raviprasad TN (1994) African marigold as a trap crop for the management of the fruit borer *Helicoverpa armigera* on tomato. Int J Pest Mang 40:56–63

Swaminathan MS (2000) For an evergreen revolution. The Hindu Survey of Indian Agriculture 2000:9–15

15 Pathogenesis-Related Proteins

Abstract

Pathogenesis-related proteins (PRs) (initially named 'b' proteins) have focused an increasing research interest in view of their possible involvement in plant resistance to pathogens. This assumption flowed from initial findings that these proteins are commonly induced in resistant plants, expressing a hypersensitive necrotic response (HR) to pathogens of viral, fungal and bacterial origin. PRs have been defined as 'proteins encoded by the host plant but induced only in pathological or related situations', the latter implying situations of non-pathogenic origin. This research encouraged the application of PR genes in gene-engineering technologies for crop improvement.

The defence strategy of plants against stress factors involves a multitude of tools, including various types of stress proteins with putative protective functions. A group of plant-coded proteins induced by different stress stimuli, named 'pathogenesis-related proteins' (PRs), is assigned an important role in plant defence against pathogenic constraints and in general adaptation to stressful environment. A large body of experimental data has been accumulated, and changing views and concepts on this hot topic have been evolved.

15.1 Introduction

The known PR inducers of biotic origin include pathogens, insects, nematodes and herbivores (Schultheiss et al. 2003). Pathogen-derived elicitors are potent PR inducers. Well-characterised are glucan and chitin fragments derived from fungal cell walls, fungus-secreted glycoproteins, peptides and proteins of elicitin family (Edreva et al. 2002). Protein products of avirulence genes in fungi and bacteria are capable of inducing PRs (Hennin et al. 2001).

Earlier data pointed that cell-wall-splitting enzymes, such as polygalacturonases, induced PR accumulation (Pierpoint et al. 1981). Later, it has been shown that polygalacturonases release biologically active pectic fragments from plant cell walls, named endogenous elicitors (Mc Neil et al. 1984), capable of inducing a set of defence responses in plants, including PR accumulation (Boudart et al. 1998). Chemicals, such as salicylic, polyacrylic and fatty acids and inorganic salts, as well as physical stimuli (wounding, UV-B radiation, osmotic shock, low temperature, water deficit and excess), are involved in PR induction. A special class of PR inducers is

P.P. Reddy, *Recent Advances in Crop Protection*,
DOI 10.1007/978-81-322-0723-8_15, © Springer India 2013

hormones (ethylene, jasmonates, abscisic acid, kinetin, auxins) (Fujibe et al. 2000). Recently, the dissipation of the proton gradient across the plasma membrane, provoked by the fungal toxin fusicoccin, activator of the plasma membrane H^+–ATPase, was reported to induce PRs (Schaller et al. 2000). Reactive oxygen species (ROS)-mediated PR formation has largely been recognised (Schultheiss et al. 2003).

PRs in plants are coded by a small multigene family. Since their discovery, regulation of PRs has been a highly active research area. Putative plasma membrane-localised receptors of PR inducers are suggested, and secondary signals of PR induction, such as salicylic acid (SA), jasmonic acid and ethylene, are established. Many of these secondary signals are well-known inducers of PR expression (Poupard et al. 2003). Cross-talks are common between signalling pathways mediated by these secondary messengers. Thus, SA-independent/jasmonate-dependent, and vice versa, pathways of PR induction have been demonstrated (Fidantsef et al. 1999). It has been proven that PRs synthesis is regulated at transcriptional level; the exact mechanisms of transcriptional regulation have been one of the most active fields of PR gene studies. Several *cis*-regulatory elements in PR-promoters mediating PR gene expression have been identified. These include Wbox, GCC box, G box, MRE-like sequence and SA-responsive element (SARE) (Zhou 1999). New mutants are developed providing clues into the better understanding of the regulation of PRs (Delaney, 2000). PRs are synthesised following a long lag period (no less than 8 h) (Matsuoka and Ohashi 1986); the synthesis proceeds in situ, that is, PRs are not translocated from the site of their induction to other plant parts, as proven by elegant grafting experiments (Gianinazzi 1982).

15.2 Occurrence

Being first detected in tobacco, PRs have subsequently been identified in numerous monocotyledonous and dicotyledonous plants across different genera and hence can be considered as ubiquitously distributed in plant kingdom. PRs are distinguished by species specificity, thus allowing their application as genetic markers in taxonomical, phylogenetical and evolutionary studies. By using PRs patterns, the origin of *Nicotiana tabacum* from the wild progenitors *N. sylvestris* and *N. tomentosiformis* was confirmed (Ahl et al. 1982).

15.3 Functions

The assumption that PRs are devoid of enzymatic functions was challenged by Legrand et al. (1987) detecting chitinase activity in four members of group three tobacco PRs. The same research team established ß-1, 3-glucanase activity in four members of group two tobacco PRs (Kauffmann et al. 1987). Later on, chitinase activity was detected in PR-4, PR-8 and PR-11, PR-4 being referred to as chitin-binding proteins. Proteinase, peroxidase, ribonuclease and lysozyme activities were established in PR-7, PR-9, PR-10 and PR-8, respectively. PR-6 was assigned proteinase-inhibitory properties.

Membrane-permeabilising functions are characteristic of defensins, thiols and lipid-transfer proteins (LTPs), referred to as PR-12, PR-13 and PR-14, respectively, and of osmotins and thaumatin-like proteins (PR-5). Multiple enzymatic, structural and receptor functions are detected in 'do-all' germins and germin-like proteins referred to as PR-15 and PR-16, respectively (Park et al. 2004a, b.).

An important common feature of most PRs is their antifungal effect; some PRs exhibited also antibacterial, insecticidal, nematicidal and – as recently shown – antiviral action. Toxicity of PRs can be generally accounted for by their hydrolytic, proteinase-inhibitory and membrane-permeabilising ability. Thus, hydrolytic enzymes (ß-1, 3-glucanases, chitinases and proteinases) can be a tool in weakening and decomposing of fungal cell walls, containing glucans, chitin and proteins, while PR-8 can disrupt Gram-positive bacteria due to lysozyme activity (Van Loon 2001; Selitrennikoff 2001). PR-10 (named Ca PR-10), induced in hot pepper (*Capsicum annuum*) by incompatible interactions with TMVPo and *Xanthomonas campestris* pv. *vesicatoria*, was shown to function as a ribonuclease.

Data are presented that subsequent phosphorylation of Ca PR-10 increases its ribonucleolytic activity to cleave invading viral RNAs and this activity is important to its antiviral pathway in vivo. Antibiotic activity of Ca PR-10 is also exerted against oomycete fungi (Park et al. 2004a). Proteinase-inhibitory properties of PR-6 may confer anti-insect and anti-nematode effects, inactivating the proteins secreted by these parasites in the invaded plant tissues. Plasma membrane-permeabilising ability proper to PR-5, PR-12, PR-13 and PR-14 contributes to plasmolysis and damage of fungal and bacterial pathogens, inhibiting their growth and development (Van Loon 2001; Selitrennikoff 2001). This effect may be due to electrostatic interactions of PRs with membrane components, leading to conformational changes, dissipation of membrane gradient and formation of pores in membranes. Multifaceted functionality of PR-15 and PR-16, including cell wall remodelling ability, can be directed against pathogens and may have protective role (Park et al. 2004b).

Interestingly, it was recently found that among the identified plant allergens, 23% belong to the group of PRs. So far, plant-derived allergens have been identified with similarities to PR families 2, 3, 4, 5, 10 and 14 (Hoffmann-Sommergruber 2001). This property may confer defensive functions displayed by the plant in hostile environment.

Apart from the above-described mechanisms of direct breakdown or damage of pathogens, PRs can operate in a distinct pathway involving the hydrolytic release of chitin and glucan fragments from fungal cell walls. These oligosaccharides are endowed with elicitor activity and can induce a chain of defence reactions in the host plant (Kombrink et al. 2001). The peroxidase activity of PR-9 can contribute to the rigidification and strengthening of plant cell wall in response to pathogen attack (Lagrimini et al. 1987).

The defensive functions of PRs against pathogens are presumably corroborated by data about their constitutive expression in seeds and plant organs; high fungitoxicity of seed osmotins and thaumatin-like proteins has been established (Abad et al. 1996). Protective role of PRs can also be inferred from their accumulation in plant cell wall appositions formed against pathogen ingress, as well as from their release into fungal structures penetrating plant tissues (Jeun and Buchenauer 2001).

Until now, the functions of the most abundant PRs family, PR-1, remain obscure. A direct inhibitory effect of basic tomato and broad bean PR-1 family members against fungal pathogens (*Phytophthora infestans* and *Uromyces fabae*, respectively) has been demonstrated by in vitro and in vivo experiments (Rauscher et al. 1999), but the mode of action as well as the cellular and molecular targets of PR-1 proteins are still unknown. Comparative 'screening' of PR-1-type proteins from various plant taxa indicates that PR-1 family is highly conserved in plants. Moreover, as already mentioned, it is related to sequences present in yeast, insects and vertebrates. The corresponding proteins in insects are major venom allergens presumably directed to other organisms. It may be speculated that related protein family in vertebrates encodes lytic enzymatic and antimicrobial activity. The widespread occurrence of PR-1 family suggests that these proteins share an evolutionary origin and possess activity essential to the functioning and surviving of living organisms (Van Loon 2001).

15.4 Relevance of PRs to Disease Resistance

Presently, it is still difficult to assign a causative role of PRs in plant resistance to pathogens. The reason for this is that the numerous data on PRs as disease resistance factor are mostly of correlative character. Four lines of supporting evidence can be outlined.

15.4.1 Stronger Accumulation of PRs in Inoculated Resistant as Compared to Susceptible Plants

Besides previous data, substantiating this statement (Van Loon 1985), differential responses of resistant/susceptible plants were recently reported in tomato plants, inoculated with *Cladosporium fulvum* (Wubben et al. 1996), *Phytophthora infestans*-infected potato (Tonón et al. 2002), *Venturia inaequalis*-inoculated apple (Poupard et al. 2003), *Pseudomonas syringae*-infected grapevine (Robert et al. 2001), *Xanthomonas campestris* pv. *vesicatoria* and TMVPo-infected hot pepper (Park et al. 2004a, b). Additional resistance gene(s) against *Cladosporium fulvum* present on the *Cf-9* introgression segment have been shown to be associated with strong PR protein accumulation (Laugé et al. 1998). In some pathosystems, mRNAs for certain PR members accumulate to similar levels in compatible and incompatible interactions, but the maximum level of expression is reached much faster in the latter (Van Kan et al. 1992).

15.4.2 Important Constitutive Expression of PRs in Plants with High Level of Natural Disease Resistance

This correlation was observed in several pathosystems, such as apple – *Venturia inaequalis* (Gau et al. 2004), tomato – *Alternaria solani* (Lawrence et al. 2000) and potato – *Phytophthora infestans* (Vleeshouwers et al. 2000), the last authors proposing PR mRNAs as molecular marker in potato breeding programmes.

15.4.3 Significant Constitutive Expression of PRs in Transgenic Plants Overexpressing PR Genes Accompanied by Increased Resistance to Pathogens

Thus, increased resistance to *Peronospora tabacina* and *Phytophthora parasitica* var. *nicotianae* was demonstrated in tobacco overexpressing PR1a gene (Alexander et al. 1993). Transgenic rice and orange plants overexpressing thaumatin-like PR-5 possessed increased tolerance to *Rhizoctonia solani* and *Phytophthora citrophthora*, respectively (Fagoaga et al. 2001), while transgenic potato overexpressing PR-2 and PR-3 had improved resistance to *Phytophthora infestans* (Bachmann et al. 1998).

PR-2 and PR-3 genes, coding for ß-1, 3-glucanase and chitinase, respectively, confer resistance of carrot to several fungal pathogens. The simultaneous expression of tobacco ß-1, 3-glucanase and chitinase genes in tomato plants results in increased resistance to fungal pathogens (Melchers et al. 1998). On the contrary, silencing of PR-1b gene in barley facilitates the penetration of the fungal pathogen *Blumeria graminis* f. sp. *hordei* in the leaves (Schultheiss et al. 2003).

15.4.4 Accumulation of PRs in Plants in Which Resistance Is Locally or Systemically Induced

Generalising this broad research area, it can be stated that PRs are recognised as markers of the systemic acquired resistance (SAR) and PR genes are involved in the list of the so-called SAR-genes (Ward et al. 1991). It has largely been demonstrated that SAR and the accompanying set of PRs are induced by different pathogens, as well as by a range of chemicals predominantly in a salicylic acid-dependent pathway; SAR is active against a broad spectrum of pathogens. Some SAR-inducing chemicals, such as benzothiadiazole (BTH), ß-aminobutyric acid (BABA) or 2,6-dichloroisonicotinic acid (DCINA), are harmless commercially supplied compounds and have promising practical application as novel tools in plant protection.

It is essential to underline that PR members induced in resistant or SAR-expressing plants, as well as PRs from transgenic resistant plants, exhibit high antimicrobial activity (Anand et al. 2004), thus suggesting their direct role in disease resistance.

In contrast to the above data, a view has been developed that PRs are rather related to the severity of symptom expression than to resistance. This assumption flows from experiments with PVY-infected potato (Naderi and Berger 1997), PVY-, PVX- and CMV-infected tobacco (Röhring 1998), viroid-infected tomato (Camacho Henrique and Sänger 1982), etc. The strong induction of PRs in tobacco leaves by PVY and necrosis-inducing abiotic factors is in line with this idea. It may be speculated that the induction of PRs accompanying the symptom expression could also have a protective role of 'last barrier', impeding the full destroyment of stressed plants by both internal and environmental constraints.

In the last years, it was surprisingly established that PRs are not synthesised during the expression of a newly reported 'induced systemic resistance' (ISR) in *Arabidopsis* (Pieterse et al. 1996), as well as in crop plants (Siddiqui and Shaukat 2004). ISR is induced during root colonisation by non-pathogenic rhizobacteria of *Pseudomonas* spp. and is effective against a broad range of pathogens. A novel signalling pathway, salicylic acid-independent but jasmonate- and ethylene-dependent, is engaged in the control of ISR (Pieterse et al. 1998). Besides by non-pathogenic *Pseudomonas* bacteria, ISR is also elicited by growth-promoting *Bacillus* spp. (Kloepper et al. 2004; Silva et al. 2004), with this approach having a beneficial effect in field conditions. The non-involvement of PRs in ISR is indicative of the diversity of plant defence strategies, pointing that PRs are only one of the multiple means employed by plants against environmental cues.

15.5 Applications: Brief Overview

Experimental evidences substantiated the utility of PR genes to develop disease resistance in transgenic plants. This practical aspect of PR gene research resulted in the release of agronomically important crops resistant to various diseases of economical interest. One promising strategy is based on the exploitation of the genes encoding antifungal hydrolases, such as ß-1, 3-glucanase and chitinase, which are associated with SAR response in plants. Increased resistance of tomato against fungal pathogens was achieved by simultaneous expression of a class I chitinase and ß-1, 3-glucanase (PR-3 and PR-2 family, respectively) from tobacco. Transgenic tomato plants expressing either of these genes alone were less protected. Field evaluation of transgenic carrot plants containing the same genes has shown a high level of resistance against major fungal pathogens of carrots. An important feature is that the majority of the transgenic lines which had resistance to one pathogen exhibited significant resistance to the other pathogens (Melchers et al. 1998). The constitutive overexpression of tobacco class I PR-2 and PR-3 transgenes in potato plants enhanced their resistance to *Phytophthora infestans*, the causal agent of late blight (Bachmann et al. 1998). Similar results about the effectiveness of the co-expression of chitinase and ß-1, 3-glucanase in plant disease resistance are reported by Kombrink et al. (2001). *Brassica napus* transgenic plants, constitutively expressing a chimeric chitinase gene, display field tolerance to fungal pathogens (Grison et al. 1996).

Increased resistance to crown rust disease in transgenic Italian ryegrass expressing the rice chitinase gene was demonstrated (Takahashi et al. 2005). Gene engineering of PR-5 is another promising strategy for improvement of crop disease resistance, based on the potent plasmolysing and antifungal effect of this PR family. Thus, overexpression of the cloned rice thaumatin-like (PR-5) gene in transgenic rice plants enhanced the environmental friendly resistance to *Rhizoctonia solani* causing sheath blight disease (Datta et al. 1999). Overexpression of a pepper basic pathogenesis-related protein one gene in tobacco plants enhances resistance to heavy metal and pathogen stresses (Sarowar et al. 2005). For biotechnological purposes, PR genes are transferred from novel sources, such as the insectivorous sundew (*Drosera rotundifolia* L) (Matušikova et al. 2004). Cautions, however, must be taken before releasing of PR transgenic crops, by assuming that some PR members display allergenic properties (Hoffmann-Sommergruber 2001).

15.6 Conclusions

Increasing amount of data enlarged the knowledge on the relevance of PRs to important plant performances, such as development, disease resistance and general adaptation to stressful environment. This research encouraged the application of PR genes in gene-engineering technologies for crop improvement. However, fundamental aspects of PR gene studies remain little understood, particularly the exact mechanisms of gene regulation; thus, the receptors, signal transducing cascades and molecular targets involved in PR induction are a challenge for both fundamental and applied studies.

References

Abad LR, Paino-D'Urzo M, Lin D, Narasimhan ML, Reuveni M, Zhu JK, Niu X, Singh NK, Hasegawa PM, Bressan RA (1996) Antifungal activity of tobacco osmotin has specificity and involves plasma permeabilization. Plant Sci 11:11–23

Ahl P, Cornu A, Gianinazzi S (1982) Soluble proteins as genetic markers in studies of resistance and phylogeny in *Nicotiana*. Phytopathology 72:80–85

Alexander D, Goodman RM, Gut-Rella M, Glascock C, Weyman K, Friedrich L, Maddox D, Ahl-Goy P, Luntz T, Ward E, Ryals J (1993) Increased tolerance to two oomycete pathogens in transgenic tobacco expressing pathogen-related protein 1a. Proc Natl Acad Sci USA 90:7327–7331

Anand A, Lei ZT, Summer LW, Mysore KS, Arakane Y, Backus W, Muthukrishnan S (2004) Apoplastic extracts from a transgenic wheat line exhibiting lesion-mimic phenotype have multiple pathogenesis-related proteins that are antifungal. Plant-Microbe Interact 17:1306–1317

Bachmann D, Rezzonico E, Retelska D, Chételat A, Schaerer S, Beffa R (1998) Improvement of potato resistance to Phytophthora infestans by overexpressing antifungal hydrolases (Abstract). In: 5th international workshop on pathogenesis-related proteins. Signalling pathways and biological activities. Aussois, 29 Mar–2 Apr 1998, p 57

Boudart G, Lafitte C, Barthe JP, Fraser D, Esquerré-Tugayé M (1998) Differential elicitation of defense responses by pectic fragments in bean seedlings. Planta 206:86–94

Camacho Henrique A, Sänger HL (1982) Analysis of acid-extractable tomato leaf proteins after infection with a viroid, two viruses and a fungus and partial purification of the'pathogenesis-related' protein P14. Arch Virol 74:181–196

Datta K, Velazhahan R, Oliva N, Ona I, Mew T, Khush GS, Muthukrishnan S, Datta SK (1999) Over-expression of the cloned rice thaumatin-like protein (PR-5) gene in transgenic rice plants enhances environmental friendly resistance to *Rhizoctonia solani* causing sheath blight disease. Theor Appl Genet 98:1138–1145

Delaney TP (2000) New mutants provide clues into regulation of systemic acquired resistance. Trends Plant Sci 5:49–51

Edreva A, Blancard D, Delon R, Bonnet P, Ricci P (2002) Biochemical changes in cryptogein-elicited tobacco: a possible basis of acquired resistance. Beitr Tabakforsch Internat 20:53–59

Fagoaga C, Rodrigo I, Conejero V, Hinarejos C, Tuset JJ, Arnau J, Pina JA, Navarro L, Peña L (2001) Increased tolerance to *Phytophthora citrophthora* in transgenic orange plants constitutively expressing a tomato pathogenesis related protein PR-5. Mol Breed 7:175–185

Fidantsef AL, Stout MJ, Thaler JS, Duffey SS, Bostock RM (1999) Signal interactions in pathogen and insect attack: Expression of lipoxygenase, proteinase inhibitor II, and pathogenesis-related protein P4 in tomato, *Lycopersicon esculentum*. Physiol Mol Plant Pathol 54:97–114

Fujibe T, Watanabe K, Nakajima N, Ohashi Y, Mitsuhara I, Yamamoto KT, Takeuchi Y (2000) Accumulation of pathogenesis-related proteins in tobacco leaves irradiated with UV-B. J Plant Res 113:387–394

Gau AE, Koutb M, Piotrowski M, Kloppstech K (2004) Accumulation of pathogenesis related proteins in apoplast of a susceptible cultivar of apple (*Malus domestica* cv Elstar) after infection by *Venturia inaequalis* and constitutive expression of PR genes in the resistant cultivar Remo. Eur J Plant Pathol 110:703–711

Gianinazzi S (1982) Antiviral agents and inducers of virus resistance: analogies with interferon. In: Wood RKS (ed) Active defence mechanisms in plants. Plenum Press, New York, pp 275–298

Grison R, Besset-Grezes B, Schneider M, Lucante N, Olsen L, Leguay JJ, Toppan A (1996) Field tolerance to fungal pathogens of *Brassica napus* constitutively expressing a chimeric chitinase gene. Nature Biotechnol 14:643–646

Hennin C, Diederichsen E, Höfte M (2001) Local and systemic resistance to fungal pathogens triggered by an AVR9-mediated hypersensitive response in tomato and oilseed rape carrying the *Cf-9* resistance gene. Physiol Mol Plant Pathol 59:287–295

Hoffmann-Sommergruber K (2001) Plant allergens and pathogenesis-related proteins (Abstract). In: 6th international workshop on PR-proteins, Spain, Belgium, 20–24 May 2001, p 35

Jeun YCh, Buchenauer H (2001) Infection structures and localization of the pathogenesis related protein AP24 in leaves of tomato plant exhibiting systemic acquired resistance against *Phytophthora infestans* after pretreatment with 3-aminobutyric acid or Tobacco necrosis virus. J Phytopathol 149:141–153

Kauffmann S, Legrand M, Geoffroy P, Fritig B (1987) Biological function of "pathogenesis- related" proteins: four PR proteins of tobacco have ß-1, 3-glucanase activity. EMBO J 6:3209–3212

Kloepper JW, Ryu CM, Zhang SA (2004) Induced systemic resistance and promotion of plant growth by *Bacillus* spp. Phytopathology 94:1259–1266

Kombrink E, Ancillo G, Büchter R, Dietrich J, Hoegen E, Ponath Y, Schmelzer E, Strömberg A, Wegener S (2001) The role of chitinases in plant defense and plant development (abstract). In: 6th international workshop on PR-proteins. Spain, Belgium, 20–24 May 2001, p 11

Lagrimini LM, Burkhart W, Moyer M, Rothstein S (1987) Molecular cloning of complementary DNA encoding the lignin-forming peroxidase from tobacco: molecular analysis and tissue-specific expression. Proc Natl Acad Sci USA 84:7542–7546

Laugé R, Dmitriev AP, Joosten MHAJ, de Wit PJGM (1998) Additional resistance gene(s) against *Cladosporium fulvum* present on the *Cf-9* introgression segment are associated with strong PR protein accumulation. Mol Plant-Microbe Inter 11:301–308

Lawrence CB, Singh NP, Qiu J, Gardner RG, Tuzun S (2000) Constitutive hydrolytic enzymes are associated with polygenic resistance of tomato to *Alternaria solani* and may function as an elicitor release mechanism. Physiol Mol Plant Pathol 57:211–220

Legrand M, Kauffmann S, Geoffroy P, Fritig B (1987) Biological function of pathogenesis-related proteins: four tobacco pathogenesis-related proteins are chitinases. Proc Natl Acad Sci USA 84:6750–6754

Matsuoka M, Ohashi Y (1986) Induction of pathogenesis-related proteins in tobacco leaves. Plant Physiol 80: 505–510

Matušikova I, Libantova J, Moravikova J, Mlynarova L, Nap JP (2004) The insectivorous sundew (*Drosera rotundifolia* L) might be a novel source of PR genes for biotechnology. Biologia (Bratislava) 59: 719–725

Mc Neil M, Darvill AG, Fry SC, Albershein P (1984) Structure and function of primary cell walls of plants. Annu Rev Biochem 53:625–663

Melchers LS, Lageweg W, Stuiver MH (1998) The utility of PR genes to develop disease resistance in transgenic crops (abstract). In: 5th international workshop on pathogenesis-related proteins. Signalling pathways and biological activities, Aussois, p 46

Naderi M, Berger PH (1997) Pathogenesis-related protein 1a is induced in potato virus Y infected plants as well as by coat protein targeted to chloroplasts. Physiol Mol Plant Pathol 51:41–44

Park CJ, An JM, Shin YC, Kim KJ, Lee BJ, Paek KH (2004a) Molecular characterization of pepper germin-like protein as the novel PR-16 family of pathogenesis-related proteins isolated during the resistance response to viral and bacterial infection. Planta 219:797–806

Park CJ, Kim KJ, Shin R, Park JM, Shin YC, Peak KH (2004b) Pathogenesis-related protein 10 isolated from hot pepper functions as a ribonuclease in an antiviral pathway. Plant J 37:186–198

Pierpoint WS, Robinson NP, Leason MB (1981) The pathogenesis-related proteins of tobacco: their induction by viruses in intact plants and their induction by chemicals in detached leaves. Physiol Plant Pathol 19:85–97

Pieterse CMJ, Van Wees SCM, Hoffland E, Van Pelt JA, Van Loon LC (1996) Systemic resistance in *Arabidopsis* induced by biocontrol bacteria is independent of salicylic acid accumulation and pathogenesis-related gene expression. Plant Cell 8:1225–1237

Pieterse CMJ, Van Wees SCM, Van Pelt JA, Knoester M, Laan R, Gerrits H, Weisbeek PJ, Van Loon LC (1998) A novel signaling pathway controlling induced systemic resistance in Arabidopsis. Plant Cell 10:1571–1580

Poupard P, Parisi L, Campion C, Ziadi S, Simoneau P (2003) A wound- and ethephon inducible *PR-10* gene subclass from apple is differentially expressed during infection with a compatible and incompatible race of *Venturia inaequalis*. Physiol Mol Plant Pathol 62:3–12

Rauscher M, Ádám AL, Wirtz S, Guggenheim R, Mendgen K, Deising HB (1999) PR-1 protein inhibits the differentiation of rust infection hyphae in leaves of acquired resistant broad bean. Plant J 19:625–633

Robert N, Ferran J, Breda C, Coutos-Thévenot P, Boulay M, Buffard D, Esnault R (2001) Molecular characterization of the incompatible interactions of *Vitis vinifera* leaves with *Pseudomonas syringae* pv *pisi*: expression of genes coding for stilbene synthase and class 10 PR protein. Eur J Plant Pathol 107:249–261

Röhring C (1998) Induction of pathogenesis-related proteins of group 1 by systemic virus infections of *Nicotiana tabacum* L. Beitr Tabakforsch Int 18:63–67

Sarowar S, Kim YJ, Kim EN, Kim KD, Hwang BK, Islam R, Shin JS (2005) Overexpression of a pepper basic pathogenesis-related protein 1 gene in tobacco plants enhances resistance to heavy metal and pathogen stresses. Plant Cell Rep 24:216–224

Schaller A, Roy P, Amrhein N (2000) Salicylic acid-independent induction of pathogenesis- related gene expression by fusicoccin. Planta 210:599–606

Schultheiss H, Dechert C, Király L, Fodor J, Michel K, Kogel KH, Hückelhoven R (2003) Functional assessment of the pathogenesis-related protein PR-1b in barley. Plant Sci 165:1275–1280

Selitrennikoff CP (2001) Antifungal proteins. Appl Env Microbiol 67:2883–2894

Siddiqui IA, Shaukat SS (2004) Systemic resistance in tomato induced by biocontrol bacteria against the root-knot nematode, *Meloidogyne javanica* is independent of salicylic acid production. J Phytopathol 152:48–54

Silva HSA, Romeiro RS, Carrer Filho R, Pereira JLA, Mizubuti ESG, Mounteer A (2004) Induction of systemic resistance by *Bacillus cereus* against tomato foliar diseases under field conditions. J Phytopathol 152:371–375

Takahashi W, Fujimori M, Miura Y, Komatsu T, Nishizawa Y, Hibi T, Takamizo T (2005) Increased resistance to crown rust disease in transgenic Italian ryegrass (*Lolium multiflorum* Lam) expressing the rice chitinase gene. Plant Cell Rep 23:811–818

Tonón C, Guevara G, Oliva C, Daleo G (2002) Isolation of a potato acidic 39 kDa ß-1, 3-glucanase with antifungal activity against *Phytophthora infestans* and analysis of its expression in potato cultivars differing in their degrees of field resistance. J Phytopathol 150:189–195

Van Kan JAL, Joosten MHAJ, Wagemakers GAM, Van den Berg-Velthuis GCM, De Wit PJGM (1992)

Differential accumulation of mRNAs encoding extracellular and intracellular PR proteins in tomato induced by virulent and avirulent races of *Cladosporium fulvum*. Plant Mol Biol 20:513–527

Van Loon LC (1985) Pathogenesis-related proteins. Plant Mol Biol 4:111–116

Van Loon LC (2001) The families of pathogenesis-related proteins (abstract). In: 6th international workshop on PR-proteins, Spain, 20–24 May 2001, p 9

Vleeshouwers VGAA, Van Dooijeweert W, Govers F, Kamoun S, Colon LT (2000) Does basal PR gene expression in *Solanum* species contribute to non-specific resistance to *Phytophthora infestans*? Physiol Mol Plant Pathol 57:35–42

Ward ER, Uknes SJ, Williams SC, Dincher SS, Widerhold DL, Alexander DC, Ahl-Goy P, Métraux JP, Ryals JA (1991) Coordinate gene activity in response to agents that induce systemic acquired resistance. Plant Cell 3:1085–1094

Wubben JP, Lawrence CB, de Wit PJGM (1996) Differential induction of chitinase and ß-1, 3-glucanase gene expression in tomato by *Cladosporium fulvum* and its racespecific elicitors. Physiol Mol Plant Pathol 48:105–116

Zhou F (1999) Signal transduction and pathogenesis-induced PR gene expression. In: Datta SK, Muthukrishnan S (eds) Pathogenesis-related proteins in plants. CRC Press LLC, Boca Raton, pp 195–207

16 Other Recent Advances

Abstract

RNA interference (RNAi) is a technology that allows for the specific down-regulation of genes and is a powerful tool for the identification of new targets for crop protection compounds. Whilst a crop protection compound inhibits its target protein, often an enzyme catalysing a specific metabolic step, RNAi targets the messenger RNA which encodes this enzyme and in consequence reduces the amount of the enzyme itself. Thus, RNAi is capable of mimicking the action of crop protection compounds. The availability of genome sequences for several model organisms made gene sequences available and provided the necessary information to target any gene of interest by using RNAi.

The biopesticide is based on fusion protein technology. This allows selected toxins from arthropods, which have no toxicity towards higher animals, to be combined with a carrier protein that makes them orally toxic to invertebrates, whereas they would normally only be effective when injected into a prey organism by a predator. The fusion protein, containing both the toxin and the carrier, is produced as a recombinant protein in a microbial expression system, which can be scaled up for industrial production.

Seed mat technology is 'advancing crop technology' by replacing agrochemicals for weed, pest and disease control by the use of advanced seed mat systems that also reduce water and labour requirement whilst improving food safety, quality and shelf life.

There are many ways the environment can be altered or managed to reduce plant diseases. Some of them include temperature, irrigation, humidity and host nutrition (fertiliser).

16.1 RNA Interference (RNAi)

Crop protection products – fungicides, herbicides and insecticides – act by inhibiting specific metabolic targets resulting in lethality of the targeted organism. Currently, only a limited number of such targets are addressed by commercially successful crop protection products. Over the last one to two decades, innovation – especially in herbicides – has been mainly restricted to improved chemicals acting on these very same targets. However, the increasing demand for

P.P. Reddy, *Recent Advances in Crop Protection*,
DOI 10.1007/978-81-322-0723-8_16, © Springer India 2013

improved crop protection products to guarantee effective and sustainable food production resulted in rising hurdles for the development of active compounds.

To overcome these obstacles and supply novel and improved compounds to farmers, the identification of new unknown targets offers new opportunities for crop protection research – exploring the potential for new compound properties like spectrum, selectivity and profiles in regard to safety and activity.

RNA interference (RNAi) is a technology that allows for the specific down-regulation of genes and is a powerful tool for the identification of new targets for crop protection compounds. Whilst a crop protection compound inhibits its target protein, often an enzyme catalysing a specific metabolic step, RNAi targets the messenger RNA which encodes this enzyme and in consequence reduces the amount of the enzyme itself. Thus, RNAi is capable of mimicking the action of crop protection compounds. The availability of genome sequences for several model organisms made gene sequences available and provided the necessary information to target any gene of interest by using RNAi.

RNA interference was introduced in crop protection research in the mid-1990s by antisense silencing of genes encoding herbicide targets and candidate target genes. In the late 1990s, the approach was widened with the goal to identify new targets by *Arabidopsis thaliana* genome-wide down-regulation of genes based on controlled antisense expression of target gene sequences.

Over the last years, significant progress has been made regarding the technology, efficacy and reliability of RNA interference which allowed for the implementation of further advanced methods such as an inducible silencing system using RNA hairpins and the satellite virus-induced silencing system (SVISS). SVISS has proven to be an excellent herbicide mimic and was used to silence various candidate herbicide target genes. This allowed for the discrimination of interesting targets that showed both silencing of the candidate gene and a strong phenotype and others that showed strong silencing of the candidate gene but did not show any impact on plant growth. Similar results were obtained with the inducible gene silencing system applied in thale cress, *A. thaliana*. For one of the candidate genes, over 90% of silenced lines showed very strong phenotypes reaching from stunted plants with bleaching of the leaf tips to total bleaching and lethality. The lines showing these phenotypes were analysed in detail and showed strong correlation of the observed phenotype with the degree of silencing of the candidate gene.

In fungicide target research, similar approaches are being followed to validate target candidate genes in rice blast (*Magnaporthe grisea*) as one of several model organisms for fungicide research. The system used is glutamate repressible and allows for down-regulation of genes by hairpin RNAs upon growth on glutamate-free medium. Applying this system, silencing an essential gene was shown to significantly reduce growth of the fungus correlating with strongly reduced messenger RNA of the gene.

In insecticide target research, a different approach is being followed which is based on direct delivery of chimeric double-stranded RNAs into larvae of the fruit fly (*Drosophila melanogaster*) as model organism.

RNAi has proven to be both a formidable tool in identifying gene function and to be a very valuable technology in identifying and validating target genes used for the discovery of novel crop protection products.

16.2 Fusion Protein-Based Biopesticides

This research and development programme is intended to result in the introduction of a novel, environmentally beneficial biopesticide to market, as a replacement for older pesticides being removed from use under legislation on pesticide safety introduced by the EC. The biopesticide is based on fusion protein technology invented and developed collaboratively by the Food and Environment Research Agency (FERA),

York and Durham University. This allows selected toxins from arthropods, which have no toxicity towards higher animals, to be combined with a carrier protein that makes them orally toxic to invertebrates, whereas they would normally only be effective when injected into a prey organism by a predator. The fusion protein, containing both the toxin and the carrier, is produced as a recombinant protein in a microbial expression system, which can be scaled up for industrial production.

Research previously funded under LINK programmes has resulted in a best candidate fusion protein (FP4) which has been evaluated for commercial production. However, in the present programme, it is intended to use a 2010 best candidate fusion protein with improved insecticidal activity (FP5), which will be more likely to result in a commercially viable product. In order to bring this technology to market, the laboratory-scale process for protein production must be optimised, to decrease production costs. Yield of recombinant protein per volume of microbial culture is a major factor in the final cost of product, and, therefore, the initial stage of the programme will be to improve the existing expression clones. Experience with an earlier fusion protein has shown that improvements in yield of product can be made by the introduction of extra copies of gene sequence into the yeast expression host, and this will be carried out for the 2010 best candidate. The resulting yeast clones will be screened for protein production, and the best clones will be selected and taken forward in the programme. The aim will be to produce recombinant protein on a 1 g/l of microbial culture scale.

If a fusion protein-based biopesticide is to be introduced as a commercial product, testing will be necessary to assess its effects on non-target organisms. The toxin selected for this programme has been shown to be non-toxic to higher animals. However, further testing of the fusion protein will need to be carried out by other partners in the programme. A more significant potential problem is effects on beneficial invertebrates, such as the predators and parasites that attack crop pests, pollinators such as bees and earthworms. Earlier experiments with the best candidate fusion protein, carried out under a previously funded LINK programme, have suggested that fusion proteins do not cause adverse effects on beneficial organisms. However, more extensive testing of the 2010 best candidate fusion protein will be required to confirm absence of any adverse environmental impact.

The academic partners will carry out preliminary work in this area, using a service provided by Newcastle University, prior to more extensive testing required to meet regulatory requirements, which will be addressed by industrial partners. Finally, further development of fusion proteins will be carried out by Durham and FERA, separately from the present programme. If fusion proteins with higher insecticidal activity or other desirable properties are developed, these will be made available for development by industrial partners in the TSB programme.

16.3 Seed Mat Technology

Seed mat technology is 'advancing crop technology' by replacing agrochemicals for weed, pest and disease control by the use of advanced seed mat systems that also reduce water and labour requirement whilst improving food safety, quality and shelf life.

It is a novel, innovative seed mat technology that has been developed by Terraseed Ltd. (commercial product 'Terraseed') primarily as a non-chemical approach to weed control in salad and vegetable production. It has taken several years and significant investment to develop the system, and the product is now being licensed to commercial growers in the UK and overseas as an alternative for effective weed control in situations where key herbicides have been withdrawn or are due to be revoked in the near future.

It is intended to further develop and extend the technology for the control of important pests and pathogens in horticultural crops and hence further reduce, and potentially eliminate, insecticide and fungicide inputs. The technology provides the opportunity to improve yield and quality and also reduce fertiliser, water and labour inputs

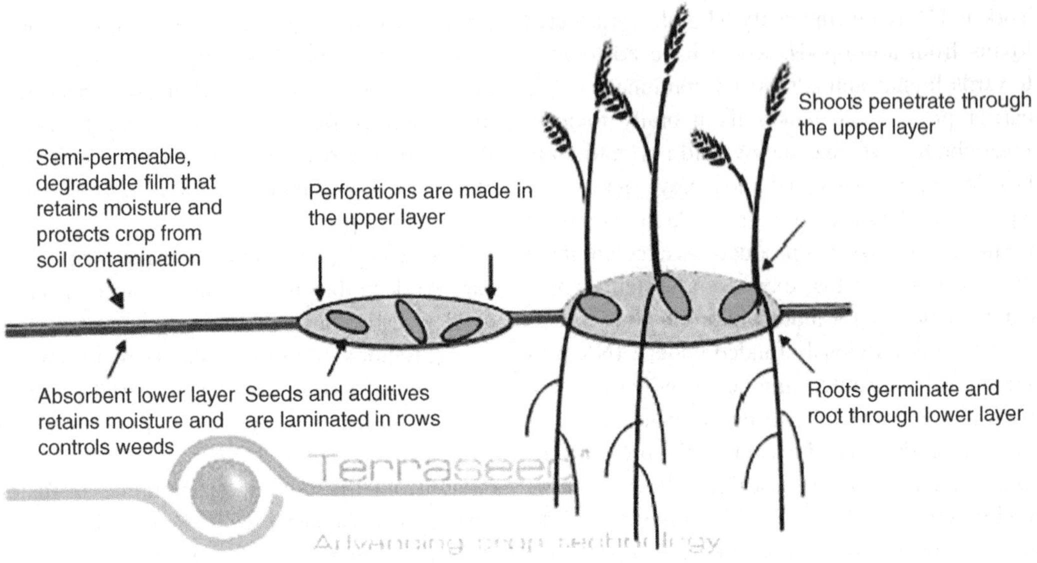

Fig. 16.1 How seed mat technology works

through improved mechanisation. The focus will be on the control of economically important pathogens and pests, for example, *Bremia, Sclerotinia, Rhizoctonia, Pythium*, carrot fly and aphids.

Studies will be primarily on lettuce and carrot crops as these are considered to have the greatest and most immediate commercial potential.

16.3.1 Terraseed: How the Technology Works

Seed mat technology consists of a physical barrier made up of an absorbent lower layer and an upper layer made from a semi-permeable, degradable, plastic film. Seeds are strategically placed between these two layers in rows (Fig. 16.1).

The absorbent lower layer retains moisture and controls weeds. The plant roots then germinate and penetrate through the lower layer into the soil.

The upper layer retains moisture and protects the crop from soil contamination, as well as conserving moisture due to reduced evaporation. The shoots germinate and grow through perforations in the plastic film.

16.3.2 Seed Mat Technology for Pest and Disease Control

Seed mat technology reduces pressure from a number of commercially important pests and diseases.

16.3.2.1 Diseases

The presence of a physical barrier on the soil surface interrupts the life cycle of soil-borne diseases such as *Bremia, Rhizoctonia* and *Sclerotinia*. Resting spores in the soil (sclerotia) can still germinate under the seed mat and the resulting fruiting bodies (apothecia) still produce spores. But, the spores do not get splashed onto the leaves, and so the life cycle is disrupted.

16.3.2.2 Insect Pests

The physical barrier can also disrupt the life cycle of soil-borne pests that emerge to feed on leaves, such as flea beetle.

The blue colour reflected from the seed mat is believed to 'look like water' to a number of flying insect pests as well as acting as a physical barrier to egg laying at the base of the plants, giving some degree of pest control for both root and leaf crops.

Growers have reported reduced levels of pests and disease damage on crops that have passed through the pack houses. It is intended to investigate these effects to quantify the level of suppression for a range of pest and disease species in a range of crops.

16.4 Environmental Methods

There are many ways the environment can be altered or managed to reduce plant diseases. Some of them include temperature, irrigation, humidity and host nutrition (fertiliser).

16.4.1 Temperature

All diseases have a specific range of temperatures under which they are the worst. If you grow plants in a greenhouse, you can alter the temperature of that structure to levels that are not optimal for the pathogen. This can greatly reduce disease severity. Sometimes, the range of temperatures that is best for the disease is different than that required for crop production. At other times, the temperatures are identical, and they cannot be changed without causing damage to the crop or perhaps lengthening production time. Other problems that exist are the added expense of changing the greenhouse temperature and the fact that many ornamentals are produced in the field where temperatures cannot be changed.

16.4.2 Irrigation and Humidity

Other elements, which greatly affect disease severity, are the irrigation method and the humidity of the growing environment. Many diseases are less severe under lower humidities. Unless you are producing the crop in an enclosed structure such as a greenhouse, you are limited in your ability to change the humidity. One of the best ways to alter the humidity around the plants is to space them farther apart. This can increase air movement between plants and thus reduce relative humidity and disease severity. It also reduces the chance of spread from infected plants to adjacent healthy plants simply because they are touching. Other methods, which reduce relative humidity in a greenhouse, are venting and heating which remove the moisture-laden air from the greenhouse just prior to sunset.

Elimination of overhead irrigation and exposure to rainfall also reduces the chance for disease development and spread since many of the most common plant pathogens require free water on the leaf surface to allow germination and infection. In addition, splashing irrigation water can easily spread spores from one infected plant or leaf throughout the entire planting.

16.4.3 Host Nutrition

Plant nutrition can influence the feeding, longevity and fecundity of phytophagous pests; the common fertiliser elements (nitrogen, phosphorous and potassium) can have direct and indirect effects on pest suppression. In general, nitrogen in high concentrations has the reputation of increasing pest incidence, particularly of sucking pests such as mites and aphids. On the other hand, phosphorous and potassium additions are known to reduce the incidence of certain pests, for example, in low phosphorous soils, wireworm populations often tend to increase.

Fertilisation promotes rapid growth and shortens the susceptible stages. It gives better tolerance to, and opportunity to compensate for, pest damage. Trace mineral and plant hormone sprays (e.g. from seaweed extracts) have been found to reduce damage by certain pests, particularly sucking pests such as some aphids and mites.

16.4.3.1 Nitrogen

Nitrogen fertiliser applied above the recommended rates can result in increased disease incidence and lesion area. This has been demonstrated for biotrophic fungal pathogens such as powdery mildews and rusts (Hoffland et al. 2000). The application of nitrogen fertiliser can increase disease severity due to increased crop canopy development. Nitrogen increased susceptibility of tomato to powdery mildew pathogen

Oidium lycopersicum and the bacterial pathogen *Pseudomonas syringae* pv. *tomato*, whilst it had no effect on the susceptibility to the vascular wilt pathogen *Fusarium oxysporum* f. sp. *lycopersici* (Hoffland et al. 2000).

16.4.3.2 Phosphorus

In general, phosphorus fertilisation tended to improve plant health, with reductions in disease incidence. Foliar application of phosphate salts has been shown to induce resistance to pathogens in cucumber (Mucharromah and Kuc 1991), broad bean (Walters and Maurray 1992) and grapevine (Reuveni and Reuveni 1995). The effects of phosphorus on plant disease may be the result of direct effects on the pathogen and host plant metabolism, leading to effects on pathogens' food supply, and effects on plant defences.

16.4.3.3 Potassium

In general, potassium is associated with disease reductions. The effects of potassium, applied as potassium chloride fertiliser, might be due to the chloride ion rather than potassium, since chloride fertilisation has been shown to suppress disease incidence. Foliar-applied K is associated with disease reductions. The effects of potassium, applied as potassium chloride, have been shown to control *Blumeria graminis* and *Septoria tritici* on wheat in field studies (Mann et al. 2004). Increase in plant resistance by application of potassium might be due to increase in epidermal cell wall thickness or disease escape as a result of vigorous crop growth (Prabhu et al. 2007).

Kirkpatrik et al. (1959a) showed that populations of *Xiphinema* and *Pratylenchus* were significantly lower under cherry trees receiving high rates of potash (2 and 6 kg of K_2SO_4 per tree). They also showed that the numbers of these nematodes were correlated negatively with leaf potash (Kirkpatrick et al. 1959b).

Increased levels of potash have significantly reduced the number of galls by *M. javanica* in tomato (Gupta and Mukhopadhyaya 1971). The growth of tomato increased (35%) and root galling due to *M. incognita* (root-knot index 1 as against 4 in control), nematode population in soil (80% reduction) and egg mass production (egg mass index 1 as against 4 in control) reduced in soil amended with 20% fly ash. The increase in plant growth may be ascribed to increased nutrient availability (N, K, Ca, Mg, Na, B, SO_4), and the reduction in nematode population might be due to toxic compounds (polycyclic aromatic hydrocarbons, dibenzofuran and dibenzo-p-dioxin mixtures) present in fly ash.

Application of potash in combination with phosphorus or nitrogen or potash alone checks the reniform nematode multiplication on okra to a great extent (Sivakumar and Meerazainuddin 1974).

16.4.3.4 Calcium

Application of calcium to soils, foliage and fruit reduces the incidence and severity of a range of diseases of crops, including cereals, vegetables, legumes, fruits as well as post-harvest diseases of tubers and fruits (Rahman and Punja 2007). Calcium has been shown to inhibit anthracnose (*Colletotrichum gloeosporoides* or *C. acutatum*) in apples (Biggs 1999) and to decrease post-harvest disease development on strawberry (Cheour et al. 1990), whilst treatment of tomato with calcium carbonate reduced *Fusarium* crown rot disease (Woltz et al. 1992). Increased calcium concentrations in storage organs lead to enhanced resistance to pathogens because calcium increases resistance of plant cell membranes and cell walls to microbial resistance to pathogens.

Application of $CaCN_2$, NaCN and urea cynamide is effective in controlling root-knot nematodes. This may be due to the release of HCN or NH_3 which are toxic to the nematodes.

16.4.3.5 Silicon

It is well known that application of silicon reduces disease severity in crop plants. Cucumbers grown in nutrient solutions supplemented with silicon were found to have significantly less powdery mildew infection than plants not receiving silicon supplementation (Miyaki and Takahashi 1983). Silicon has been shown to suppress both foliar- and soil-borne pathogens in cucurbits (Belanger et al. 1995). Phytoalexin accumulation occurs in silicon-mediated resistance in both dicots and monocots.

16.4.3.6 Sulphur

Elemental sulphur was most efficient against rust and powdery mildew, but was also successfully used against common scab of potato. Next to its fungicidal effect, sulphur is an acaricide and used to combat mites. Soil-applied elemental sulphur reduced the infection rate and severity of potatoes infected with *Rhizoctonia solani* (Kilkocka et al. 2005).

References

Belanger RR, Bowen P, Ehret DL, Menzies JG (1995) Soluble silicon: its role in crop and disease management of greenhouse crops. Plant Dis 79:329–336

Biggs AR (1999) Effects of calcium salts on apple bitter rot caused by two *Colletotrichum* spp. Plant Dis 83:1001–1005

Cheour F, Willemot C, Arul J, Deasjaardins Y, Makhlouf J, Charest PM, Gosselin A (1990) Foliar application of calcium chloride delays post-harvest ripening of strawberry. J Am Soc Hortic Sci 115:789–792

Gupta DC, Mukhopadhyaya MC (1971) Effect of N, P and K on the root-knot nematode, *Meloidogyne javanica* (Treub) Chitwood. Sci Cult 37:246–247

Hoffland E, Jeger MJ, van Beusichem ML (2000) Effect of nitrogen supply rate on disease resistance in tomato depends on the pathogen. Plant Soil 218:239–247

Kilkocka H, Haneklaus S, Bloem E, Schnug E (2005) Influence of sulfur fertilization on infections of potato tubers (*Solanum tuberosum*) with *Rhizoctonia solani* and *Streptomyces scabies*. J Plant Nutr 28:819–833

Kirkpatrik JD, Mai WF, Fisher EC, Parker KG (1959a) Population levels of *Pratylenchus penetrans* and *Xiphinema americanum* in relation to potassium fertilization on Montmorency sour cherries on Mazzard rootstock (Abstract). Phytopathol 49:543

Kirkpatrik JD, Mai WF, Fisher EC, Parker KG (1959b) Relation of nematode population in nutrition of sour cherries (Abstract). Phytopathology 49:543

Mann RL, Kettlewell PS, Jenkinson P (2004) Effect of foliar applied potassium chloride on septoria leaf blotch on winter wheat. Plant Pathol 53:653–659

Miyaki Y, Takahashi E (1983) Effect of silicon on the growth of solution cultured cucumber plant. Soil Sci Plant Nutr 29:71–83

Mucharromah E, Kuc J (1991) Oxalates and phosphates induce systemic resistance against disease caused by fungi, bacteria and viruses in cucumber. Crop Prot 10:265–270

Prabhu AS, Fageria NK, Huber DM, Rodrihues FA (2007) Potassium and plant disease. In: Datnoff LE, Elmer WH, Huber DM (eds) Mineral nutrition and plant disease. APS Press, St. Paul, pp 57–78

Rahman M, Punja ZK (2007) Calcium and plant disease. In: Datnoff LE, Elmer WH, Huber DM (eds) Mineral nutrition and plant disease. APS Press, St. Paul, pp 79–94

Reuveni MM, Reuveni R (1995) Efficacy of foliar application of phosphates in controlling powdery mildew fungus on field grown wine grapes: effects on cluster yield and peroxidase activity. J Phytopathol 143:21–25

Sivakumar CV, Meerazainuddin M (1974) Influence of N, P and K on the reniform nematode, *Rotylenchulus reniformis* and its effect on the yield of okra. Indian J Nematol 4:243–244

Walters DR, Maurray DC (1992) Induction of systemic resistance to rust in *Vicia faba* by phosphate and EDTA: effects of calcium. Plant Pathol 41:444–448

Woltz SS, Jones JP, Scott JW (1992) Sodium chloride, nitrogen source and lime influence Fusarium crown rot severity in tomato. Hortic Sci 27:1087–1088

Index

P.P. Reddy, *Recent Advances in Crop Protection*,
DOI 10.1007/978-81-322-0723-8, © Springer India 2013

Made in the USA
Las Vegas, NV
09 May 2026

46662559R00162